D0075038

Oxford Applied Mathematics and Computing Science Series

General Editors

J. N. Buxton, R. F. Churchhouse, and A. B. Tayler

OXFORD APPLIED MATHEMATICS AND COMPUTING SCIENCE SERIES

I. Anderson: *A First Course in Combinatorial Mathematics*
D. W. Jordan and P. Smith: *Nonlinear Ordinary Differential Equations* (*Second Edition*)
D. S. Jones: *Elementary Information Theory*
B. Carré: *Graphs and Networks*
A. J. Davies: *The Finite Element Method*
W. E. Williams: *Partial Differential Equations*
R. G. Garside: *The Architecture of Digital Computers*
J. C. Newby: *Mathematics for the Biological Sciences*
G. D. Smith: *Numerical Solution of Partial Differential Equations* (*Third Edition*)
J. R. Ullmann: *A Pascal Database Book*
S. Barnett and R. G. Cameron: *Introduction to Mathematical Control Theory* (*Second Edition*)
A. B. Tayler: *Mathematical Models in Applied Mechanics*
R. Hill: *A First Course in Coding Theory*
P. Baxandall and H. Liebeck: *Vector Calculus*
D. C. Ince: *An Introduction to Discrete Mathematics and Formal System Specification*
P. Thomas, H. Robinson, and J. Emms: *Abstract Data Types: Their Specification, Representation, and Use*
D. R. Bland: *Wave Theory and Applications*

D. R. BLAND
Cranfield Institute of Technology

Wave Theory and Applications

CLARENDON PRESS · OXFORD
1988

Oxford University Press, Walton Street, Oxford OX2 6DP

Oxford New York Toronto
Delhi Bombay Calcutta Madras Karachi
Petaling Jaya Singapore Hong Kong Tokyo
Nairobi Dar es Salaam Cape Town
Melbourne Auckland

and associated companies in
Berlin Ibadan

Published in the United States
by Oxford University Press, New York

British Library Cataloguing in Publication Data

Bland, D. R.
Wave theory and applications.
1. Waves
I. Title
531'.1133

ISBN 0-19-859654-5
ISBN 0-19-859669-3 (Pbk)

Library of Congress Cataloging-in-Publication Data

Bland, D. R. (David Russell)
Wave theory and applications/D. R. Bland.
(Oxford applied mathematics and computing science series)
Includes index.
1. Wave-motion, Theory of. I. Title. II. Series.
QC157.B53 1988 530.1'4—dc19 88-470

ISBN 0-19-859654-5
ISBN 0-19-859669-3 (Pbk)

Typeset by Macmillan India Ltd, Bangalore 25.
Printed and bound in Great Britain by
Biddles Ltd, Guildford and King's Lynn

Preface

The aim of this book is to introduce the reader to the theory of waves and to bring him or her up to the level at which the advanced books and papers on the subject can be studied. Most chapters treat different examples of wave motion, generally starting with the derivation of the governing equations and then solving particular problems in detail. Consideration is restricted to waves with one independent space variable, generally cartesian. However, spherical waves and reflection in two dimensions of plane waves are treated in Chapter 7, and vibrations of water in tanks in Chapters 3 and 4.

In terms of an English undergraduate honours mathematics course, chapters 1 and 2[†] are late first-year standard. Chapters 3, 4, and 5 are second year. The chapters should be read in numerical order, except that Chapter 5 can be read immediately after Chapter 2. Chapter 6 *et seq.* are third-year standard, verging on the post-graduate by Chapter 10.

The equations of three-dimensional linear isentropic elasticity are stated at the beginning of Chapter 7. They are derived in second-year courses on the foundations of continuum mechanics. The full equations of one-dimensional thermoelasticity are derived in Chapter 8. This can be done concisely because the Eulerian and Lagrangian components of the stress tensor in the direction of motion are identical.

Like all applied mathematics, the theory of waves combines both mathematics and physics. Most applied mathematicians have a detailed physical feel for only a few branches of the subject. Since the present author has worked primarily in solid mechanics, he has chosen to set most of the more advanced chapters of this book in that field. The general ideas developed apply in other fields, but the details are different.

This book is not concerned with experimental techniques or with numerical or computational methods *per se*. Experimental measurements and numerical solutions are quoted when required.

---

† Chapters 1 and 2 are in effect a second edition of the author's previous textbook *Vibrating Strings*, now out of print.

I hope not only that this book will prove useful as a course text but also that it will be read for individual enlightenment and pleasure. Exercises are included in all chapters. Where required, answers are given at the end of the chapter.

My thanks are due to Mr I. G. Highton for checking both text and exercises and to Professor J. F. Clarke, Dr D. F. Parker, Dr T. G. Rogers, Mr E. J. Watson, and Professor L. C. Woods for their comments.

Cranfield D. R. B.
June 1987

Contents

1 Transverse waves on strings

Transverse motion of strings is one of the easiest examples of wave motion to visualize both in reality and in the interpretation of the various solutions of the governing equations. The main application is to stringed musical instruments, whether bowed, plucked, or struck, where the frequencies of the sound produced are predicted by the theory. In common with most linear theories in applied mathematics, the linear theory of transverse motions is exact only in the limit as the amplitude of the motion tends to zero. For large amplitudes, non-linear terms must be included. The value of a linear theory depends upon whether the predictions of the theory for small amplitudes agree with observation to within acceptable error†. In the case of strings this criterion is satisfied.

1.1. Derivation of the governing equations

We start by considering a string at rest stretched to tension T_0. We assume that body forces such as that due to gravity are negligible and that no surface forces act on the string except at its ends. In these circumstances the string is straight. The string is now disturbed from its rest position. The first two chapters are devoted to determining the subsequent motions of the string, i.e. to finding the position of each particle of the string at a later time.

Treatment will be limited to plane motions of the string. If the x-axis is taken to lie along the string in its rest position and the y-axis to be in the plane of motion, then the particle initially at rest at $(x, 0)$ is at $(x+u, y)$ at time t. There is no dependence on the out-of-plane coordinate z. The mathematical problem of the vibrating string is the determination of the two functions $u(x, t)$ and $y(x, t)$, called, respect-

† What is 'acceptable' error depends upon the application. In some parts of physics, quantities can be measured accurately to 1 part in 10^6 or even higher; in solid mechanics, accuracies rarely exceed 0.1 per cent and are often around only 1 per cent.

ively, the longitudinal and transverse displacements (of a particle) of the string.

Consider a small elemental length of the string whose end particles are initially at $(x, 0)$ and $(x+\delta x, 0)$. Let these particles have co-ordinates $(x+u, y)$ and $(x+\delta x+u+\delta u, y+\delta y)$ at some later time t (Fig. 1.1) Since $u=u(x, t)$ and $y=y(x, t)$ and t is temporarily fixed,

$$\delta u=\frac{\partial u}{\partial x}\delta x \quad \text{and} \quad \delta y=\frac{\partial y}{\partial x}\delta x.$$

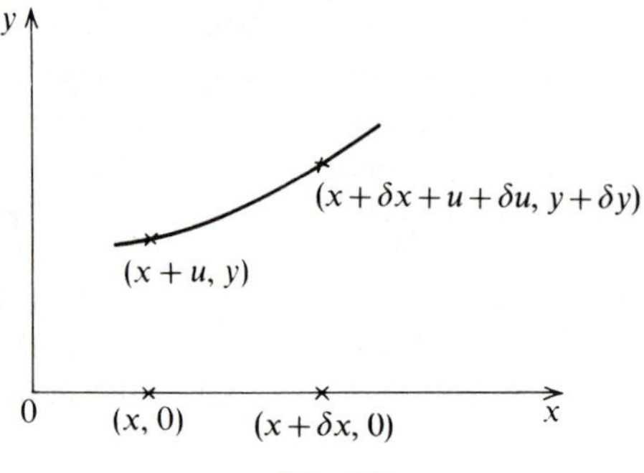

Fig. 1.1

If Ψ denotes the angle made by the element at time t with the x-axis, then $\Psi=\Psi(x, t)$ and

$$\tan\Psi=\lim_{\delta x\to 0}\frac{\delta y}{\delta x+\delta u}=\frac{\partial y}{\partial x}\left(1+\frac{\partial u}{\partial x}\right)^{-1}. \tag{1.1}$$

The strain ε of the element is defined as the increase of its length divided by its original length, i.e.

$$\begin{aligned}\varepsilon&=\lim_{\delta x\to 0}\{[(\delta x+\delta u)^2+(\delta y)^2]^{\frac{1}{2}}-\delta x\}/\delta x\\&=\left[\left(1+\frac{\partial u}{\partial x}\right)^2+\left(\frac{\partial y}{\partial x}\right)^2\right]^{\frac{1}{2}}-1.\end{aligned} \tag{1.2}$$

We first treat purely longitudinal motions of the string, i.e. $y\equiv 0$. Only strings of constant initial linear density ρ, mass per unit length, are treated. Since the tension T at any point of the string acts along the tangent at that point and the string remains parallel to the x-axis, the equation of motion† of the element is

$$\rho\,\delta x\frac{\partial^2 u}{\partial t^2}=T(x+\delta x,t)-T(x,t)=\frac{\partial T}{\partial x}\delta x,$$

or

$$\rho\frac{\partial^2 u}{\partial t^2}=\frac{\partial T}{\partial x}.$$

By Hooke's law, the increase in tensile stress or in the force per unit cross-sectional area, is equal to Young's modulus E times the strain, i.e.

$$(T-T_0)/A=E\varepsilon, \tag{1.3}$$

where A is the cross-sectional area. Use of eqn (1.2) with $y=0$ gives $\varepsilon=\partial u/\partial x$, so that

$$T-T_0=EA\frac{\partial u}{\partial x}$$

and

$$\rho\frac{\partial^2 u}{\partial t^2}=\frac{\partial T}{\partial x}=EA\frac{\partial^2 u}{\partial x^2},$$

or

$$\frac{\partial^2 u}{\partial t^2}=c_l^2\frac{\partial^2 u}{\partial x^2}, \tag{1.4}$$

where

$$c_l=\sqrt{\frac{EA}{\rho}}. \tag{1.5}$$

Equation (1.4) is the governing equation for longitudinal motions of the string; c_l is a constant. Note that ρ/A is the volume density of the string, ρ_v say, so that eqn (5) can be written

$$c_l=\sqrt{\frac{E}{\rho_v}}. \tag{1.6}$$

For a metal string, plastic yielding commences at a strain of about 0.25 per cent. Since the string is initially under tension, the additional tensile strain ε must be less than 0.25 per cent. A reasonable estimate of the greatest magnitude of ε or $\partial u/\partial x$, for which eqn (1.4) is valid, is of the order of 0.2 per cent or 0.002.

---

† Newton's second law: mass × acceleration = force.

In transverse motion the displacement y is a non-zero function of x and t varying with both and, hence, by eqn (1.1), Ψ is non-zero in general. This chapter is restricted to transverse motions in which Ψ is sufficiently small to neglect its square compared to T_0/EA. Since $T_0/(EA) < 0.0025$ for most metals, Ψ^2 can be neglected compared to unity. We assume in the theory of small transverse motions that the longitudinal displacement u is zero. The conditions for this assumption to be consistent with the rest of the theory are found at the end of this section. Because the shortest distance between two points is a straight line, the string must extend over some or all of its parts in a transverse motion; i.e. the strain $\varepsilon(x, t) \not\equiv 0$. From eqn (1.1), when $u = 0$,

$$\tan\Psi = \frac{\partial y}{\partial x}$$

and, neglecting Ψ^2 compared with 1,

$$\Psi = \frac{\partial y}{\partial x}, \quad \sin\Psi = \Psi = \frac{\partial y}{\partial x} \quad \text{and} \quad \cos\Psi = 1. \tag{1.7}$$

From eqn (1.2), since $\partial u/\partial x = 0$,

$$\varepsilon = \left[1 + \left(\frac{\partial y}{\partial x}\right)^2\right]^{\frac{1}{2}} - 1 = \left[1 + \frac{1}{2}\left(\frac{\partial y}{\partial x}\right)^2 - \frac{1}{8}\left(\frac{\partial y}{\partial x}\right)^4 + \ldots\right] - 1.$$

Therefore

$$\varepsilon = \frac{1}{2}\left(\frac{\partial y}{\partial x}\right)^2, \tag{1.8}$$

on neglecting terms of order $(\partial y/\partial x)^4$ compared to $(\partial y/\partial x)^2$. The increase in tension above the static value T_0 is given by eqns (1.3) and (1.8) as

$$T - T_0 = EA\varepsilon = \frac{1}{2}EA\left(\frac{\partial y}{\partial x}\right)^2,$$

$$T = T_0\left[1 + \frac{EA}{2T_0}\left(\frac{\partial y}{\partial x}\right)^2\right]. \tag{1.9}$$

Since $(EA/T_0)/(\partial y/\partial x)^2$ is neglected compared to unity,

$$T = T_0. \tag{1.10}$$

The tension is constant and equal to T_0 to the order of accuracy to which we are working in small transverse motions.

The equation of motion in the y-direction of an element of the string is

$$\rho\,\delta x \frac{\partial^2 y}{\partial t^2} = T_0 \sin(\Psi + \delta\Psi) - T_0 \sin\Psi.$$

(see Fig. 1.2). Use of eqns (1.7) enables $\sin(\Psi + \delta\Psi)$ to be replaced by $\Psi + \delta\Psi$ and $\sin\Psi$ by Ψ, so that

$$\rho\,\delta x \frac{\partial^2 y}{\partial t^2} = T_0\,\delta\Psi = T_0 \frac{\partial \Psi}{\partial x}\,\delta x.$$

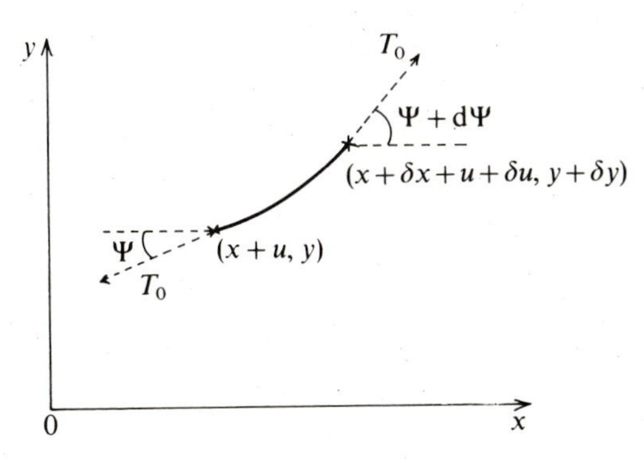

Fig. 1.2

Also by eqns (1.7) $\Psi = \partial y/\partial x$, so that

$$\rho \frac{\partial^2 y}{\partial t^2} = T_0 \frac{\partial^2 y}{\partial x^2},$$

or

$$\frac{\partial^2 y}{\partial t^2} = c^2 \frac{\partial^2 y}{\partial x^2}, \tag{1.11}$$

where the constant c is given by

$$c = \sqrt{\frac{T_0}{\rho}}. \tag{1.11a}$$

Comparison of eqns (1.11) and (1.4) show that the governing equation for small transverse motions of a string is identical to that for longitudinal motions, except that the constant has different values in the two cases. Both equations are examples of the classical wave equation which occurs in many branches of applied mathematics. In the following sections we shall be concerned with the solutions of eqn (1.11). These solutions can be taken over to any other phenomenon governed by the classical wave equation provided that (i) c is replaced by the constant appropriate to that phenomenon, and that (ii) the boundary conditions correspond.

It remains to be found under what conditions (if any) $u = 0$ is consistent with the theory of small transverse motions. The equation of motion in the x-direction of an element of the string, when $T = T_0$, is

$$\rho\, dx \frac{\partial^2 u}{\partial t^2} = T_0 \cos(\Psi + \mathrm{d}\Psi) - T_0 \cos \Psi = 0,$$

since both $\cos \Psi$ and $\cos(\Psi + \mathrm{d}\Psi)$ can be replaced by unity to the order of accuracy to which we are working. Hence

$$\frac{\partial^2 u}{\partial t^2} = 0,$$

and integrating twice:

$$\frac{\partial u}{\partial t} = f(x) \quad \text{and} \quad u = t f(x) + g(x).$$

If there exists any time t at which $u = \partial u/\partial t = 0$, then $f(x) = g(x) = 0$ and $u(x, t) \equiv 0$. It follows that u is identically zero in a small transverse motion provided that no longitudinal displacement or velocity is imparted to the string when the transverse motion is set up. These are the conditions required. In the following sections, these conditions are taken to be satisfied.

Exercise 1.1. Prove that, if $y = u(x, t)$ and $y = v(x, t)$ are any two solution of the wave equation (1.11), and if A and B are constants, then $y = Au(x, t) + Bv(x, t)$ is also a solution.

Corollary to Exercise *1*: If $y = u_i(x, t)$ are solutions, and if A_i are constants, $i = 1, 2, \ldots, n$, then $y = \sum_{i=1}^{n} A_i u_i(x, t)$ is also a solution.

1.2. D'Alembert's solution

We now look for the general solution of the classical wave equation (1.11). The equation can be written

$$\left(\frac{\partial^2}{\partial x^2}-\frac{1}{c^2}\frac{\partial^2}{\partial t^2}\right)y=0, \quad c \text{ constant.}$$

The partial differential operator

$$\frac{\partial^2}{\partial x^2}-\frac{1}{c^2}\frac{\partial^2}{\partial t^2}$$

can be factorized and the equation becomes either

$$\left(\frac{\partial}{\partial x}-\frac{1}{c}\frac{\partial}{\partial t}\right)\left(\frac{\partial}{\partial x}+\frac{1}{c}\frac{\partial}{\partial t}\right)y=0,$$

or

$$\left(\frac{\partial}{\partial x}+\frac{1}{c}\frac{\partial}{\partial t}\right)\left(\frac{\partial}{\partial x}-\frac{1}{c}\frac{\partial}{\partial t}\right)y=0.$$

THEOREM. *The equation*

$$\left(\frac{\partial}{\partial x}+\frac{1}{c}\frac{\partial}{\partial t}\right)y=0$$

has solution $y=f(x-ct)$, *where* f *is any function of its argument* $x-ct$.

Proof.

$$\frac{\partial y}{\partial x}=\frac{\partial f(x-ct)}{\partial(x-ct)}\frac{\partial(x-ct)}{\partial x}=f'(x-ct), \tag{1.12}$$

where a dash denotes a derivative of a function with respect to its argument, and

$$\frac{\partial y}{\partial t}=\frac{\partial f(x-ct)}{\partial(x-ct)}\frac{\partial(x-ct)}{\partial t}=-cf'(x-ct). \tag{1.13}$$

Hence

$$\frac{\partial y}{\partial x}+\frac{1}{c}\frac{\partial y}{\partial t}=f'(x-ct)+\frac{1}{c}[-cf'(x-ct)]=0. \qquad \text{Q.E.D.}$$

By replacing c by $-c$,

$$\left(\frac{\partial}{\partial x}-\frac{1}{c}\frac{\partial}{\partial t}\right)y=0$$

has solution $g(x+ct)$ where $g(x+ct)$ is any function of its argument $x+ct$. These two solutions suggest the next theorem.

THEOREM. *The solution of*

$$\frac{\partial^2 y}{\partial x^2}-\frac{1}{c^2}\frac{\partial^2 y}{\partial t^2}=0 \tag{1.11}$$

is

$$y=f(x-ct)+g(x+ct). \tag{1.14}$$

Proof.

$$\frac{\partial^2 y}{\partial x^2}=\frac{\partial}{\partial x}\frac{\partial y}{\partial x}=\frac{\partial}{\partial x}(f'(x-ct)+g'(x+ct)) \quad \text{on using eqn (1.12),}$$

$$=\frac{\partial f'(x-ct)}{\partial(x-ct)}\frac{\partial(x-ct)}{\partial x}+\frac{\partial g'(x+ct)}{\partial(x+ct)}\frac{\partial(x+ct)}{\partial x}$$

$$=f''(x-ct)+g''(x+ct);$$

and

$$\frac{\partial^2 y}{\partial t^2}=\frac{\partial}{\partial t}\frac{\partial y}{\partial t}=\frac{\partial}{\partial t}(-cf'(x-ct)+cg'(x+ct) \quad \text{on using eqn (1.13),}$$

$$=-c\frac{\partial f'(x-ct)}{\partial(x-ct)}\frac{\partial(x-ct)}{\partial t}+c\frac{\partial g'(x+ct)}{\partial(x+ct)}\frac{\partial(x+ct)}{\partial t}$$

$$=c^2f''(x-ct)+c^2g''(x+ct).$$

Therefore,

$$\frac{\partial^2 y}{\partial x^2}-\frac{1}{c^2}\frac{\partial^2 y}{\partial t^2}=0. \qquad \text{Q.E.D.}$$

Hence eqn (1.11) has solution eqn (1.14). Moreover, since eqn (1.14) contains two arbitrary functions, it is the general solution. Equation (1.14) is known as D'Alembert's solution of the wave equation†.

What is the physical interpretation of the transverse displacement $y=f(x-ct)$? Since y has the same value whenever $x-ct$ has the same value, consider any two times t_1 and t_2, $t_1<t_2$ (Fig. 1.3). y will have the same value at the point x_1 at time t_1 as at the point x_2 at time t_2 if

$$x_1-ct_1=x_2-ct_2, \quad \text{or} \quad x_2-x_1=c(t_2-t_1).$$

† Eqn (1.11) will frequently be referred to as the 'wave equation'. Later in the book, when other wave equations are treated, it will be distinguished by the prefix 'classical'.

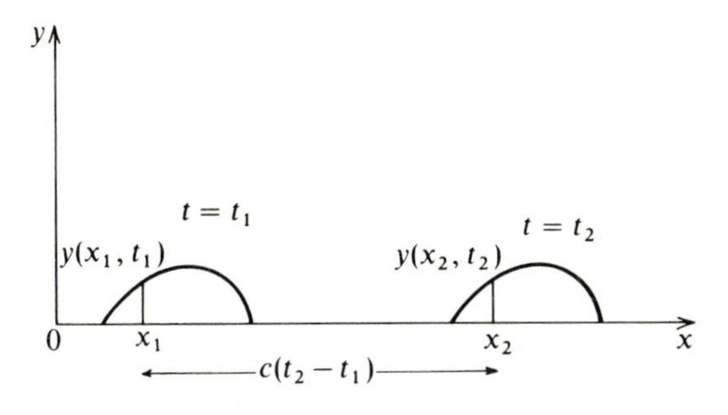

Fig. 1.3

If x_1 is the space coordinate of any point on the curve $y = f(x - ct)$ at time t_1, then the same point at time t_2 has coordinate x_2, where x_2 and x_1 are related by the above equation. The mean velocity of the point is $(x_2 - x_1)/(t_2 - t_1)$. But $(x_2 - x_1)/(t_2 - t_1) = c$, a constant independent of both x_1 and $t_2 - t_1$. Since x_1 was any point on the curve and $t_2 - t_1$ any time interval, $y = f(x - ct)$ represents a displacement of arbitrary form travelling at constant velocity c in the positive x-direction without change of shape or amplitude. Similarly, $y = g(x + ct)$ represents a displacement of arbitrary form travelling at constant velocity c in the negative x-direction without change of shape or amplitude.

It follows that eqn (1.14), D'Alembert's solution to the wave equation, represents two waves of arbitrary form each travelling with velocity c without change of shape or amplitude, one in the positive x-direction and one in the negative x-direction.

How are the functions $f(x - ct)$ and $g(x + ct)$ determined in a particular case? It is a physical fact that what happens at later times cannot alter what happens at earlier times. It is generally convenient to choose the zero of time at the instant at which the transverse displacement is set up. Since

$$y(x, 0) = f(x) + g(x) \quad \text{and} \quad \left(\frac{\partial y}{\partial t}\right)_{x,0} = -cf'(x) + cg'(x), \qquad (1.15)$$

both the functions $f(x)$ and $g(x)$ appear independently, apart from equal and opposite arbitrary additive constants, in the equations for the initial displacement and velocity of the string. It follows that $f(x)$

and $g(x)$ must be determined at the initial time for any x which is the coordinate of a point on the string†.

The form of the equations (1.15) suggests the following theorem.

THEOREM. *If*

$$y = \phi(x) \quad \text{at } t = 0, \tag{1.16}$$

$$\frac{\partial y}{\partial t} = \chi(x) \quad \text{at } t = 0, \tag{1.17}$$

and

$$y = f(x - ct) + g(x + ct) \quad \text{for } t \geqslant 0, \tag{1.14}$$

then

$$f(x) = \frac{1}{2}\phi(x) - \frac{1}{2c}\int_a^x \chi(\xi)\,\mathrm{d}\xi \tag{1.18}$$

and

$$g(x) = \frac{1}{2}\phi(x) + \frac{1}{2c}\int_a^x \chi(\xi)\,\mathrm{d}\xi, \tag{1.19}$$

where a is an arbitrary constant.

Proof. Put $t = 0$ in eqn (1.14) and use eqn (1.16) to give

$$f(x) + g(x) = y(x, 0) = \phi(x). \tag{1.20}$$

Differentiate eqn (1.14) with respect to t:

$$\frac{\partial y}{\partial t} = -cf'(x - ct) + cg'(x + ct). \tag{1.21}$$

Put $t = 0$ in eqn (1.21) and use eqn (1.17) to give

$$-cf'(x) + cg'(x) = \left(\frac{\partial y}{\partial t}\right)_{x,0} = \chi(x).$$

Integrate with respect to x and divide through by c:

$$-f(x) + g(x) = \frac{1}{c}\int_a^x \chi(\xi)\,\mathrm{d}\xi, \tag{1.22}$$

where a is an arbitrary constant. Subtraction and addition in turn of eqns (1.20) and (1.22) gives eqns (1.18) and (1.19) respectively.

---

† If the string extends from $x = a$ to $x = b$, $b > a$, then $f(x)$ and $g(x)$ must be determined at $t = 0$ for all x, $a \leqslant x \leqslant b$.

Note that $f(x)$ and $g(x)$ are determined in terms of $\phi(x)$ and $\chi(x)$, D'Alembert's solution can be expressed in terms of $\phi(x)$ and $\chi(x)$. From eqns (1.18) and (1.19),

$$f(x-ct)=\frac{1}{2}\phi(x-ct)-\frac{1}{2c}\int_a^{x-ct}\chi(\xi)\,\mathrm{d}\xi$$

and

$$g(x-ct)=\frac{1}{2}\phi(x+ct)+\frac{1}{2c}\int_a^{x+ct}\chi(\xi)\,\mathrm{d}\xi.$$

Therefore,

$$\begin{aligned} y(x,t)&=f(x-ct)+g(x+ct)\\ &=\frac{1}{2}\phi(x-ct)+\frac{1}{2}\phi(x+ct)+\frac{1}{2c}\int_{x-ct}^{x+ct}\chi(\xi)\,\mathrm{d}\xi, \end{aligned}\tag{1.23}$$

since

$$\begin{aligned} -\int_a^{x-ct}\chi(\xi)\,\mathrm{d}\xi+\int_a^{x+ct}\chi(\xi)\,\mathrm{d}\xi&=\int_{x-ct}^{a}\chi(\xi)\,\mathrm{d}\xi+\int_a^{x+ct}\chi(\xi)\,\mathrm{d}\xi\\ &=\int_{x-ct}^{x+ct}\chi(\xi)\,\mathrm{d}\xi. \end{aligned}$$

Note that the arbitrary constant a has disappeared from the equation for y, eqn (1.23).

Exercise 1.2. This is an alternative way of deriving D'Alembert's solution. Introduce new independent variables u and v, defined by $u=x-ct$ and $v=x+ct$, and show that the wave equation (1.11) becomes $\partial^2 y/\partial u\partial v=0$. Two integrations and substitution back for u and v then give D'Alembert's solution.

Exercise 1.3. Perhaps the easiest way of deriving D'Alembert's solution is to look for solutions of eqn (1.11) which represent waves travelling with velocity V without change of shape or amplitude, i.e. $y=F(x-Vt)$. V and the form of F are chosen so that eqn (1.11) is satisfied.

We now treat some commonly occurring initial conditions.

Case 1. The given initial velocity is zero for all x. In this case $\chi(x)=0$ and eqn (1.23) gives

$$y(x,t)=\frac{1}{2}\phi(x-ct)+\frac{1}{2}\phi(x+ct).\tag{1.24}$$

The physical interpretation of eqn (1.24) is that two waves, equal in shape to the original displacement but of half the amplitude, travel along the x-axis, one wave in each sense and with velocity c.

Consider a specific example in which the displacement is initially non-zero only over a finite interval, namely

$$\phi(x) = \begin{cases} 0 & \text{for} \quad x > \pi \quad \text{or} \quad x < 0 \\ \dfrac{1}{500}\sin^2 x & \text{for} \quad 0 \leqslant x \leqslant \pi, \end{cases} \tag{1.25}$$

and $\chi(x) = 0$ for all x. Find the displacement at times $\pi/4c$, $\pi/2c$ and π/c. Figure 1.4(a) shows the initial displacement (at time $t = 0$), and its division into two halves that form the two waves. At time $t = \pi/4c$, each wave has moved a distance $\pi/4c$ times c, i.e. $\pi/4$. Each wave is shown in Fig. 1.4(b) by dashed lines, and the displacement y, which is the sum of the displacements in each wave, by a solid line. At times $\pi/2c$ and π/c, the distances moved by each wave are $\pi/2$ and π respectively. At time $\pi/2c$ the two waves are just about to separate;

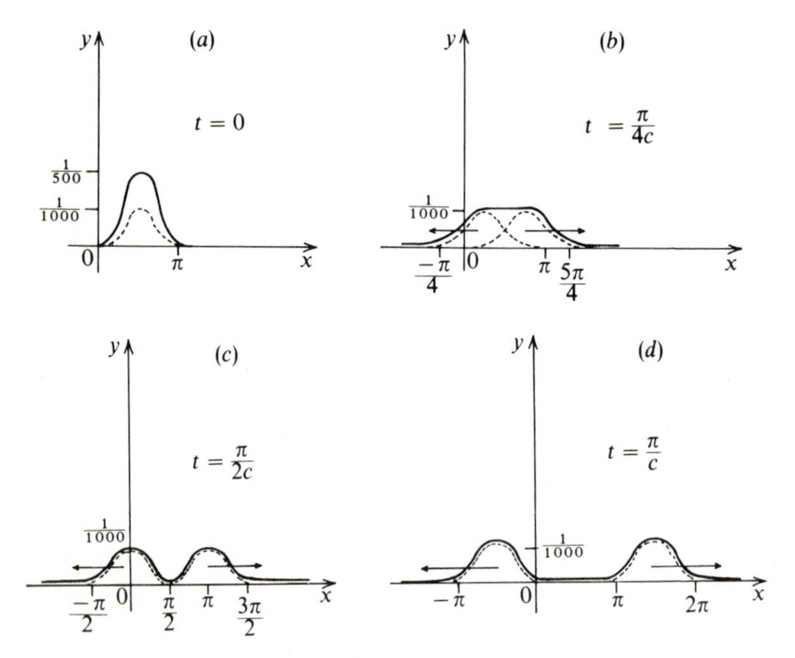

Fig. 1.4

for times greater than $\pi/2c$ the waves are separate (see Figs 1.4(c) and (d)). In the last three figures, the solid line representing the total displacement is drawn slightly above its actual position to distinguish it from the dashed lines.

The time of separation of the two waves, set up by any initial displacement which is non-zero only over a finite length L, can be found. The waves separate when the left-hand end of the positive-moving wave meets the right-hand end of the negative-moving wave. These two ends are initially a distance L apart. Since the relative velocity of the two waves is $2c$, the time of separation is $L/(2c)$. In the example just studied $L=\pi$ and the time of separation is $\pi/2c$, confirmed in Fig. 1.4(c).

An alternative method of solution is possible. First we must express $\phi(x)$ as one equation for all values of x, instead of as three equations, one for each of the three different ranges of x, as in eqns (1.25). To do this we introduce the Heaviside unit function $H(x)$, defined by

$$H(x)=\begin{cases} 1 & \text{for } x>0, \\ \frac{1}{2} & \text{for } x=0, \\ 0 & \text{for } x<0. \end{cases} \tag{1.26}$$

Equations (1.25) can now be written as

$$\phi(x)=\frac{1}{500}[H(x)-H(x-\pi)]\sin^2 x, \tag{1.27}$$

because

(i) if $x>\pi$, $H(x)=H(x-\pi)=1$, $H(x)-H(x-\pi)=0$;
(ii) if $0<x<\pi$, $H(x)=1$, $H(x-\pi)=0$, $H(x)-H(x-\pi)=1$;
(iii) if $x<0$, $H(x)=0$, $H(x-\pi)=0$, $H(x)-H(x-\pi)=0$;
(iv) if $x=0$ or $x=\pi$, $\sin x=0$.

Now that $\phi(x)$ has been expressed as one equation for all values of x, we can substitute from eqn (1.27) into eqn (1.23) to find the displacement at later times:

$$y=\frac{1}{1000}\{[H(x-ct)-H(x-ct-\pi)]\sin^2(x-ct) + [H(x+ct)-H(x+ct-\pi)]\sin^2(x+ct)\}. \tag{1.28}$$

We shall now check that eqn (1.28) gives the same results as Fig. 1.4(*b*) by substituting $t = \pi/4c$ into eqn (1.28):

$$y = \frac{1}{1000}\{[H(x-\pi/4) - H(x-5\pi/4)]\sin^2(x-\pi/4)$$

$$+[H(x+\pi/4) - H(x-3\pi/4)]\sin^2(x+\pi/4)\}.$$

It is convenient to express this equation as several different equations not involving Heaviside functions for different ranges of values of x. Since one Heaviside function changes value at each of the points $x = 5\pi/4$, $3\pi/4$, $\pi/4$ and $-\pi/4$, there are five ranges to be considered, as shown in Table 1.1. The reader can verify that the values of y given in the last column are identical with those in Fig. 1.4(*b*).

Exercise 1.4. The initial displacement of the string is

$$\phi(x) = \frac{1}{500}x^2(2-x)^2 \quad \text{for } 0 \leqslant x \leqslant 2 \quad \text{and}$$

$$\phi(x) = 0 \qquad \text{for } x > 2 \quad \text{or} \quad x < 0.$$

The initial velocity is zero for all x. Find the time τ taken for the two waves to separate and sketch the displacement at $t = \tau/4$, $\tau/2$, τ and $3\tau/2$.

Case 2. The given initial displacement is zero for all x. In this case $\phi(x) = 0$ and eqn (1.23) becomes

$$y(x, t) = \frac{1}{2c}\int_{x-ct}^{x+ct} \chi(\xi)\,\mathrm{d}\xi. \tag{1.29}$$

When $\chi(x)$ is zero outside the range $-a \leqslant x \leqslant a$, $y = 0$ if $x + ct < -a$ or $x - ct > a$. If $x + ct > a$, the upper limit of the integral can be replaced by a; and if $x - ct < -a$, the lower limit can be replaced by $-a$. The four points $x + ct = \pm a$ and $x - ct = \pm a$ divide the x-axis into five ranges and the integral on the r.h.s. of the eqn (1.29) can be simplified in most of these ranges. The ordering of the four points depends upon whether ct is $> a$ or $< a$.

Case 3. The initial conditions produce a wave in the positive sense only. Since $y = f(x-ct) + g(x+ct)$ (eqn (1.14)), $g(x+ct)$ must reduce to a constant in this case. Call the constant A and substitute in eqn (1.19):

$$\frac{1}{2}\phi(x) + \frac{1}{2c}\int_a^x \chi(\xi)\,\mathrm{d}\xi = A.$$

Table 1.1

Range	$H(x-5\pi/4)$	$H(x-3\pi/4)$	$H(x-\pi/4)$	$H(x+\pi/4)$	y
$x > 5\pi/4$	1	1	1	1	0
$5\pi/4 > x > 3\pi/4$	0	1	1	1	$\frac{1}{1000}\sin^2\left(x-\frac{\pi}{4}\right)$
$3\pi/4 > x > \pi/4$	0	0	1	1	$\frac{1}{1000}\left[\sin^2\left(x-\frac{\pi}{4}\right)+\sin^2\left(x+\frac{\pi}{4}\right)\right] = \frac{1}{1000}$
$\pi/4 > x > -\pi/4$	0	0	0	1	$\frac{1}{1000}\sin^2\left(x+\frac{\pi}{4}\right)$
$x < -\pi/4$	0	0	0	0	0

Differentiate with respect to x:

$$\chi(x) = -c\phi'(x). \tag{1.30}$$

We have shown that eqn (1.30) is a necessary condition for the initial conditions to produce a wave in the positive sense only. To prove that eqn (1.30) is also sufficient, substitute from eqn (1.30) into eqn (1.23):

$$\begin{aligned} y(x,t) &= \frac{1}{2}\phi(x-ct) + \frac{1}{2}\phi(x+ct) + \frac{1}{2c}\int_{x-ct}^{x+ct} (-c\phi'(\xi))\,\mathrm{d}\xi \\ &= \frac{1}{2}\phi(x-ct) + \frac{1}{2}\phi(x+ct) - \frac{1}{2}\phi(\xi)\Big|_{x-ct}^{x+ct}, \end{aligned}$$

where $\phi(\xi)|_a^b$ denotes $\phi(b)-\phi(a)$. Hence

$$y(x,t) = \phi(x-ct).$$

This last equation represents a wave moving in the positive sense.

Exercise 1.5. What initial conditions produce a displacement $y = (1/500)\sin(x-ct)$ for all x and for all $t \geqslant 0$?

Exercise 1.6. Show that a necessary and sufficient condition for a wave in the negative sense only is

$$\chi(x) = c\phi'(x). \tag{1.31}$$

1.3. Kinetic and potential energies of travelling transverse waves

The kinetic energy of an element of string of length δx in a small transverse vibration is equal to $\frac{1}{2}\rho\delta x(\partial y/\partial t)^2$, since the longitudinal displacement $u = 0$. The kinetic energy per unit length, K, is given by

$$K = \frac{1}{2}\rho\left(\frac{\partial y}{\partial t}\right)^2. \tag{1.32}$$

The potential energy of the element is equal to the work done by the tension in extending the element, i.e. force $T_0 \times$ extension $\varepsilon\delta x = T_0\varepsilon\,\delta x$. The potential energy per unit length, V, is given by

$$V = T_0\varepsilon = \frac{1}{2}T_0\left(\frac{\partial y}{\partial x}\right)^2, \tag{1.33}$$

on use of eqn (1.8). The total energy per unit length is $K+V$.

Substitute the general solution for travelling waves, eqn (1.23), into eqns (1.32) and (1.33) to give

$$K = \tfrac{1}{2}\rho c^2 [f'(x-ct) - g'(x+ct)]^2$$
$$= \tfrac{1}{2}\rho c^2 [(f'(x-ct))^2 - 2f'(x-ct)\,g'(x+ct)$$
$$+ (g'(x+ct))^2] \qquad (1.34)$$

and

$$V = \tfrac{1}{2} T_0 [f'(x-ct) + g'(x+ct)]^2$$
$$= \tfrac{1}{2} T_0 [(f'(x-ct))^2 - 2f'(x-ct)\,g'(x+ct)$$
$$+ (g'(x+ct))^2]. \qquad (1.35)$$

Since $T_0 = \rho c^2$,

$$K + V = T_0[(f'(x-ct))^2 + (g'(x+ct))^2]. \qquad (1.36)$$

From eqns (1.34) to (1.36), the following conclusions can be drawn. (i) The kinetic and potential energies of an element are equal if and only if just one wave, either in the positive or negative sense, is passing through that element at the time considered. (ii) The total energy, but neither the kinetic nor the potential energy, when two non-zero waves are passing through the element is the sum of the total energies which would obtain if each wave passed through separately. In the above conclusions, the two waves considered are travelling in opposite directions.

1.4. Harmonic waves

We shall see later that arbitrary functions either of $x-ct$ or of $x+ct$ are not solutions of most wave equations. The classical wave equation (1.11) is one of the exceptions. However functions of the form

$$y = a\cos(\alpha(x-ct)+\varepsilon) \qquad (1.37)$$

or

$$y = b\cos(\beta(x+ct)-\gamma) \qquad (1.38)$$

satisfy a much wider group of wave equations and are of great practical importance. Waves with the form of eqn (1.37) or (1.38) are known as harmonic waves: c, a, α, ε, b, β and γ are real constants, with c, a, α, b and β positive. Clearly the r.h.s. of eqns (1.37) and (1.38) are special cases of $f(x-ct)$ and $g(x+ct)$ respectively and therefore satisfy the classical wave equation (1.11).

The waves represented by eqns (1.37) and (1.38) can be generated as transverse vibrations of strings by the initial conditions

$$y = a\cos(\alpha x + \varepsilon) \quad \text{and} \quad \frac{\partial y}{\partial t} = \alpha c a \sin(\alpha x + \varepsilon) \qquad \text{for eqn (1.37),}$$

and

$$y = b\cos(\beta x - \gamma) \quad \text{and} \quad \frac{\partial y}{\partial t} = -\beta c b \sin(\beta x - \gamma) \quad \text{for eqn (1.38).}$$

The reader can verify that eqns (1.30) and (1.31) respectively are satisfied in these two cases.

The maximum value of the displacement y in a harmonic wave is known as the 'amplitude'. In eqns (1.37) and (1.38), it is a and b respectively. At any given time, the displacement is periodic in x. The minimum distance after which the displacement repeats its values is known as the 'wavelength' and is denoted by λ. In eqns (1.37) and (1.38),

$$\lambda = \frac{2\pi}{\alpha} \quad \text{and} \quad \lambda = \frac{2\pi}{\beta}, \quad \text{respectively.} \tag{1.39}$$

The time for a complete wavelength to pass a given point is called the 'period' of the wave and denoted by τ. Since the waves in both senses travel with speed c,

$$\tau = \frac{\lambda}{c}. \tag{1.40}$$

The 'frequency' of the wave, denoted by n, is the number of wavelengths passing a fixed point in unit time. Since a length c of the wave passes a fixed point in unit time,

$$n = \frac{c}{\lambda} = \frac{1}{\tau}. \tag{1.41}$$

The number of waves in unit distance, denoted by k, is given by

$$k = 1/\lambda. \tag{1.42}$$

Equations (1.37) and (1.38) can now be written in the form

$$y = a\cos(2\pi(kx - nt) + \varepsilon) \tag{1.43}$$

and

$$y = b\cos(2\pi(\bar{k}x + nt) - \gamma) \tag{1.44}$$

respectively. These equations can be written more concisely if we introduce k and ω through $k = 2\pi\bar{k}$ and $\omega = 2\pi n$. k is called the 'wavenumber' and ω the 'angular or radian frequency'. It follows that

$$k = 2\pi\bar{k} = 2\pi/\lambda \tag{1.45}$$

and

$$\omega = 2\pi n = 2\pi/\tau. \tag{1.46}$$

Equations (1.43) and (1.44) become

$$y = a\cos(kx - \omega t + \varepsilon) \tag{1.47}$$

and

$$y = b\cos(kx + \omega t - \gamma). \tag{1.48}$$

From eqns (1.41), (1.45) and (1.46),

$$c = \frac{\lambda}{\tau} = \frac{\omega}{k}. \tag{1.49}$$

The equation $\omega = ck$ occurs again and again in wave theory.

The constant ε in eqn (1.47), or γ in eqn (48), is called the 'phase constant' of the wave. The significance of the phase constant is that at any fixed time, compared to the wave which has zero phase constant but which is otherwise identical, the original wave is shifted an amount equal to the phase constant divided by k along the direction of propagation but in the opposite sense. If the difference between the phase constants of two harmonic waves of the same frequency is equal to 0 or an even integral multiple of π, then the waves are said to be 'in phase'; otherwise they are 'out of phase'. If the difference is an odd integral multiple of π, then the waves are 'exactly out of phase'.

The reader must be familiar with yet another formulation of harmonic waves. This formulation depends upon (i) the equation

$$e^{i\theta} = \cos\theta + i\sin\theta$$

and (ii) the convention that, whenever a real physical quantity is represented in an equation as equal to a complex quantity, it is understood that the real part of the complex quantity is to be taken.

Equation (1.47) can be written as[†]

$$\begin{aligned} y &= a\,R[e^{i(kx-\omega t+\varepsilon)}] \\ &= R[ae^{i\varepsilon}\,e^{i(kx-\omega t)}] \\ &= R[Ae^{i(kx-\omega t)}], \end{aligned} \tag{1.50}$$

where

$$A = ae^{i\varepsilon}. \tag{1.51}$$

A is called the complex amplitude. Its modulus $|A|$ is equal to the amplitude a and its argument $\arg A$ is equal to the phase constant ε. By means of the convention, eqn (1.50) can be written

$$y = Ae^{i(kx-\omega t)}. \tag{1.52}$$

This is the complex form for a harmonic wave travelling in the positive sense. From eqn (1.48), for a harmonic wave travelling in the negative sense, the complex form is

$$y = Be^{i(kx+\omega t)}, \tag{1.53}$$

$$B = be^{-i\gamma}. \tag{1.54}$$

It is worth repeating that, for harmonic waves, k and ω are real but A and B may be complex.

The complex form of y can be used in any operation for which the real part of the result of the operation on the complex form of y is equal to the result of the operation on the real form of y. Such operations include addition, subtraction, multiplication or division by a real quantity (but not a complex quantity), and any combination of these operations such as integration or differentiation with respect to a real variable. However, it is not possible to multiply two complex forms together (or to square the same form of y) and then to interpret the real part of the product as the product of the real part; the real part of $(a+ib)$ times $(c+id)$ is $ac-bd$, which is only equal to ac if either b or d is zero, i.e. if one of the two factors of the product is real.

The concepts of amplitude, wavelength, period, frequency, wave-number and phase constant can be applied to any periodic wave, not only to harmonic waves.

The energies of a harmonic wave are now found. For a wave given

---

[†] $R[z]$ denotes the real part of z.

by eqn (1.47) with $\omega = ck$,

$$y = f(x - ct) = a\cos(k(x - ct) + \varepsilon)$$

and

$$y' = f'(x - ct) = -ka\sin(k(x - ct) + \varepsilon).$$

Use of eqns (1.34) and (1.35) with $g(x + ct) = 0$ shows that

$$K = V = \tfrac{1}{2}\rho c^2 k^2 a^2 \sin^2(k(x - ct) + \varepsilon).$$

The kinetic and potential energies per wavelength are obtained by integrating from any fixed value of x, say X, to $X + \lambda$. Now

$$\int_X^{X+\lambda} \sin^2(k(x - ct) + \varepsilon)\,\mathrm{d}x = \frac{1}{2}\int_X^{X+\lambda} (1 - \cos 2(k(x - ct) + \varepsilon))\,\mathrm{d}x$$

$$= \frac{\lambda}{2} - \frac{1}{4k}\sin 2(k(x - ct) + \varepsilon)\Bigg|_{x=X}^{x=X+\lambda} = \frac{\lambda}{2},$$

since

$$\sin 2(k(x - ct) + \varepsilon)\Bigg|_X^{X+\lambda} = \sin\left(\frac{4\pi}{\lambda}(x - ct) + 2\varepsilon\right)\Bigg|_X^{X+\lambda} = 0,$$

because $\sin(4\pi/\lambda)(x - ct)$ is periodic in x with period $\lambda/2$. Therefore, the kinetic and potential energies per wavelength are each equal to

$$\tfrac{1}{4}\rho c^2 k^2 a^2 \lambda = \tfrac{1}{4}\rho\omega^2\lambda a^2, \tag{1.55}$$

with various other alternative expressions.

Note that the complex forms of y, eqns (1.52) and (1.53), cannot be used for direct substitution into eqns (1.34) and (1.35) to calculate the energies of a travelling wave, because this would involve the product of two complex quantities.

What is the effect of superimposing two harmonic waves of equal amplitude and frequency but travelling in opposite directions? The displacement is given by summing the right-hand sides of eqns (1.47) and (1.48) with $b = a$:

$$y = a[\cos(kx - \omega t + \varepsilon) + \cos(kx + \omega t - \gamma)]$$

$$= 2a\cos(kx + \tfrac{1}{2}(\varepsilon - \gamma))\cos(\omega t - \tfrac{1}{2}(\varepsilon + \gamma)). \tag{1.56}$$

The displacements of all particles of the string are in phase. No wave moves along the string. The motion is called a 'standing' wave in contradistinction to the 'travelling' wave considered hereforeto. The points at which y is zero for all times are called 'nodes', points at which the amplitude of oscillation is a maximum are called 'antinodes'. The distance apart of two neighbouring nodes is π/k or $\lambda/2$, half a wavelength. The motion of the string between any two nodes would be unaffected if the two nodal points were fixed to rigid supports. The string is effectively one of finite length fixed at its two ends. Also, eqn (1.56) is of the form $y = X(x)\,\theta(t)$, i.e. a product of a function of x times a function of t. This suggests that if we look for solutions of the wave equation (1.11) by the method of separation of variables and impose the boundary conditions that $y=0$ at $x=0$ and at $x=l$, $l>0$, then we should arrive at solutions similar to eqn (1.56). We shall do this in Chapter 2.

Exercise 1.7. A harmonic wave of frequency 16 Hz (cycles per second) and of 2 mm amplitude travels along a string of density 4 g cm^{-1}, which is stretched to a tension of 25,600 dynes. Find the wavelength, velocity, period, wavenumber and average kinetic and potential energies per centimetre of the wave.

Exercise 1.8. Show that the expression (1.55) is valid when the harmonic wave is given by eqn (1.48), i.e. when the wave travels in the negative sense.

1.5. Reflection and transmission at a discontinuity of the string

So far the string has been assumed to be uniform along its length, i.e. ρ constant. In this section we consider the effect of a discontinuity at just one point, chosen as the origin. We take $\rho=\rho_1$ for $x<0$ and $\rho=\rho_2$ for $x>0$, where ρ_1 and ρ_2 are constants.

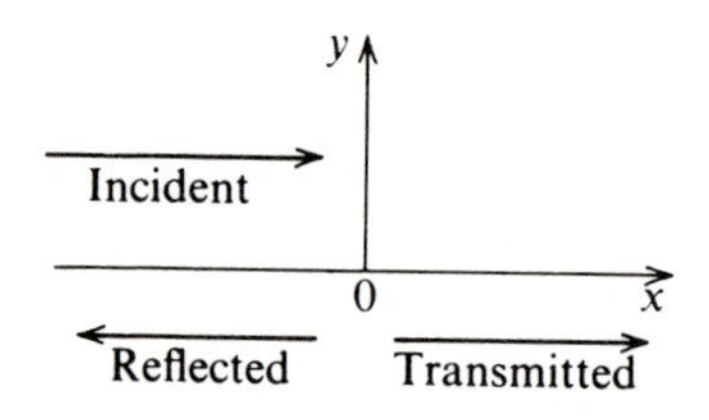

Fig. 1.5

When a wave is incident on a discontinuity (Fig. 1.5), one of three different effects may occur: (i) a wave may be transmitted in the string beyond the discontinuity; (ii) a wave may be reflected back from the discontinuity; (iii) both transmitted and reflected waves may be generated. However, it is not necessary to consider all three possibilities separately. It is sufficient to consider possibility (iii) because, if either (i) or (ii) is in fact true, then either the reflected or the transmitted wave respectively will be zero.

First consider the conditions to be satisfied at the origin. Since the string has no break, the value of the displacement at both sides of $x=0$ must be the same, i.e. y is continuous at $x=0$ for all t.

Since no mass is concentrated at the origin, by Newton's second law the force vectors due to the tension acting on either side of $x=0$ must have the same magnitude and direction but opposite senses. This requires both the tension T and the slope of the string to be the same on either side of the origin. This identity of slope is equivalent to requiring $\partial y/\partial x$ to be continuous at $x=0$ for all t.

Since the tension is the same on both sides of the origin but the density different, the wave velocity c changes at the origin:

$$c=c_1=\sqrt{\frac{T}{\rho_1}} \quad \text{for } x<0 \quad \text{and} \quad c=c_2=\sqrt{\frac{T}{\rho_2}} \quad \text{for } x>0. \tag{1.57}$$

Since the incident wave travels in the positive sense along that part of the string where the density is ρ_1 it is of the form†

$$y=f^*(t-x/c_1).$$

The reflected wave travels along the same part of the string but in the negative sense and has form

$$y=g^*(t+x/c_1).$$

The total displacement for $x<0$ is the sum of the displacements caused by the incident and reflected waves, i.e.

$$y=f^*(t-x/c_1)+g^*(t+x/c_1) \quad \text{for } x<0. \tag{1.58}$$

The transmitted wave travels in the positive sense along that part of the string where the density is ρ_2 and is the only wave on the string for $x>0$. Hence the displacement is of the form

$$y=h^*(t-x/c_2) \quad \text{for } x>0. \tag{1.59}$$

The continuity of y at $x=0$ requires that the values of y calculated from eqns (1.58) and (1.59) should be equal at $x=0$ for all t, i.e.

$$f^*(t)+g^*(t)=h^*(t). \tag{1.60}$$

† It is convenient when evaluating the conditions at $x=0$ to have replaced $f(x-ct)=f(-c(t-x/c))$ by $f^*(t-x/c)$, and $g(x+ct)$ by $g^*(t+x/c)$.

Similarly, the two values of $\partial y/\partial x$ calculated from eqns (1.58) and (1.59) should be equal at $x=0$ for all t, i.e.

$$\frac{\mathrm{d}f^*(t-x/c_1)}{\mathrm{d}(t-x/c_1)}\frac{\partial(t-x/c_1)}{\partial x}+\frac{\mathrm{d}g^*(t+x/c_1)}{\mathrm{d}(t+x/c_1)}\frac{\partial(t+x/c_1)}{\partial x}$$
$$=\frac{\mathrm{d}h^*(t-x/c_2)}{\mathrm{d}(t-x/c_2)}\frac{\partial(t-x/c_2)}{\partial x}\quad \text{at } x=0,$$

or

$$-\frac{1}{c_1}f^{*\prime}(t)+\frac{1}{c_1}g^{*\prime}(t)=-\frac{1}{c_2}h^{*\prime}(t),$$

where a prime denotes the derivative of a function with respect to its argument. Integrating this last equation with respect to t,

$$-\frac{1}{c_1}f^*(t)+\frac{1}{c_1}g^*(t)=-\frac{1}{c_2}h^*(t)+k. \tag{1.61}$$

We shall assume that there exists a time t' such that $f^*(t')$, $g^*(t')$ and $h^*(t')$ are all zero; with this physical assumption, $k=0$. Solving eqns (1.60) and (1.61) with $k=0$ for $g^*(t)$ and $h^*(t)$ in terms of $f^*(t)$,

$$g^*(t)=\frac{c_2-c_1}{c_2+c_1}f^*(t)\quad\text{and}\quad h^*(t)=\frac{2c_2}{c_2+c_1}f^*(t).$$

The incident wave $f^*(t-x/c_1)$ will be given. The reflected and transmitted waves are found from the above equations as

$$g^*(t+x/c_1)=\frac{c_2-c_1}{c_2+c_1}f^*(t+x/c_1) \tag{1.62}$$

and

$$h^*(t-x/c_2)=\frac{2c_2}{c_2+c_1}f^*(t-x/c_2). \tag{1.63}$$

If the incident wave is harmonic with radian frequency ω,

$$f^*(t-x/c_1)=A_1\mathrm{e}^{\mathrm{i}\omega(t-x/c_1)}$$
$$=A_1\mathrm{e}^{\mathrm{i}(\omega t-k_1x)}, \tag{1.64}$$

where

$$k_1=\omega/c_1. \tag{1.65}$$

Substitute $f^*(t) = A_1 e^{i\omega t}$ into eqns (1.62) and (1.63):

$$g^*(t + x/c_1) = \frac{c_2 - c_1}{c_2 + c_1} A_1 e^{i\omega(t + x/c_1)}$$

$$= B_1 e^{i(\omega t + k_1 x)} \quad \text{where } B_1 = \frac{c_2 - c_1}{c_2 + c_1} A_1, \tag{1.66}$$

and

$$h^*(t - x/c_2) = \frac{2c_2}{c_2 + c_1} A_1 e^{i\omega(t - x/c_2)}$$

$$= A_2 e^{i(\omega t - k_2 x)} \quad \text{where } A_2 = \frac{2c_2}{c_2 + c_1} A_1 \tag{1.67}$$

and

$$k_2 = \omega / c_2. \tag{1.68}$$

Equations (1.66) and (1.67) are the equations of the reflected and transmitted harmonic waves, respectively.

The reflected and transmitted waves have the same frequency as the incident wave. If the zero of time is chosen so that the coefficient of the incident wave, A_1, is real and positive, then the coefficient of the transmitted wave, A_2, is real and positive. The coefficient of the transmitted wave, B_1, is real; it is positive if $c_2 > c_1$, i.e. if $\rho_1 > \rho_2$, and negative if $\rho_2 > \rho_1$. Hence, if the incident wave is on the denser part of the string, the transmitted and reflected waves are both in phase with the incident wave; otherwise the transmitted wave is in phase and the reflected wave is exactly out of phase with the incident wave. Note that A_2 is only zero if $c_2 = 0$† and that B_1 is only zero if $\rho_1 = \rho_2$. But, if $\rho_1 = \rho_2$, there is no discontinuity and the incident wave travels on unaltered.

The 'coefficient of reflection' R is defined as the ratio $|B_1/A_1|^2$. Since the total energy of a progressive wave is proportional to the square of its amplitude at constant density, eqn (1.55), the coefficient of reflection is the ratio of the reflected to the incident total energy per unit length.

Substituting from eqns (1.66) and (1.57),

$$R = |B_1/A_1|^2 = \left(\frac{c_2 - c_1}{c_2 + c_1}\right)^2 = \left(\frac{\sqrt{\rho_1} - \sqrt{\rho_2}}{\sqrt{\rho_1} + \sqrt{\rho_2}}\right)^2. \tag{1.69}$$

† $c_2 \to 0$ as $\rho_2 \to \infty$. This is equivalent to a semi-infinite string $x < 0$ with the end $x = 0$ fixed.

The 'coefficient of transmission' T^* is defined as $1-R$. Since no energy is dissipated, T^* must be equal to the ratio of the energies carried by the transmitted and incident waves. Note that we have talked of total energies, rather than of kinetic and potential energies. The reason derives from the conclusions in the last paragraph of Section 1.3. It is only the total energy of a wave that is unaltered by the presence of another wave travelling in the opposite sense on the same part of the string.

An example will now be treated in detail. The density of a string changes from $4\rho'$ $(x<0)$ to ρ' $(x>0)$ at the origin. At time $t=0$, a wave is travelling in the positive sense with displacement given by

$$y=\begin{cases} 0 & \text{for } x>-\pi, \\ \dfrac{1}{100}\sin^2 x & \text{for } -\pi>x>-2\pi, \\ 0 & \text{for } x<-2\pi. \end{cases}$$

If $c'=\sqrt{(T/\rho')}$, where T is the tension of the string, find the displacement of the string at time $t=6\pi/c'$. Sketch the displacement and indicate the waves present at this time.

The initial displacement is shown in Fig. 1.6. Since the wave is moving in the positive sense, at $t=0$ there is an incident wave but there are no reflected or transmitted waves. At $t=0$, the incident wave formula $y=f^*(t-(x/c_1))$ reduces to $y=f^*(-x/c_1)$. Using the initial data and the Heaviside unit function,

$$f^*\left(-\frac{x}{c_1}\right)=\frac{1}{100}\sin^2 x\,[H(x+2\pi)-H(x+\pi)].$$

Expressing the right-hand side as a function of $-x/c_1$,

$$f^*\left(-\frac{x}{c_1}\right)=-\frac{1}{100}\sin^2\left(c_1\left(-\frac{x}{c_1}\right)\right)\left[H\left(2\pi-c_1\left(-\frac{x}{c_1}\right)\right)\right.$$
$$\left.-H\left(\pi-c_1\left(-\frac{x}{c_1}\right)\right)\right]. \tag{1.70}$$

The incident wave is therefore

$$f^*\left(t-\frac{x}{c_1}\right)=-\frac{1}{100}\sin^2\left(c_1\left(t-\frac{x}{c_1}\right)\right)\left[H\left(2\pi-c_1\left(t-\frac{x}{c_1}\right)\right)\right.$$
$$\left.-H\left(\pi-c_1\left(t-\frac{x}{c_1}\right)\right)\right]. \tag{1.71}$$

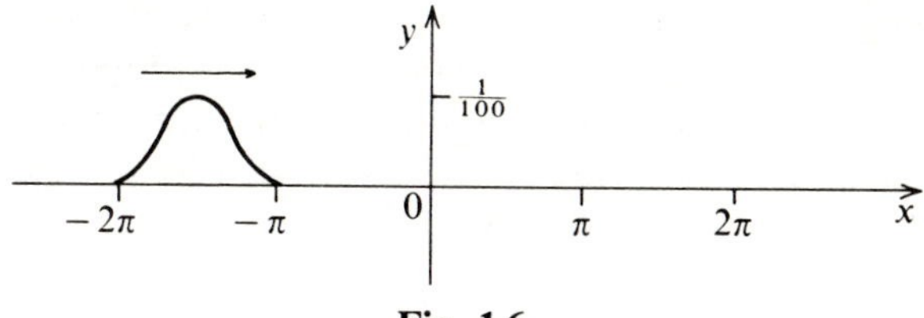

Fig. 1.6

The reflected and transmitted waves are given by eqns (1.62), (1.63) and (1.70) as

$$g^*\left(t+\frac{x}{c_1}\right)=\frac{c_2-c_1}{c_2+c_1}\,f^*\left(t+\frac{x}{c_1}\right)$$

$$=-\frac{1}{100}\frac{c_2-c_1}{c_2+c_1}\sin^2\left(c_1\left(t+\frac{x}{c_1}\right)\right)\left[H\left(2\pi-c_1\left(t+\frac{x}{c_1}\right)\right)\right.$$

$$\left.-H\left(\pi-c_1\left(t+\frac{x}{c_1}\right)\right)\right] \tag{1.72}$$

and

$$h^*\left(t-\frac{x}{c_2}\right)=\frac{2c_2}{c_2+c_1}\,f^*\left(t-\frac{x}{c_2}\right)$$

$$=-\frac{1}{100}\frac{2c_2}{c_2+c_1}\sin^2\left(c_1\left(t-\frac{x}{c_2}\right)\right)\left[H\left(2\pi-c_1\left(t-\frac{x}{c_2}\right)\right)\right.$$

$$\left.-H\left(\pi-c_1\left(t-\frac{x}{c_2}\right)\right)\right]. \tag{1.73}$$

Now $c_1=\sqrt{(T/\rho_1)}=\sqrt{(T/4\rho')}=\frac{1}{2}c'$ and $c_2=\sqrt{(T/\rho_2)}=\sqrt{(T/\rho')}=c'$. Substitute $c_1=\frac{1}{2}c'$, $c_2=c'$ and $t=6\pi/c'$ into eqns (1.71), (1.72) and (1.73) to find the waves at time $6\pi/c'$:

$$f^*\left(\frac{6\pi-2x}{c'}\right)=-\frac{1}{100}\sin^2(3\pi-x)[H(x-\pi)-H(x-2\pi)],$$

$$g^*\left(\frac{6\pi+2x}{c'}\right)=-\frac{1}{300}\sin^2(3\pi+x)[\{H(-x-\pi)-H(-x-2\pi)\}]$$

and

$$h^*\left(\frac{6\pi-x}{c'}\right)=-\frac{4}{300}\sin^2\left(3\pi-\frac{x}{2}\right)\left(H\left(\frac{x}{2}-\pi\right)-H\left(\frac{x}{2}-2\pi\right)\right). \tag{1.74}$$

The incident wave travels on that part of the string for which $x<0$. It is seen from the first of eqns (1.74) that $f^*((6\pi-2x)/c')$ is zero for all $x<0$ and therefore there is no incident wave at time $t = 6\pi/c'$. The reflected wave also travels on the part $x<0$, and it can be seen from the second of eqns (1.74) that

$$g^*\left(\frac{6\pi+2x}{c'}\right)=\begin{cases}-\dfrac{1}{300}\sin^2(3\pi+x) & \text{if } -2\pi<x<-\pi,\\ 0 & \text{if } x>-\pi \text{ or if } x<-2\pi.\end{cases}$$

Since $f^*((6\pi-2x)/c')=0$, the displacement of the string for $x<0$ is equal to $g^*((6\pi+2x)/c')$. The transmitted wave travels on that part of the string for which $x>0$. It is seen from the last of eqns (1.74) that

$$h^*\left(\frac{6\pi-x}{c'}\right)=\begin{cases}-\dfrac{4}{300}\sin^2\left(3\pi-\dfrac{x}{2}\right) & \text{if } 2\pi<x<4\pi,\\ 0 & \text{if } x<2\pi \text{ or } x>4\pi.\end{cases}$$

$h^*((6-x)/c')$ is also the displacement of the string for $x>0$.

Figure 1.7 shows the displacement of the string and the direction of motion of the waves at time $t=6\pi/c'$. Since the velocity of the transmitted wave is twice that of the incident wave, the length of the 'pulse' is doubled. Since the ratio of the amplitudes of the reflected wave and of the initial incident wave is 1/3, the coefficient of reflection R is 1/9 and the coefficient of transmission T^* is 8/9. The fact that the transmitted wave travels on a portion of the string that is four times less dense than that on which the incident wave travels, more than counterbalances the additional pulse length and greater amplitude of the transmitted wave compared with the incident wave, when the total energy is calculated.

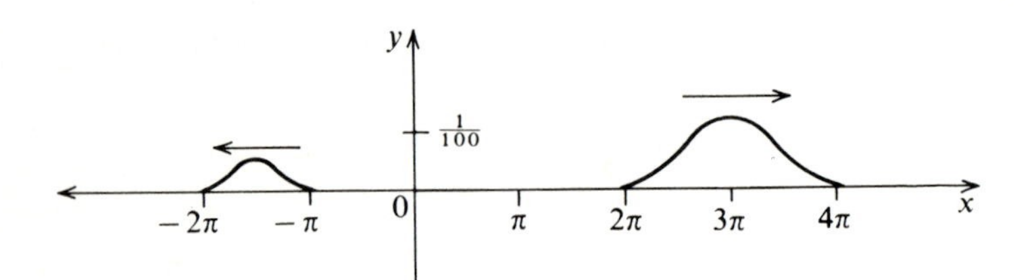

Fig. 1.7

Exercise 1.9. The density of a string changes from $(9/4)\rho'$ to ρ' at the origin. At time $t=0$ a wave is travelling in the positive sense with displacement given by

$$y=\begin{cases} 0 & \text{for } x>-1, \\ -\dfrac{1}{50}(x+1)^2(x+2)^2 & \text{for } -2<x<-1, \\ 0 & \text{for } x<-2. \end{cases}$$

If $c'=\sqrt{(T/\rho')}$, find the displacement of the string at time $t=6/c'$. Sketch the displacement and indicate the waves present at this time.

1.6. Reflection and transmission when a mass is attached to the string

In this section the density of the string is constant everywhere but a mass M is attached at the origin. Since there is no break in the string, y is continuous at $x=0$. Assuming the mass and string are at rest under tension T_0 before the arrival of the incident wave, the analysis of the first section shows that the tension on either side remain equal to T_0 (Fig. 1.8). The equation of motion in the x-direction of the mass gives

$$M(\text{acc})_x = T_0\cos\Psi_+ - T_0\cos\Psi_- = T_0 - T_0 = 0$$

on neglecting Ψ_+^2 and Ψ_-^2 compared to 1. Hence, the x-component of acceleration of the mass is zero and, since its displacement and velocity are initially zero, its x-component of displacement remains

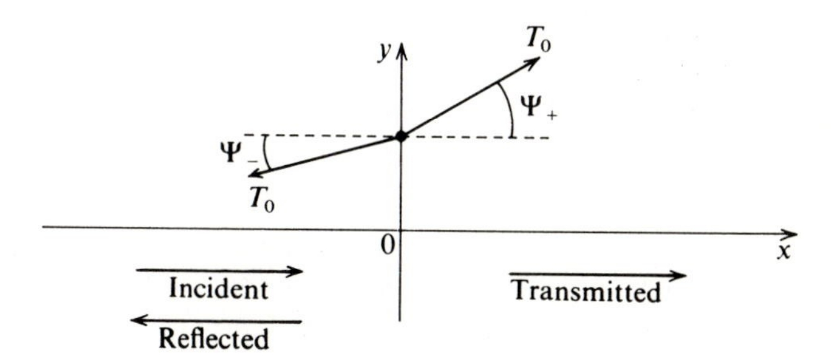

Fig. 1.8

zero. The equation of motion in the y-direction of the mass is

$$M\left(\frac{\partial^2 y}{\partial t^2}\right)_{y=0} = T_0 \sin \Psi_+ - T_0 \sin \Psi_-$$

$$= T_0\left(\frac{\partial y}{\partial x}\right)_+ - T_0\left(\frac{\partial y}{\partial x}\right)_-. \tag{1.75}$$

A suffix + denotes a quantity just on the positive side of the origin and a suffix − denotes one just on the negative side.

Since there is no change of density or tension at the origin, we take the incident, reflected and transmitted waves to be

$$f^*\left(t - \frac{x}{c}\right), \quad g^*\left(t + \frac{x}{c}\right) \quad \text{and} \quad h^*\left(t - \frac{x}{c}\right)$$

respectively. The displacement of the string is given by

$$y = f^*\left(t - \frac{x}{c}\right) + g^*\left(t + \frac{x}{c}\right) \quad \text{for } x < 0, \tag{1.76}$$

and

$$y = h^*\left(t - \frac{x}{c}\right) \quad \text{for } x > 0. \tag{1.77}$$

Since y is continuous at $x = 0$,

$$f^*(t) + g^*(t) = h^*(t). \tag{1.78}$$

Substituting from eqns (1.76) and (1.77) into eqn (1.75),

$$-\frac{T}{c}h^{*\prime}(t) + \frac{T}{c}f^{*\prime}(t) - \frac{T}{c}g^{*\prime}(t) = M h^{*\prime\prime}(t).$$

Integrating once with respect to t,

$$-h^*(t) + f^*(t) - g^*(t) = \frac{cM}{T}h^{*\prime}(t) + K. \tag{1.79}$$

As after eqn (1.61), we make the physical assumption that there exists a time t' for which $h^*(t') = h^{*\prime}(t') = g^*(t') = f^*(t') = 0$. Hence $K = 0$. Eliminating $g^*(t)$ between eqns (1.78) and (1.79),

$$\frac{cM}{T}h^{*\prime}(t) + 2h^*(t) = 2f^*(t).$$

The integrating factor for this equation is $e^{\alpha t}$, where

$$\alpha = 2T/cM; \tag{1.80}$$

the solution is $h^*(t) = \alpha e^{-\alpha t} \int e^{\alpha t} f^*(t)\, dt$.

The constant of integration is determined by the fact that $h^*(t)$ is zero for all times prior to that time, t'' say, at which the incident wave first strikes the mass. But $f^*(t)$ is also zero for such times. The condition on $h^*(t)$ will therefore be satisfied if the constant of integration is chosen so that

$$h^*(t) = \alpha e^{-\alpha t} \int_{-\infty}^{t} e^{\alpha \tau} f^*(\tau)\, d\tau,$$

because $f^*(\tau)$ is zero for $\tau < t''$.

Substituting for $h^*(t)$ into eqn (1.78),

$$g^*(t) = h^*(t) - f^*(t) = \alpha e^{-\alpha t} \int_{-\infty}^{t} e^{\alpha \tau} f^*(\tau)\, d\tau - f^*(t).$$

The transmitted and reflected waves are

$$h^*\left(t - \frac{x}{c}\right) = \alpha e^{-\alpha(t - x/c)} \int_{-\infty}^{t - x/c} e^{\alpha \tau} f^*(\tau)\, d\tau \tag{1.81}$$

and

$$g^*\left(t + \frac{x}{c}\right) = \alpha e^{-\alpha(t + x/c)} \int_{-\infty}^{t + x/c} e^{\alpha \tau} f^*(\tau)\, d\tau - f^*(t + x/c). \tag{1.82}$$

If the incidental wave is harmonic with angular frequency ω,

$$f^*(t - x/c) = A_1 e^{i\omega(t - x/c)}. \tag{1.83}$$

The zero of time is chosen so that A_1 is real and positive.

Substitute $f^*(t) = A_1 e^{i\omega t}$ in eqn (1.81):

$$h^*\left(t - \frac{x}{c}\right) = \alpha e^{-\alpha(t - x/c)} \int_{-\infty}^{t - x/c} A_1 e^{(\alpha + i\omega)\tau}\, d\tau$$

$$= A_2 e^{i\omega(t - x/c)} \quad \text{where } A_2 = \frac{A_1}{1 + i\omega/\alpha}; \tag{1.84}$$

since $\alpha > 0$, $e^{(\alpha + i\omega)t} \to 0$ as $t \to -\infty$.

$$g^*\left(t + \frac{x}{c}\right) = h^*(t + x/c) - f^*(t + x/c)$$

$$= B_1 e^{i\omega(t + x/c)} \quad \text{where } B_1 = A_2 - A_1 = \frac{-i\omega/\alpha}{1 + i\omega/\alpha} A_1. \tag{1.85}$$

The reflected and transmitted waves both have angular frequency ω. A_2 and B_1 are both complex and therefore both reflected and transmitted waves are out of phase with the incident wave. Use of eqns (1.54) and (1.51) shows that their amplitudes are $|B_1|$ and $|A_2|$ and their phase angles $-\arg B_1$ and $\arg A_2$, both pairs respectively. The coefficients of reflection and transmission, R and T^* are given by

$$R = \left|\frac{B_1}{A_1}\right|^2 = \frac{\omega^2/\alpha^2}{1 + \omega^2/\alpha^2} \quad \text{and} \quad T^* = 1 - R = \frac{1}{1 + \omega^2/\alpha^2}. \tag{1.86}$$

Exercise 1.10. A string of density $4\,\mathrm{g\,cm^{-1}}$, stretched to a tension of 57,600 dynes, has a mass of $24/\pi$ g attached at the origin. A harmonic wave of frequency 20 Hz and amplitude 0.5 mm travels along the string in the positive sense and is incident on the mass. Find the transmitted and reflected waves and the coefficient of reflection.

1.7. Effect of air resistance on travelling waves

Almost all terrestrial waves die out unless maintained by external force. The energy of the wave is dissipated by various mechanisms. In the case of transverse waves on strings, the mechanisms are internal friction (viscoelasticity) of the string, damping of the supports, and air resistance. Easiest to treat mathematically is linearized air resistance, and we therefore commence this section by considering a string that is subject at each point to a resistive force per unit mass that is proportional to its transverse velocity at that point.

In Fig. 1.2, the element of string of mass $\rho\,\delta x$ has transverse velocity $\partial y/\partial t$. The resistive force is therefore $-\kappa\rho\,\delta x(\partial y/\partial t)$, where κ is the positive constant of proportionality; the minus sign is inserted because the resistive force acts in the opposite sense to the transverse velocity. The equation of motion in the y-direction of the element becomes

$$\rho\,\delta x \frac{\partial^2 y}{\partial t^2} = T_0 \sin(\Psi + \delta\Psi) - T_0 \sin\Psi - \kappa\rho\,\delta x \frac{\partial y}{\partial t},$$

leading, by the same argument as before, to

$$\frac{\partial^2 y}{\partial t^2}+\kappa\frac{\partial y}{\partial t}=c^2\frac{\partial^2 y}{\partial x^2}. \tag{1.87}$$

In the absence of air resistance, $\kappa=0$, and the equation reduces to the wave equation. It has already been shown that harmonic waves of the form of eqns (1.52) and (1.53) travel along the string, whose transverse motion is governed by the wave equation. The right-hand sides of both these equations are the form $e^{i\omega t}$ times a function of x. This suggests that if we substitute

$$y=X(x)e^{i\omega t}, \tag{1.88}$$

in eqn (1.87), solve for $X(x)$, and substitute back into eqn (1.88), we might obtain a solution which can be interpreted in terms of transverse waves travelling along a string in the presence of air resistance. A wave whose time dependence is of the form $e^{i\omega t}$ (or $\cos\omega t$ or $\sin\omega t$) is said to be sinusoidal.

Substituting from eqn (1.88) into eqn (1.87):

$$(i\omega)^2e^{i\omega t}X(x)+\kappa(i\omega)e^{i\omega t}X(x)=c^2e^{i\omega t}\frac{d^2X(x)}{dx^2},$$

or

$$\frac{d^2X}{dx^2}+\frac{\omega^2-i\kappa\omega}{c^2}X=0. \tag{1.89}$$

The solution of this equation can be written

$$X=Ae^{\lambda x}+Be^{-\lambda x}, \tag{1.90}$$

where A and B are the constants of integration and

$$\lambda^2+\frac{\omega^2-i\kappa\omega}{c^2}=0.$$

Put $\lambda=a+ib$ where a and b are real. Then

$$a^2-b^2=-\frac{\omega^2}{c^2}\quad\text{and}\quad 2ab=\frac{\kappa\omega}{c^2}.$$

Solving for a and b and substituting back in $\lambda=a+ib$,

$$\lambda=\pm(\mu+i\omega/V), \tag{1.91}$$

where

$$\mu = \frac{\omega}{\sqrt{2}c}\left[\left(1+\frac{\kappa^2}{\omega^2}\right)^{\frac{1}{2}} - 1\right]^{\frac{1}{2}} \tag{1.92}$$

and

$$\frac{1}{V} = \frac{1}{\sqrt{2}c}\left[\left(1+\frac{\kappa^2}{\omega^2}\right)^{\frac{1}{2}} + 1\right]^{\frac{1}{2}}. \tag{1.93}$$

The positive sign is taken in all square roots. Substitute back for λ into eqn (1.90) and then for X in eqn (1.88):

$$y = A\mathrm{e}^{-\mu x}\mathrm{e}^{\mathrm{i}\omega(t - x/V)} + B\mathrm{e}^{\mu x}\mathrm{e}^{\mathrm{i}\omega(t + x/V)}. \tag{1.94}$$

Compare the first term of eqn (1.94) with eqn (1.52). It is seen that the first term represents a harmonic wave of angular frequency ω travelling with velocity V in the positive sense, except that its amplitude decreases exponentially with the distance travelled. Comparison of the second term of eqn (1.94) with eqn (1.53) shows that the second term represents the same type of wave as the first term but travelling in the negative sense.

A wave is said to be attenuated if its amplitude decreases with distance travelled. If the decrease is exponential, then the reciprocal of the distance to be travelled for the wave amplitude to be decreased by a factor e^{-1} is called the attenuation. It has dimensions $(\text{length})^{-1}$. For the waves in eqn (1.94), x must change by an amount $1/\mu$ in the direction of travel for the amplitude to decrease by a factor e^{-1}. Since $(1/\mu)^{-1} = \mu$, μ is the attenuation. Equation (1.92) shows that the attenuation is frequency dependent.

Equation (1.93) shows that the velocity of the waves is also frequency dependent. Such waves are said to be dispersed. The dependence of velocity on frequency is known as dispersion. The reader should note that, from eqns (1.52) and (1.53), the waves governed by the classical wave equation (1.11) are both non-attenuated and non-dispersed.

If the air resistance is sufficiently small for κ^2/ω^2 to be neglected compared to 1, then eqns (1.93) and (1.94) can be simplified:

$$\mu = \frac{\omega}{\sqrt{2}c}\left(1 + \frac{1}{2}\frac{\kappa^2}{\omega^2} + \mathrm{O}\left(\frac{\kappa^4}{\omega^4}\right) - 1\right)^{\frac{1}{2}} = \frac{\kappa}{2c}$$

and

$$V = \sqrt{2}c\left(1 + \mathrm{O}\left(\frac{\kappa^2}{\omega^2}\right) + 1\right)^{-\frac{1}{2}} = c.$$

To the accuracy of the approximation used ($\kappa^2/\omega^2 \ll 1$), both the attenuation μ and the velocity V are independent of frequency. Furthermore, the wave velocity is the same as for the corresponding harmonic wave. The difference is that the waves are still attenuated, albeit with a small attenuation constant $\kappa/2c$.

Suppose a signal is sent along a string by feeding in a periodic transverse vibration at one end and that this signal is the sum of a number of sinusoidal components of different frequencies. If waves of different frequencies travel at the same velocity and with the same attenuation, then the signal received at the far end will be the same shape as that transmitted, though possibly attenuated. If, however, the waves are dispersed, i.e. if each component travels at a different speed, then when the components are recombined at the far end the original shape will in general not be reproduced; the signal is distorted. The situation is summarized in Table 1.2.

Table 1.2

Air resistance, k	Attenuation	Velocity	Distortion
None	None	Independent of frequency	None
$k \neq 0$ but $k^2 \ll 1$	Present but independent of frequency	Independent of frequency	None
$k \neq 0$ and k^2 not $\ll 1$	Present and dependent on frequency	Dependent on frequency	Present

Further exercises

Exercise 1.11. This exercise is very difficult.

Suppose that $(\partial y/\partial x)^2$ can be neglected compared to 1, except where $(\partial y/\partial x)^2$ is multiplied by AE/T_0 and that $\partial u/\partial x$ is of the same order as $(\partial y/\partial x)^2$. Show that the governing equations for y and u are

$$\frac{1}{c^2}\frac{\partial^2 y}{\partial t^2} = \frac{\partial^2 y}{\partial x^2} + \frac{EA}{T_0}\frac{\partial}{\partial x}\left(\frac{\partial u}{\partial x}\frac{\partial y}{\partial x} + \frac{1}{2}\left(\frac{\partial y}{\partial x}\right)^3\right)$$

and

$$\frac{1}{c_l^2}\frac{\partial^2 u}{\partial t^2} = \frac{\partial^2 u}{\partial x^2} + \frac{\partial y}{\partial x}\frac{\partial^2 y}{\partial x^2}$$

N.B. The price paid for allowing $(\partial y/\partial x)^2$ to be larger is that T is no longer effectively constant.

Exercise 1.12. By letting $\rho_2 \to \infty$, show that the wave reflected back along a string at a fixed end has the same shape and amplitude as the incident wave but is opposite in sign.

Exercise 1.13. Verify that if $\rho_2 = \rho_1$ in Section 1.5, the reflected wave is zero and the transmitted wave is a continuation of the incident wave.

Exercise 1.14. From eqns (1.4) and (1.5), write down D'Alembert's solution for longitudinal waves and determine the wave velocity. What are the initial conditions required to find the solution at later times?

Answers to exercises

1.4. Time at separation $= 1/c$. Figure 1.9 is drawn for $t = 1/c$.

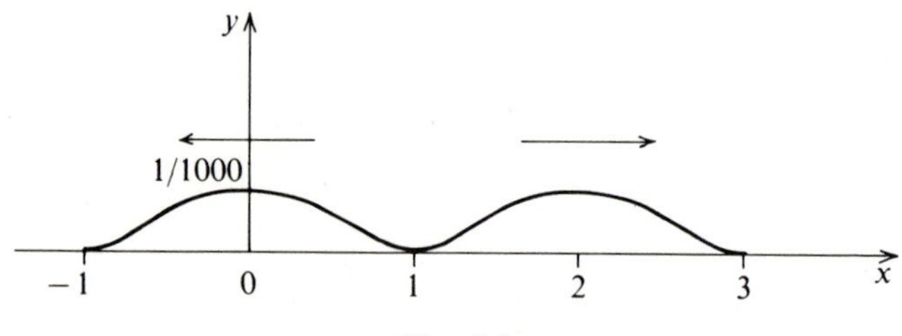

Fig. 1.9

1.5. $\phi(x) = \dfrac{1}{500}\sin x$; $\chi(x) = -\dfrac{c}{500}\cos x$.

1.7. $\lambda = 5$ cm; $c = 80$ cm s^{-1}; $\tau = 1/16$ s; $k = 2\pi/5$ cm^{-1}; average kinetic and potential energies per cm each equal 404.3 ergs cm^{-1}.

1.9. $y = 0$ for $x \leqslant -3$, $-2 \leqslant x \leqslant 3$ and $x \geqslant 9/2$;

$$y = -\frac{1}{250}(x+3)(x+2) \qquad \text{for } -3 \leqslant x \leqslant -2;$$

$$y = -\frac{4}{375}(x-3)(x-9/2) \qquad \text{for } 3 \leqslant x \leqslant 9/2.$$

1.10. Taking the incident wave as $y = 0.05e^{2\pi i(20t - x/6)}$, the reflected wave is $y = -0.025(1+i)e^{2\pi i(20t + x/6)}$ and the transmitted wave is $y = 0.025(1 - i)e^{2\pi i(20t - x/6)}$, $R = \frac{1}{2}$.

1.14. Longitudinal displacement and velocity given at the initial time.

2 Transverse vibrations on strings

In this chapter we shall look for solutions of the wave equation (1.11) by the method of separation of variables, when the string of length l is fixed at both ends. The origin is chosen so that the boundary conditions are

$$y=0 \quad \text{at} \quad x=0 \quad \text{and} \quad x=l. \tag{2.1}$$

The results will be interpreted in terms of the transverse vibrations of strings, the major application, but they also apply to longitudinal vibrations provided the constant $c=\sqrt{(T_0/\rho)}$ is replaced by $c_1=\sqrt{(E/\rho)}$.

2.1. Solution by separation of variables with boundary conditions eqn (2.1)

Look for solutions to the wave equation (1.11) of the form

$$y=X(x)\,\theta(t), \tag{2.2}$$

where X is a function of x only and θ is a function of t only. On substitution into

$$\frac{\partial^2 y}{\partial x^2}=\frac{1}{c^2}\frac{\partial^2 y}{\partial t^2}, \tag{1.11}$$

$$\theta\frac{\mathrm{d}^2 X}{\mathrm{d}x^2}=\frac{1}{c^2}X\frac{\mathrm{d}^2\theta}{\mathrm{d}t^2}, \tag{2.3}$$

because $\partial/\partial x$ operates only on X and $\partial/\partial t$ only on θ; since X is a function of x only, $\partial X/\partial x$ can be written $\mathrm{d}X/\mathrm{d}x$ and similarly $\partial\theta/\partial t$ can be written $\mathrm{d}\theta/\mathrm{d}t$. Divide eqn (2.3) through by $X\theta$:

$$\frac{1}{X}\frac{\mathrm{d}^2 X}{\mathrm{d}x^2}=\frac{1}{c^2\theta}\frac{\mathrm{d}^2\theta}{\mathrm{d}t^2}. \tag{2.4}$$

In eqn (2.4), since the l.h.s. is a function of x only and the r.h.s. of t only, the variables have been separated. Since the two sides are equal, they must depend on the same variables. But the l.h.s. does not depend on θ and the r.h.s. does not depend on x; the only possibility remaining is that both sides are constant. It is convenient to choose the constant as $-p^2$, i.e.

$$\frac{1}{X}\frac{d^2X}{dx^2}=\frac{1}{c^2\theta}\frac{d^2\theta}{dt^2}=-p^2, \tag{2.5}$$

or

$$\frac{d^2X}{dx^2}+p^2X=0$$

and

$$\frac{d^2\theta}{dt^2}+c^2p^2\theta=0.$$

The solutions are

$$X=A'\cos px+B'\sin px$$

and

$$\theta=C'\cos pct+D'\sin pct,$$

where A', B', C' and D' are constants. Substituting back into eqn (2.2),

$$y=A\sin px\cos pct+B\sin px\sin pct+C\cos px\cos pct$$
$$+D\cos px\sin pct \tag{2.6}$$

Equation (2.6) is the solution of wave equation (1.11) by separation of variables: p, A, B, C and D are arbitrary constants. Since the sum of solutions is also a solution, a more general solution is

$$y=\sum_p(A_p\sin px\cos pct+B_p\sin px\sin pct$$
$$+C_p\cos px\cos pct+D_p\cos px\sin pct), \tag{2.7}$$

where $\sum_p$ denotes a sum over any finite or infinite set of constants p and a different set of constants A_p, B_p, C_p and D_p exists for each value of p.

We now impose the boundary conditions $y=0$ at $x=0$ and at $x=l$ for all t on eqn (2.6). For $y=0$ at $x=0$ for all t,

$$0=C\cos pct+D\sin pct.$$

For this equation to hold for all t, $C=D=0$. Hence

$$y=\sin px(A\cos pct+B\sin pct). \tag{2.8}$$

For $y=0$ at $x=l$ for all t,

$$0=\sin pl\,(A\cos pct+B\sin pct).$$

Hence either $A\cos pct+B\sin pct=0$ for all t or $\sin pl=0$.

If $A\cos pct+B\sin pct=0$ for all t, $A=B=0$. Substitution back into eqn (2.8) gives $y=0$ for all x and for all t. This solution does not represent a deformation, static or dynamic, of the string and so it is rejected. There remains $\sin pl=0$ or

$$pl=n\pi, \quad \text{where } n \text{ is an integer.}$$

If $n=0$, $p=0$, and from eqn (2.8) $y\equiv 0$. This solution is again rejected. If n is a positive integer,

$$p=\frac{n\pi}{l}, \quad n+\text{ve integer}, \tag{2.9}$$

and from eqn (2.8),

$$y=\sin\frac{n\pi x}{l}\left(A_n\cos\frac{n\pi ct}{l}+B_n\sin\frac{n\pi ct}{l}\right), \quad n+\text{ve integer}, \tag{2.10}$$

where the A_n and B_n are arbitrary constants. If n is a negative integer, put $n=-m$ so that $p=-m\pi/l$, $\sin(-(m\pi/l)x)=-\sin(m\pi x/l)$ and

$$y=-\sin\frac{m\pi x}{l}\left(A_{-m}\cos\frac{m\pi ct}{l}-B_{-m}\sin\frac{m\pi ct}{l}\right);$$

but this solution is identical to eqn (2.10) since A_{-m} and B_{-m} are arbitrary constants. Nothing new is introduced by taking n a negative integer. We therefore restrict n to being a positive integer. Since the sum of solutions is also a solution, the most general solution obtained by separation of variables to the wave equation when $y=0$ at $x=0$ and at $x=l$ is

$$y=\sum_{n=1}^{\infty}\sin\frac{n\pi x}{l}\left(A_n\cos\frac{n\pi ct}{l}+B_n\sin\frac{n\pi ct}{l}\right). \tag{2.11}$$

It will be shown in Section 2.5 that this solution is both general and unique.

2.2. Normal modes

Consider the displacement in the particular case of all pairs of A_n and B_n zero except for one pair; then, from eqn (2.11),

$$y = \sin\frac{n\pi x}{l}\left(A_n \cos\frac{n\pi ct}{l} + B_n \sin\frac{n\pi ct}{l}\right), \tag{2.12}$$

where n is any positive integer. Let us interpret (A_n, B_n) temporarily as cartesian coordinates of a point in a plane. Let (C_n, ε_n) be the polar coordinates of this point.

Then

$$C_n = +\sqrt{A_n^2 + B_n^2}, \tag{2.13}$$

and

$$\varepsilon_n = \tan^{-1} B_n/A_n. \tag{2.14}$$

If $B_n > 0$, $0 < \varepsilon_n < \pi$; if $B_n < 0$, $\pi < \varepsilon_n < 2\pi$; if $B_n = 0$ and $A_n > 0$, $\varepsilon_n = 0$; and if $B_n = 0$ and $A_n < 0$, $\varepsilon_n = \pi$. Conversely

$$A_n = C_n \cos\varepsilon_n \tag{2.15}$$

and

$$B_n = C_n \sin\varepsilon_n. \tag{2.16}$$

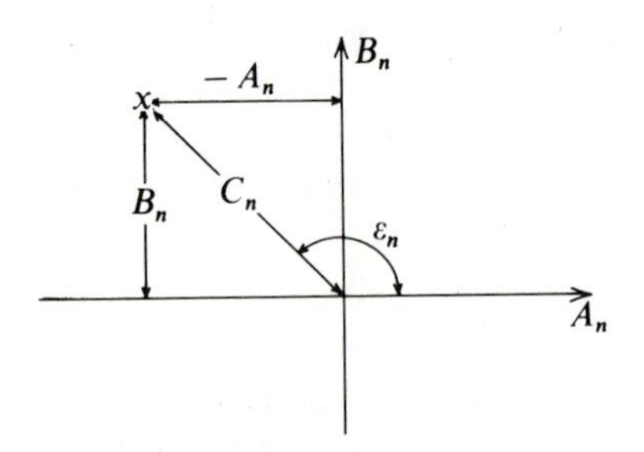

Fig. 2.1

Substituting into eqn (2.12),

$$y = C_n \sin\frac{n\pi x}{l}\cos\left(\frac{n\pi ct}{l} - \varepsilon_n\right). \tag{2.17}$$

Equation (2.17) states the transverse displacement at any point x on the string varies sinusoidally with time and has amplitude $C_n \sin(n\pi x/l)$ and cycle frequency $nc/2l$. Furthermore, the transverse displacements at different points are in phase, they are all zero when $n\pi ct/l = \varepsilon_n + [(2N+1)/2]\,\pi$, N integral, and all have their greatest absolute value when $n\pi ct/l = \varepsilon_n + N\pi$. When a string oscillates in such a way that the transverse displacements at all points along its length are in phase, then it is said to be oscillating in a ‘normal mode'. The mode corresponding to a particular value of n is known as the 'nth normal mode'. The mode corresponding to $n=1$ is called the 'fundamental mode'. C_n and ε_n are known as the 'normal amplitude' and the 'phase constant' of the nth normal mode respectively.

The cyclic frequency ν_1 of the fundamental mode is given by $\nu_1 = c/2l$. Substituting from eqn (1.11a),

$$\nu_1 = \frac{1}{2l}\sqrt{\frac{T}{\rho}}. \qquad (2.18)$$

In 1636, Mersenne discovered experimentally what is now known as Mersenne's law: 'The cyclic frequency of the fundamental mode of transverse vibration of a string varies as the square root of the tension, inversely as the length and inversely as the square root of the density.'

ν_1 is referred to as the 'fundamental frequency' or the 'ground note'. All three parts of Mersenne's law can be deduced from eqn (2.18). The predictions of the linear theory of the transverse vibrations of strings agree with Mersenne's observations. The period of the fundamental mode, equal to $2l/c$, is known as the 'fundamental period' of the string. The frequencies ν_n, $n>1$, are known as the 'harmonics' or 'overtones' of the fundamental. The fundamental frequency is known in music as the 'pitch' of a note. The 'tone' of a note is not precisely defined but it depends on the magnitude of the various harmonics present in a particular vibration. It is the presence of these harmonics in differing degrees which distinguishes the notes of different musical instruments.

The reader should observe that eqn (2.11) can now be interpreted to mean: 'The displacement, produced by summing the displacements of arbitrary amplitude and phase constant in any or all of the normal modes, is itself a possible displacement for a vibrating string.'

Example (i). A string is held in the position $y = \alpha \sin(\pi x/l)$ for $t<0$ and released at time $t=0$. Find the displacement at later times.

When the string is released, the tension in the string causes it to be displaced from its initial position and to oscillate. The tension, however, is finite and can only produce finite acceleration on finite mass, i.e. there is no instantaneous change in velocity. Since the velocity is zero for $t<0$, the velocity is zero at $t=0$. The initial conditions are therefore

$$y = \alpha \sin\frac{\pi x}{l} \quad \text{at } t=0, \quad \text{and} \quad \frac{\partial y}{\partial t} = 0 \quad \text{for all } x \text{ at } t=0. \qquad (2.19)$$

We wish to find which terms in the general solution, eqn (2.11), are appropriate to this particular problem. Since only terms corresponding to $n=1$ appear in the initial conditions, it is reasonable to try a solution corresponding to the terms $n=1$ in the general solution, i.e. we try

$$y=\sin\frac{\pi x}{l}\left(A_1\cos\frac{\pi ct}{l}+B_1\sin\frac{\pi ct}{l}\right). \tag{2.20}$$

The values of y and $\partial y/\partial t$ at $t=0$, calculated from eqn (2.20), are

$$y=A_1\sin\frac{\pi x}{l} \quad \text{and} \quad \frac{\partial y}{\partial t}=B_1\frac{\pi c}{l}\sin\frac{\pi x}{l}, \quad \text{respectively.} \tag{2.21}$$

Comparing eqns (2.19) and (2.21), it can be seen that the initial conditions are satisfied if $A_1=\alpha$ and $B_1=0$. Substituting back into eqn (2.20),

$$y=\alpha\sin\frac{\pi x}{l}\cos\frac{\pi ct}{l}. \tag{2.22}$$

Equation (2.22) is the solution of this problem because it satisfies the governing differential equation (1.11), the spatial boundary conditions $y=0$ at $x=0$ and at $x=l$ and the initial conditions, eqns (2.19). Comparing eqn (2.22) with eqn (2.19), we see that the string is oscillating in its first normal or fundamental mode with normal amplitude equal to α and phase constant equal to zero.

The velocity of the particle of the string with coordinate x in this example is given by

$$\frac{\partial y}{\partial t}=-\frac{\alpha\pi c}{l}\sin\frac{\pi x}{l}\sin\frac{\pi ct}{l}. \tag{2.23}$$

The displacement is a maximum at each value of x when $\cos(\pi ct/l)=\pm 1$, i.e when $\pi ct/l=N\pi$ with N integral, $t=Nl/c$. The displacement is zero for all values of x when $\cos(\pi ct/l)=0$, i.e.

$$\frac{\pi ct}{l}=\frac{2N+1}{2}\pi, \quad t=\frac{(2N+1)l}{2c}.$$

The velocity is a maximum at each value of x when $\sin(\pi ct/l)=\pm 1$, i.e. when

$$\frac{\pi ct}{l}=\frac{2N+1}{2}\pi, \quad t=\frac{(2N+1)l}{2c}.$$

The velocity is zero for all values of x when $\sin(\pi ct/l)=0$, i.e. when $\pi ct/l=N\pi$, $t=Nl/c$. Hence the displacement is a maximum when the velocity is zero and vice versa.

We can now sketch the motion of the string (Fig. 2.2). At time $t=0$, it is in position (1) with zero velocity at all points. At time $t=l/2c$, it lies along OL with velocity at a maximum 'downwards', position (2). At time $t=l/c$, it is in position (3) with zero velocity at all points. At time $t=3l/2c$ it lies along OL again with velocity at a maximum but this time 'upwards', position (4). At time $t=2l/c$, the string is again in position (1) with zero velocity. The motion now repeats itself, in fact it will continue to repeat itself with period $2l/c$ until some external force alters or stops the motion. In practice, air resistance and internal friction will gradually damp out the vibration.

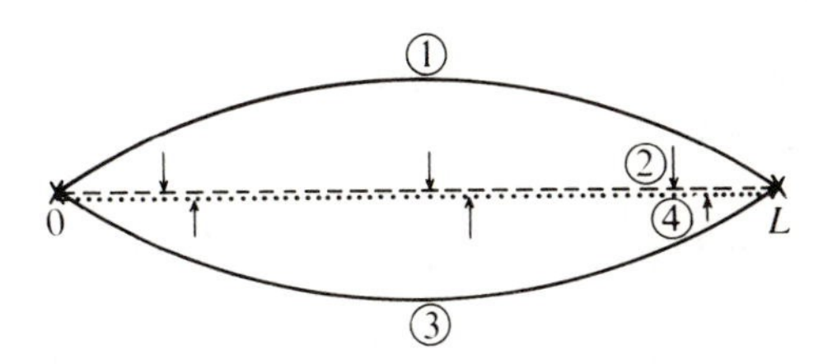

Fig. 2.2

Example (ii). A string is held in the position $y=\alpha\sin(n\pi x/l)$ for $t<0$ and released at time $t=0$. Find the displacement at later times.

The method of solution of this problem is identical to that of Example (i), except that it is the terms in the general solution, given by eqn (2.12), that are picked out. The final solution is

$$y=\alpha\sin\frac{n\pi x}{l}\cos\frac{n\pi ct}{l}. \tag{2.24}$$

This represents a vibration in the nth normal mode with normal amplitude equal to α and phase constant equal to zero. One of the two positions of maximum displacement is plotted in Fig. 2.3 below for $n=1$, 2, 3 and 4.

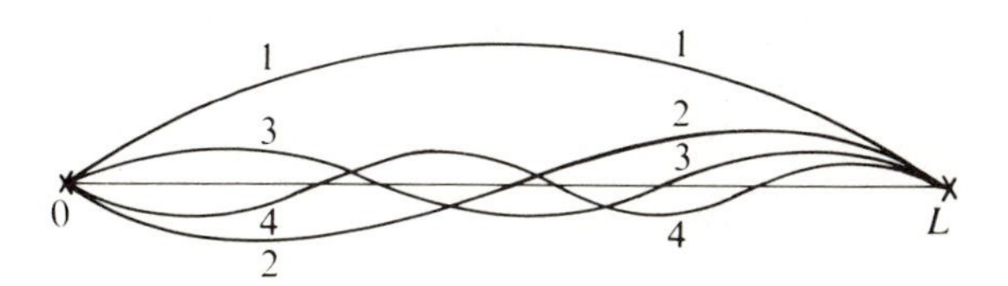

Fig. 2.3

A point at which the displacement is zero for all times for a given mode is called a 'node' of that mode. A point at which the amplitude of oscillation is a maximum is called an 'antinode'. It can be seen from the above diagram that the third normal mode has two nodes at $l/3$ and $2l/3$ and three antinodes at $l/6$, $l/2$ and $5l/6$. All modes have nodes at 0 and l. The reader will note the identity of the definitions of this paragraph with those given on pp. 21 and 22 in the discussion of standing waves.

Example (iii). The initial conditions are

$$y=\varepsilon\sin\frac{\pi x}{l} \quad \text{and} \quad \frac{\partial y}{\partial t}=\eta\sin\frac{2\pi x}{l} \quad \text{at } t=0.$$

The values of y and $\partial y/\partial t$ at $t=0$, obtained from the general solution, equation (2.11), are

$$y=\sum_{n=1}^{\infty} A_n\sin\frac{n\pi x}{l} \quad \text{and} \quad \frac{\partial y}{\partial t}=\sum_{n=1}^{\infty}\frac{n\pi c}{l}B_n\sin\frac{n\pi x}{l}. \tag{2.25}$$

The initial conditions of this problem can be satisfied if we choose

$$A_1=\varepsilon, \quad \text{all other } A_n=0,$$

and

$$\frac{2\pi c}{l}B_2=\eta, \quad \text{all other } B_n=0.$$

Substituting back for the A_n and B_n into eqn (2.11), the solution of this problem is

$$y=\varepsilon\sin\frac{\pi x}{l}\cos\frac{\pi ct}{l}+\frac{\eta l}{2\pi c}\sin\frac{2\pi x}{l}\sin\frac{2\pi ct}{l}. \tag{2.26}$$

We now investigate whether the properties of y and $\partial y/\partial t$ being zero for all x at (different) particular times, true for any normal mode, carry over to the vibration described by eqn (2.26). Amplitude y can only be zero for all x if the coefficient of each term in x in eqn (2.26) is zero at a particular time. This requires, since ε and η are non-zero,

$$\cos\frac{\pi ct}{l}=0 \quad \text{and} \quad \sin\frac{2\pi ct}{l}=0.$$

For $\cos(\pi ct/l)=0$,

$$\frac{\pi ct}{l}=\frac{\pi}{2}, \frac{3\pi}{2}, \frac{5\pi}{2}, \ldots$$

or

$$t=\frac{l}{2c}, \frac{3l}{2c}, \frac{5l}{2c}, \ldots$$

For $\sin(2\pi ct/l)=0$,

$$\frac{2\pi ct}{l}=0, \pi, 2\pi, 3\pi, \ldots$$

or

$$t=0,\ l/2c,\ l/c,\ 3l/2c, \ldots$$

We see that both terms are zero when $t=l/2c, 3l/2c, 5l/2c, \ldots$. At these times the displacement is zero for all x.

The transverse velocity is given by differentiating eqn (2.26) with respect to t:

$$\frac{\partial y}{\partial t}=-\frac{\varepsilon\pi c}{l}\sin\frac{\pi x}{l}\sin\frac{\pi ct}{l}+\eta\sin\frac{2\pi x}{l}\cos\frac{2\pi ct}{l}. \tag{2.27}$$

For $\partial y/\partial t$ to be zero for all x at any one particular time,

$$\sin\frac{\pi ct}{l}=0 \quad \text{and} \quad \cos\frac{2\pi ct}{l}=0.$$

For $\sin(\pi ct/l)=0$,

$$\frac{\pi ct}{l}=0, \pi, 2\pi, 3\pi, \ldots$$

or

$$t=0, l/c, 2l/c, 3l/c, \ldots$$

For $\cos(2\pi ct/l)=0$,

$$\frac{2\pi ct}{l}=\frac{\pi}{2}, \frac{3\pi}{2}, \frac{5\pi}{2}, \ldots$$

or

$$t=l/4c, 3l/4c, 5l/4c, \ldots$$

There is no time at which both $\sin(\pi ct/l)$ and $\cos(2\pi ct/l)$ are zero and therefore $\partial y/\partial t$ cannot be zero for all x at any one time.

Exercise 2.1. The initial conditions are $y=\varepsilon\sin(2\pi x/l)$, $\partial y/\partial t=\eta\sin(\pi x/l)$ at $t=0$. Find the displacement at later times and show that $\partial y/\partial t$ but not y can be zero for all x at certain times.

Exercise 2.2. The initial conditions are $y=\varepsilon_1\sin(\pi x/l)+\varepsilon_3\sin(3\pi x/l)$ and $\partial y/\partial t=0$ at $t=0$. Find the displacement at later times and show that either y or $\partial y/\partial t$ can be zero for all x at certain times.

2.3. Initial conditions and Fourier sine series

The wave equation, equation (1.11), contains second-order partial derivatives with respect to t. To determine the arbitrary functions of x which are introduced on integrating the differential equation with respect to t, it will be sufficient to specify both the dependent variable and its first time derivative for all values of x from 0 to l at the initial time, which is nearly always taken as $t=0$. However, it is seen from eqns (2.25) that the general solution, eqn (2.11), gives values of y and $\partial y/\partial t$ at $t=0$ each in the form of an infinite series in $\sin(n\pi x/l)$ with constant coefficients and with n summed over all integral values from 1 to ∞. Then, if we are going to use the general solution in the form of eqn (2.11), we must be able to express the initial given values of y and $\partial y/\partial t$ as infinite series of the same form as eqns (2.25).

Recall the problem of equation fitting to a curve. If we have four points (x_1, y_1), (x_2, y_2), (x_3, y_3) and (x_4, y_4) on the curve C (Fig. 2.4),

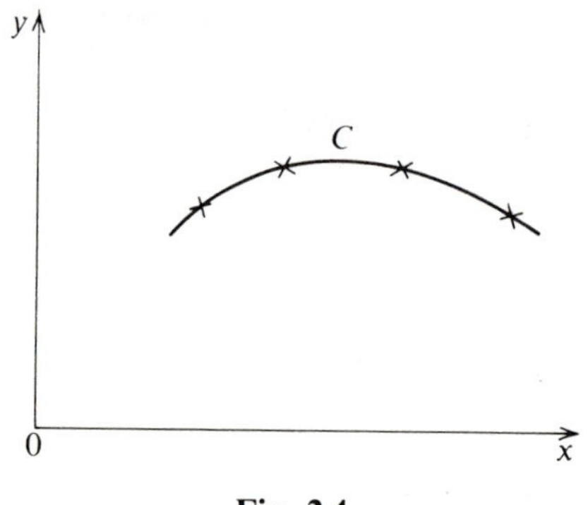

Fig. 2.4

we can fit an equation of the form

$$y = a + bx + cx^2 + dx^3$$

through these points. Constants a, b, c and d are determined by

$$y_1 = a + bx_1 + cx_1^2 + dx_1^3,$$

$$y_2 = a + bx_2 + cx_2^2 + dx_2^3,$$

etc.,

i.e. four linear equations for a, b, c and d. There is no reason why the equation should be a polynomial. We could take

$$y = a\sin\frac{\pi x}{l} + b\sin\frac{2\pi x}{l} + c\sin\frac{3\pi x}{l} + d\sin\frac{4\pi x}{l};$$

substitution of each point in turn gives four equations for a, b, c and d. Since $\sin(n\pi x/l)$ is zero at $x=0, l$, we can clearly fit an equation of the form

$$y = \sum_{1}^{n} A_n \sin\frac{n\pi x}{l}$$

through any n points x_i, $i=1, 2, \ldots, n$ where $0 < x_1 < x_2 < \ldots < x_n < l$, and any such equation gives a y that is zero for $x=0, l$. If the points x_i are a distance $l/(n+1)$ apart, they are equally spaced. What happens if we let $n \to \infty$, so that we try to fit an infinite number of points on a curve? That this is not a simple question is clear if we consider a curve with a discontinuity. Each $\sin(n\pi x/l)$ is continuous, so the sum to a finite number n is continuous. The answer to the question is given by Fourier's theorem for sine series in the interval $(0, l)$: 'Any function $f(x)$, single-valued and continuous, except possibly for a finite number of discontinuities, in the interval $0 \leqslant x \leqslant l$ and which has only a finite number of maxima and minima in that interval, can be represented uniquely in that interval by a series of the form

$$f(x) = \sum_{1}^{\infty} C_n \sin\frac{n\pi x}{l}, \tag{2.28}$$

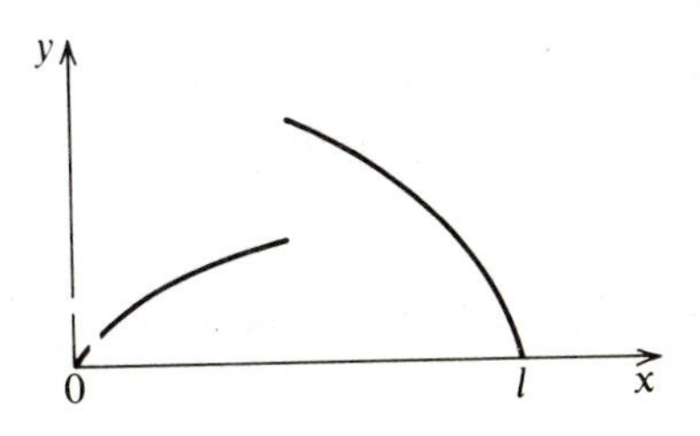

Fig. 2.5

where the C_n are constants. The series converges to zero at $x=0, l$ and to the mean of the two values of the function of either side of the discontinuity at points where the function has a discontinuity.' The r.h.s. of eqn (2.28) is known as a Fourier sine series.

Note that if the function is such that it has no discontinuities and $f(0)=f(l)=0$, then the Fourier sine series converges to the original function at all points in the interval, including 0 and l.

The general Fourier series contains cosine as well as sine terms, but we shall not need such series in this chapter.

How are the coefficients C_n determined? First prove the following.

LEMMA.

$$\int_0^l \sin\frac{n\pi x}{l}\sin\frac{m\pi x}{l}\,\mathrm{d}x=\begin{cases}0 & \text{if } m\neq n,\\ l/2 & \text{if } m=n,\end{cases}$$

where m and n are positive integers.

Proof.

$$\int_0^l \sin\frac{n\pi x}{l}\sin\frac{m\pi x}{l}\,\mathrm{d}x=\frac{1}{2}\int_0^l\left(\cos\frac{(n-m)\pi x}{l}-\cos\frac{(n+m)\pi x}{l}\right)\mathrm{d}x$$

$$\int_0^l \cos\frac{(n+m)\pi x}{l}\,\mathrm{d}x=-\frac{l}{(n+m)\pi}\sin\frac{(n+m)\pi x}{l}\bigg|_0^l=0$$

For $m\neq n$,

$$\int_0^l \cos\frac{(n-m)\pi x}{l}\,\mathrm{d}x=-\frac{l}{(n-m)\pi}\sin\frac{(n-m)\pi x}{l}\bigg|_0^l=0.$$

For $m=n$,

$$\int_0^l \cos\frac{(n-m)\pi x}{l}\,\mathrm{d}x=\int_0^l 1\cdot\mathrm{d}x=l.$$

The result follows on back-substitution for the cosine integrals. Multiply eqn (2.28) through by $\sin(m\pi x/l)$ and integrate with respect to x from 0 to l:

$$\int_0^l f(x)\sin\frac{m\pi x}{l}\,\mathrm{d}x=\sum_{n=1}^{\infty}C_n\int_0^l \sin\frac{n\pi x}{l}\sin\frac{m\pi x}{l}\,\mathrm{d}x=\frac{l}{2}C_m,$$

since the integral on the r.h.s. is zero unless $n=m$ and equal to $l/2$ when $n=m$. Therefore,

$$C_m=\frac{2}{l}\int_0^l f(x)\sin\frac{m\pi x}{l}\,\mathrm{d}x.$$

Replacing m by n,

$$C_n=\frac{2}{l}\int_0^l f(x)\sin\frac{n\pi x}{l}\,\mathrm{d}x. \tag{2.29}$$

Let the given values of y and $\partial y/\partial t$ at $t=0$ be $u(x)$ and $v(x)$ respectively. Let the Fourier sine series expansions of $u(x)$ and $v(x)$ be

$$u(x)=\sum_{n=1}^{\infty} a_n \sin\frac{n\pi x}{l} \quad \text{and} \quad v(x)=\sum_{n=1}^{\infty} b_n \sin\frac{n\pi x}{l}. \tag{2.30}$$

Comparing eqns (2.25) and (2.30),

$$A_n=a_n \quad \text{and} \quad B_n=\frac{l}{n\pi c} b_n. \tag{2.31}$$

Determination of the A_n and B_n from eqns (2.31) and substitution into eqn (2.11) gives the solution to any problem in which the initial displacement and velocity are given. Note that the A_n are determined solely by the displacement and the B_n solely by the velocity. The method of solution will now be illustrated by examples.

Example (iv). The midpoint of the string is held a distance h from its equilibrium position for $t<0$ and released at time $t=0$ (Fig. 2.6). Find the displacement at later times.

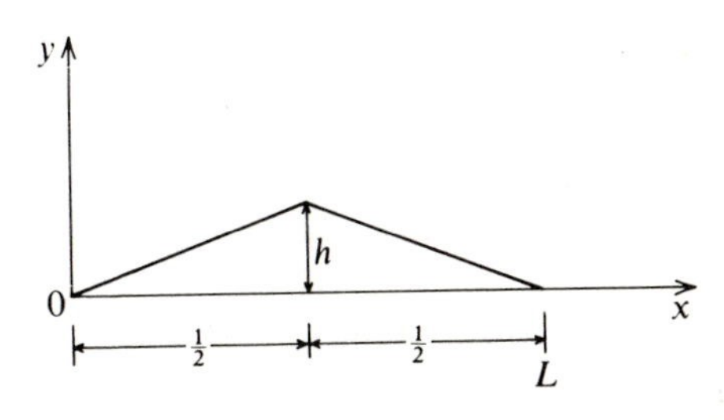

Fig. 2.6

The given initial conditions are

$$u(x)=\begin{cases} \dfrac{2hx}{l} & \text{for } 0\leqslant x\leqslant \tfrac{1}{2}l, \\[2ex] \dfrac{2h(l-x)}{l} & \text{for } \tfrac{1}{2}l\leqslant x\leqslant l, \end{cases} \tag{2.32}$$

and $v(x)=0$ for $0\leqslant x\leqslant l$.

First obtain the coefficients a_n and b_n in the Fourier sine series representations of $u(x)$ and $v(x)$. From eqns (2.29) and (2.30), we see that $b_n=0$ and

$$a_n=\frac{2}{l}\int_0^l u(x)\sin\frac{n\pi x}{l}\,\mathrm{d}x$$

$$=\frac{2}{l}\int_0^{\frac{1}{2}l}\frac{2hx}{l}\sin\frac{n\pi x}{l}\,\mathrm{d}x$$

$$+\frac{2}{l}\int_{\frac{1}{2}l}^{l}\frac{2h(l-x)}{l}\sin\frac{n\pi x}{l}\,\mathrm{d}x.$$

Integrating by parts,[†]

$$a_n=\frac{4h}{l^2}\left(-\frac{l}{n\pi}\right)\left(x\cos\frac{n\pi x}{l}\bigg|_0^{\frac{1}{2}l}-\int_0^{\frac{1}{2}l}\cos\frac{n\pi x}{l}\,\mathrm{d}x+(l-x)\cos\frac{n\pi x}{l}\bigg|_{\frac{1}{2}l}^{l}\right.$$

$$\left.+\int_{\frac{1}{2}l}^{l}\cos\frac{n\pi x}{l}\,\mathrm{d}x\right)=\frac{8h}{n^2\pi^2}\sin\frac{n\pi}{2},$$

on evaluation of the integrals.

From eqn (2.31),

$$A_n=a_n=\begin{cases}\dfrac{8h}{n^2\pi^2} & \text{if } n=1, 5, 9, 13, \ldots\\[2ex] -\dfrac{8h}{n^2\pi^2} & \text{if } n=3, 7, 11, 15, \ldots\\[2ex] 0 & \text{if } n \text{ is even.}\end{cases}$$

$$B=\frac{l}{n\pi c}b_n=0.$$

Substituting for A_n and B_n in eqn (2.11), the displacement at later times is given by

$$y=\frac{8h}{\pi^2}\left(\sin\frac{\pi x}{l}\cos\frac{\pi ct}{l}-\frac{1}{3^2}\sin\frac{3\pi x}{l}\cos\frac{3\pi ct}{l}+\frac{1}{5^2}\sin\frac{5\pi x}{l}\cos\frac{5\pi ct}{l}\right.$$

$$\left.-\frac{1}{7^2}\sin\frac{7\pi x}{l}\cos\frac{7\pi ct}{l}+\ldots\right). \tag{2.33}$$

[†] $f(x)|_a^b$ means the difference of the values of $f(x)$ evaluated at b and a, i.e. $f(b)-f(a)$.

The displacement y must satisfy the differential equation

$$\frac{\partial^2 y}{\partial t^2} = c^2 \frac{\partial^2 y}{\partial x^2}.$$

If the series for y, eqn (2.33), is differentiated twice with respect to t,

$$\frac{\partial^2 y}{\partial t^2} = -\frac{8hc^2}{l^2}\left(\sin\frac{\pi x}{l}\cos\frac{\pi ct}{l} - \sin\frac{3\pi x}{l}\cos\frac{3\pi ct}{l}\right.$$

$$\left. + \sin\frac{5\pi x}{l}\cos\frac{5\pi ct}{l} - \dots\right);$$

$c^2(\partial^2 y/\partial x^2)$ gives the same series. The series are equal term-by-term but they are not convergent for general values of x and t.

It is impossible to say that the differential equation is satisfied. However, the equation would be satisfied if it were possible to terminate the series after N terms, where N can be very large but not infinite, because two finite series which are equal term by term, have equal sums.

In practice, it is never possible to measure any physical quantity exactly; for example, when we say that the length of a rod is 10 cm, we mean that it is in the range $(10 \pm \varepsilon)$ cm, where ε may be 10^{-2} or 10^{-5} or 10^{-8} cm, according to the accuracy of the measuring instrument; ε can never be zero. If we wish the initial displacement of the string to be given by eqns (2.32) and the accuracy to which the displacement can be measured is $\pm\varepsilon$ $(\varepsilon > 0)$, any initial displacement which differs by less than ε at all points from eqns (2.32) is physically indistinguishable from the required displacement. It can be shown that it is always possible to find finite N so that the sum of the first N terms of the Fourier series for eqn (2.32) differs by less than ε from eqns (2.32) for all values of x, $0 \leqslant x \leqslant l$. This series of N terms leads to a series for the displacement at later times identical to eqn (2.33) but cut off after N terms, and the latter series, as we have seen, satisfies the wave equation. The eqn (2.33) can therefore be interpreted as the solution to the physical problem, for which the initial conditions are nearly eqns (2.32), the solution being obtained by retaining only a finite number of terms on the right-hand side of eqn (2.33).

Example (v). The string is undisplaced for $t < 0$. At time $t = 0$ it is subject to an impulse that gives it a parabolic velocity distribution of the form

$$v(x)=\frac{4V}{l^2}(lx-x^2);$$

V is the velocity of the mid-point $x=l/2$. Since $u(x)=0$, $a_n=0$. Using the second of eqns (2.30),

$$b_n=\frac{2}{l}\int_0^l \frac{4V}{l^2}(lx-x^2)\sin\frac{n\pi x}{l}\,\mathrm{d}x.$$

Integrating twice by parts,

$$b_n=\frac{8V}{l^3}\left(-\frac{l}{n\pi}\right)\left((lx-x^2)\cos\frac{n\pi x}{l}\Big|_0^l-\int_0^l(l-2x)\cos\frac{n\pi x}{l}\,\mathrm{d}x\right)$$

$$=\frac{8V}{l^3}\left(\frac{l}{n\pi}\right)^2\left((l-2x)\sin\frac{n\pi x}{l}\Big|_0^l+2\int_0^l\sin\frac{n\pi x}{l}\,\mathrm{d}x\right)$$

$$=\frac{16V}{n^3\pi^3}(1-(-)^n).$$

From eqns (2.31):

$$A_n=a_n=0$$

and

$$B_n=\frac{l}{n\pi c}b_n=\begin{cases}\dfrac{32lV}{n^4\pi^4 c} & \text{if } n \text{ is odd,}\\[2ex] 0 & \text{if } n \text{ is even.}\end{cases}$$

Substituting in eqn (2.11), the solution is

$$y=\frac{32lV}{\pi^4 c}\left(\sin\frac{\pi x}{l}\sin\frac{\pi ct}{l}+\frac{1}{3^4}\sin\frac{3\pi x}{l}\sin\frac{3\pi ct}{l}\right.$$
$$\left.+\frac{1}{5^4}\sin\frac{5\pi x}{l}\sin\frac{5\pi ct}{l}+\dots\right).$$

Exercise 2.3. A portion of the string from $x=l/4$ to $x=3l/4$ is held at constant displacement h for $t<0$ and released at $t=0$; see Fig. 2.7. Find the displacement at later times.

Exercise 2.4. A portion of the string from $x=l/3$ to $x=2l/3$ is held at constant displacement h for $t<0$ and released at $t=0$. Find the displacement at later times.

Exercise 2.5. The initial displacement and velocity are given by

$$u(x)=\alpha(lx-x^2)/l^2$$

and

$$v(x)=\begin{cases}\beta x/l & \text{for } 0\leqslant x\leqslant l/2,\\ \beta(l-x)/l & \text{for } l/2\leqslant x\leqslant l.\end{cases}$$

Find the displacement at later times (α and β are constants).

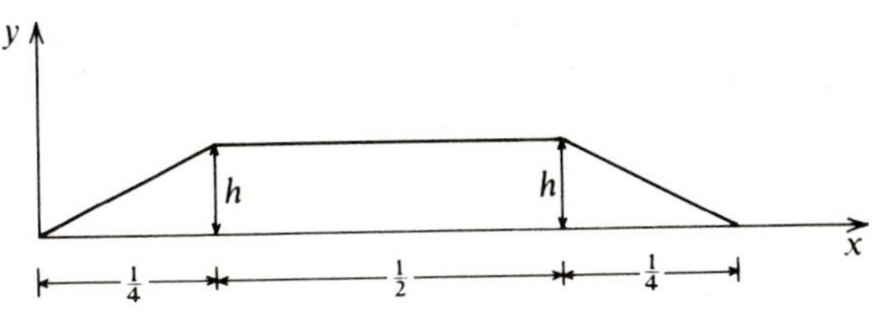

Fig. 2.7

2.4. Kinetic and potential energies of transverse vibrations

The kinetic and potential energies per unit length are given by eqns (1.32) and (1.33). When the string is vibrating in a normal mode, the kinetic and potential energies of the whole string, denoted by K_l and V_l respectively, are obtained by substitution for y from eqn (2.17) into eqns (1.32) and (1.33) and by integration with respect to x from 0 to l:

$$K_l=\frac{1}{2}\rho\left(\frac{n\pi cC_n}{l}\right)^2\sin^2\left(\frac{n\pi ct}{l}-\varepsilon_n\right)\int_0^l\sin^2\frac{n\pi x}{l}\,\mathrm{d}x$$

and

$$V_l=\frac{1}{2}T\left(\frac{n\pi C_n}{l}\right)^2\cos^2\left(\frac{n\pi ct}{l}-\varepsilon_n\right)\int_0^l\cos^2\frac{n\pi x}{l}\,\mathrm{d}x.$$

Now

$$\int_0^l\sin^2\frac{n\pi x}{l}\,\mathrm{d}x=\frac{1}{2}\int_0^l\left(1-\cos\frac{2n\pi x}{l}\right)\mathrm{d}x$$
$$=\frac{1}{2}\left(x-\frac{l}{2n\pi}\sin\frac{2n\pi x}{l}\right)\bigg|_0^l=\frac{l}{2} \tag{2.34}$$

and

$$\int_0^l\cos^2\frac{n\pi x}{l}\,\mathrm{d}x=\int_0^l\left(1-\sin^2\frac{n\pi x}{l}\right)\mathrm{d}x=l-\frac{l}{2}=\frac{l}{2}. \tag{2.35}$$

Hence

$$K_l = \frac{\rho c^2 \pi^2}{4l} (nC_n)^2 \sin^2\left(\frac{n\pi ct}{l} - \varepsilon_n\right),$$

$$V_l = \frac{T\pi^2}{4l} (nC_n)^2 \cos^2\left(\frac{n\pi ct}{l} - \varepsilon_n\right).$$

These expressions can be simplified by introducing the normal radian frequency of the nth mode, denoted by ω_n. Since $\omega_n = 2\pi\nu_n$ and $\nu_n = nc/2l$,

$$\omega_n = \frac{n\pi c}{l}, \tag{2.36}$$

$$K_l = \tfrac{1}{4}\rho l \omega_n^2 C_n^2 \sin^2(\omega_n t - \varepsilon_n), \tag{2.37}$$

$$V_l = \tfrac{1}{4}\rho l \omega_n^2 C_n^2 \cos^2(\omega_n t - \varepsilon_n), \tag{2.38}$$

where $T = \rho c^2$ has been used in deriving eqn 2.38. The total energy is

$$K_l + V_l = \tfrac{1}{4}\rho l \omega_n^2 C_n^2, \tag{2.39}$$

which is constant. For a given string oscillating in a normal mode, all the energies are proportional to the square of the frequency and to the square of the amplitude.

Next we find the energies in the general case. At the beginning of Section 2.2 it was shown that the displacement

$$y = \sin\frac{n\pi x}{l}\left(A_n \cos\frac{n\pi ct}{l} + B_n \sin\frac{n\pi ct}{l}\right)$$

can be expressed in the form

$$y = C_n \sin\frac{n\pi x}{l} \cos\left(\frac{n\pi ct}{l} - \varepsilon_n\right).$$

Hence the displacement in the general case can be expressed either in the form of eqn (2.11) or as

$$y = \sum_{n=1}^{\infty} C_n \sin\frac{n\pi x}{l} \cos\left(\frac{n\pi ct}{l} - \varepsilon_n\right). \tag{2.40}$$

Differentiating eqn (2.40) with respect to t,

$$\frac{\partial y}{\partial t}=-\frac{\pi c}{l}\sum_{n=1}^{\infty} nC_n \sin\frac{n\pi x}{l}\sin\left(\frac{n\pi ct}{l}-\varepsilon_n\right)$$

$$=-\sum_{n=1}^{\infty}\omega_n C_n \sin\frac{n\pi x}{l}\sin(\omega_n t-\varepsilon_n). \tag{2.41}$$

Hence

$$K_l=\frac{1}{2}\rho\int_0^l\left(\frac{\partial y}{\partial t}\right)^2 \mathrm{d}x=\frac{1}{2}\rho\int_0^l\left(\sum_{n=1}^{\infty}\omega_n C_n \sin\frac{n\pi x}{l}\sin(\omega_n t-\varepsilon_n)\right)^2 \mathrm{d}x. \tag{2.42}$$

Squaring the infinite series on the right-hand side of eqn (2.42) will produce terms of two types, namely square and product terms, which must then be integrated with respect to x from 0 to l. A typical square term is

$$\omega_n^2 C_n^2 \sin^2\frac{n\pi x}{l}\sin^2(\omega_n t-\varepsilon_n).$$

Using eqn (2.34),

$$\int_0^l \omega_n^2 C_n^2 \sin^2\frac{n\pi x}{l}\sin^2(\omega_n t-\varepsilon_n)\,\mathrm{d}x=\frac{\omega_n^2 C_n^2 l}{2}\sin^2(\omega_n t-\varepsilon_n). \tag{2.43}$$

A typical product term is

$$\omega_n\omega_m C_n C_m \sin\frac{n\pi x}{l}\sin\frac{m\pi x}{l}\sin(\omega_n t-\varepsilon_n)\sin(\omega_m t-\varepsilon_m) \quad n\neq m.$$

Since

$$\int_0^l \sin\frac{n\pi x}{l}\sin\frac{m\pi x}{l}\,\mathrm{d}x=0, \quad n\neq m, \tag{2.44}$$

the product terms make no contribution to the total kinetic energy of the string. Summing the square terms over all values of n and using eqn (2.43),

$$K_l=\frac{\rho l}{4}\sum_{n=1}^{\infty}\omega_n^2 C_n^2 \sin^2(\omega_n t-\varepsilon_n). \tag{2.45}$$

This is the kinetic energy in the general case.

To find the potential energy, differentiate eqn (2.40) with respect to x:

$$\frac{\partial y}{\partial x}=\frac{\pi}{l}\sum_{n=1}^{\infty} nC_n \cos\frac{n\pi x}{l}\cos(\omega_n t-\varepsilon_n);$$

$$V_l=\frac{1}{2}T\int_0^l\left(\frac{\partial y}{\partial x}\right)^2 \mathrm{d}x$$

$$=\frac{T\pi^2}{2l^2}\int_0^l\left(\sum_{n=1}^{\infty} nC_n\cos\frac{n\pi x}{l}\cos(\omega_n t-\varepsilon_n)\right)^2 \mathrm{d}x. \quad (2.46)$$

A typical square term gives on integration, using eqn (2.35),

$$\int_0^l n^2C_n^2\cos^2\frac{n\pi x}{l}\cos^2(\omega_n t-\varepsilon_n)\,\mathrm{d}x=\frac{n^2C_n^2 l}{2}\cos^2(\omega_n t-\varepsilon_n). \quad (2.47)$$

A typical product term is

$$nmC_nC_m\cos\frac{n\pi x}{l}\cos\frac{m\pi x}{l}\cos(\omega_n t-\varepsilon_n)\cos(\omega_m t-\varepsilon_m)$$

with $n\neq m$. But with $n\neq m$

$$\int_0^l\cos\frac{n\pi x}{l}\cos\frac{m\pi x}{l}\mathrm{d}x=\frac{1}{2}\int_0^l\left(\cos\frac{(n+m)\pi x}{l}+\cos\frac{(n-m)\pi x}{l}\right)\mathrm{d}x$$

$$=\frac{1}{2}\left(\frac{l}{(n+m)\pi}\sin\frac{(n+m)\pi x}{l}\right.$$

$$\left.+\frac{l}{(n-m)\pi}\sin\frac{(n-m)\pi x}{l}\right)\Bigg|_0^{\pi}=0. \quad (2.48)$$

The product terms make no contribution to the total potential energy of the string. Summing the square terms over all values of n and using eqn (2.47),

$$V_l=\frac{T\pi^2}{2l^2}\sum_{n=1}^{\infty}\frac{n^2C_n^2 l}{2}\cos^2(\omega_n t-\varepsilon_n).$$

Using $T=\rho c^2$ and eqn (2.36),

$$V_l=\frac{\rho l}{4}\sum_{n=1}^{\infty}\omega_n^2C_n^2\cos^2\left(\frac{n\pi ct}{l}-\varepsilon_n\right). \quad (2.49)$$

This is the potential energy in the general case. Summing eqns (2.45) and (2.49), the total energy equals

$$K_l+V_l=\frac{\rho l}{4}\sum_{n=1}^{\infty}\omega_n^2C_n^2, \quad (2.50)$$

which is constant.

Equations (2.45), (2.49) and (2.50) state the following:

> The kinetic, potential or total energy of any transverse vibration of a string is equal to the sum of the kinetic, potential or total energies respectively in each of the normal modes into which the vibration can be resolved.

The kinetic, potential and total energies are found for the vibrations of Examples (iv) and (v) of Section 2.3. In example (iv) the displacement was given by

$$y=\frac{8h}{\pi^2}\left(\sin\frac{\pi x}{l}\cos\frac{\pi ct}{l}-\frac{1}{3^2}\sin\frac{3\pi x}{l}\cos\frac{3\pi ct}{l}\right.$$
$$\left.+\frac{1}{5^2}\sin\frac{5\pi x}{l}\cos\frac{5\pi ct}{l}-\ \ldots\right).$$

Comparison with eqn (2.40) shows that the constants C_n and ε_n in this vibration are

$$C_n=\frac{8h}{n^2\pi^2} \qquad \text{if } n \text{ odd,}$$
$$=0 \qquad \text{if } n \text{ even;}$$

and

$$\varepsilon_n=0 \qquad \text{unless } n=4N+3, \quad N \text{ integral,}$$
$$=\pi \qquad \text{if } n=4N+3.$$

Substituting into eqn (2.45) and using eqn (2.36), the kinetic energy is

$$K_l=\frac{16\rho c^2h^2}{l\pi^2}\left(\sin^2\frac{\pi ct}{l}+\frac{1}{3^2}\sin^2\frac{3\pi ct}{l}+\frac{1}{5^2}\sin^2\frac{5\pi ct}{l}+\ \ldots\right);$$

and substituting into eqn (2.49), the potential energy is

$$V_l=\frac{16\rho c^2h^2}{l\pi^2}\left(\cos^2\frac{\pi ct}{l}+\frac{1}{3^2}\cos^2\frac{3\pi ct}{l}+\frac{1}{5^2}\cos^2\frac{5\pi ct}{l}+\ \ldots\right).$$

The total energy is

$$K_l+V_l=\frac{16\rho c^2h^2}{l\pi^2}\left(1+\frac{1}{3^2}+\frac{1}{5^2}+\ \ldots\right). \tag{2.51}$$

But, as will be proved at the end of this section,

$$1+\frac{1}{3^2}+\frac{1}{5^2}+\ \ldots\ =\frac{\pi^2}{8}.$$

Therefore,

$$K_l + V_l = \frac{2\rho c^2 h^2}{l} = \frac{2Th^2}{l}.$$

This result can be checked from the initial conditions (Fig. 2.8). All the energy in the initial position is potential energy because the initial velocity is zero. The potential energy is equal to the tension times the increase in length of the string. Therefore initial energy is

$$2T\left\{\left[\left(\frac{l}{2}\right)^2 + h^2\right]^{\frac{1}{2}} - \frac{l}{2}\right\} = Tl\left[\left(1 + \frac{4h^2}{l^2}\right)^{\frac{1}{2}} - 1\right]$$

$$= \frac{2h^2 T}{l}\left[1 + O\left(\frac{2h}{l}\right)^2\right].$$

But $2h/l = \tan\alpha = \alpha + O(\alpha^3)$, where $\tan\alpha$ is the initial slope of the string. Since squares of α can be neglected compared to 1, squares of $2h/l$ can be neglected compared to 1 and the initial energy is $2Th^2/l$, which agrees with the value calculated for the total energy.

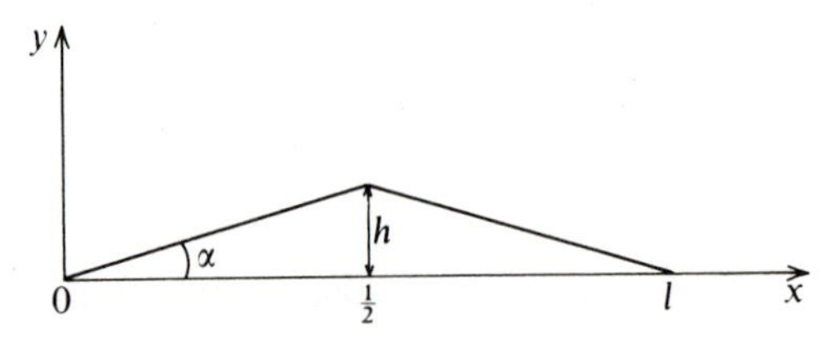

Fig. 2.8

The fraction of the total energy in the fundamental mode is the ratio of the first term in eqn (2.51) to the sum of the series, i.e. $1 : \pi^2/8$ or $8/\pi^2$. Since $\pi^2 \simeq 9.87$, over 80 per cent of the total energy is in the fundamental mode.

In Example (v) of Section (2.3),

$$y = \frac{32lV}{\pi^4 c}\left(\sin\frac{\pi x}{l}\sin\frac{\pi ct}{l} + \frac{1}{3^4}\sin\frac{3\pi x}{l}\sin\frac{3\pi ct}{l} + \frac{1}{5^4}\sin\frac{5\pi x}{l}\sin\frac{5\pi ct}{l} + \ldots\right).$$

Comparison with eqn (2.40) gives

$$C_n = \frac{32lV}{n^4\pi^4 c} \quad \text{if } n \text{ is odd,}$$

$$= 0 \qquad \text{if } n \text{ is even,}$$

and

$$\varepsilon_n = \frac{\pi}{2}.$$

From eqns (2.45) and (2.49), the kinetic and potential energies are

$$K_l = \frac{256\rho l V^2}{\pi^6} \sum_{n=1}^{\infty} \frac{1}{n^6} \cos^2 \frac{n\pi ct}{l}, \quad n \text{ odd,}$$

and

$$V_l = \frac{25\rho l V^2}{\pi^6} \sum_{n=1}^{\infty} \frac{1}{n^6} \sin^2 \frac{n\pi ct}{l}, \quad n \text{ odd.}$$

The total energy is

$$K_l + V_l = \frac{256\rho l V^2}{\pi^6} \sum_{n=1}^{\infty} \frac{1}{n^6}, \quad n \text{ odd.}$$

But, as we shall prove,

$$\sum_{n=1}^{\infty} \frac{1}{n^6} = \frac{\pi^6}{960}, \quad n \text{ odd,}$$

therefore $K_l + V_l = \frac{4}{15}\rho l V^2$.

The initial conditions in this problem were displacement zero and velocity equal to $(4V/l^2)(lx - x^2)$. Initially the energy is kinetic only and of amount

$$\frac{1}{2}\rho \int_0^l \left(\frac{4V}{l^2}(lx - x^2)\right)^2 \mathrm{d}x = \frac{8\rho V^2}{l^4} \int_0^l (l^2x^2 - 2lx^3 + x^4)\,\mathrm{d}x = \frac{4}{15}\rho l V^2,$$

which agrees with the total energy previously calculated.

The fraction of the total energy in the fundamental mode is $1:(\pi^6/960)$ or 99.86 per cent. All but 0.14 per cent of the total energy is in the fundamental mode.

We shall now prove the results, stated earlier in this section, of the summation of certain series. The Fourier sine series in the interval

$0<x<\pi$ for the function $u(x)$ of eqn (2.32) with $h=1$ and $l=\pi$ is given by

$$u(x)=\frac{8}{\pi^2}\left(\sin x-\frac{1}{3^2}\sin 3x+\frac{1}{5^2}\sin 5x-\ \ldots\right);$$

or, for $0<x<\pi/2$,

$$x=\frac{4}{\pi}\left(\sin x-\frac{1}{3^2}\sin 3x+\frac{1}{5^2}\sin 5x-\ \ldots\right).$$

Since $u(x)$ was continuous at $x=\pi/2$, the series for x converges to the value of x at $x=\pi/2$, i.e to $\pi/2$. Substituting $x=\pi/2$,

$$1+\frac{1}{3^2}+\frac{1}{5^2}+\ \ldots=\frac{\pi^2}{8}.$$

Fourier series may be integrated term-by-term. In particular, if the series for x is integrated from x to $\pi/2$,

$$\frac{1}{2}\left[\left(\frac{\pi}{2}\right)^2-x^2\right]=\frac{4}{\pi}\left(\cos x-\frac{1}{3^3}\cos 3x+\frac{1}{5^3}\cos 5x-\ \ldots\right).$$

Integrated again from 0 to x,

$$\frac{1}{2}\left[\left(\frac{\pi}{2}\right)^2 x-\frac{x^3}{3}\right]=\frac{4}{\pi}\left(\sin x-\frac{1}{3^4}\sin 3x+\frac{1}{5^4}\sin 5x-\ \ldots\right).$$

Substituting $x=\pi/2$,

$$1+\frac{1}{3^4}+\frac{1}{5^4}\ \ldots\ =\frac{\pi}{8}\cdot\frac{\pi^3}{12}=\frac{\pi^4}{96}.$$

Integrating again from x to $\pi/2$,

$$\left(\frac{\pi}{2}\right)^2\left[\frac{1}{4}\left(\frac{\pi}{2}\right)^2-\frac{1}{4}x^2\right]-\frac{1}{24}\left[\left(\frac{\pi}{2}\right)^4-x^4\right]=\frac{4}{\pi}\left(\cos x-\frac{1}{3^5}\cos 3x\right.$$
$$\left.+\frac{1}{5^5}\cos 5x-\ \ldots\right).$$

And integrating once again from 0 to x,

$$\frac{\pi^4}{64}x-\frac{\pi^2}{48}x^3-\frac{\pi^4}{384}x+\frac{1}{120}x^5=\frac{4}{\pi}\left(\sin x-\frac{1}{3^6}\sin 3x+\frac{1}{5^6}\sin 5x-\ \ldots\right);$$

putting $x=\pi/2$,

$$1+\frac{1}{3^6}+\frac{1}{5^6}+\ \ldots\ =\frac{\pi^6}{960}.$$

Sums of other infinite series involving powers of odd integers can be obtained by substituting $x=0$ into those Fourier series above which consist of cosine terms.

Exercise 2.6. Find the kinetic, potential and total energies for the vibration of Exercises 2.3. Check the result for the total energy from the initial conditions and find what proportion of the total energy is in the fundamental mode.

Exercise 2.7. As Exercise 2.6 but for the vibration of Exercise 2.5 with $\beta=2c\alpha/l$.

2.5. External force and air resistance

In this section we shall consider the effect of transverse external force and of air resistance on the motion of the string. A more sophisticated approach will be adopted in the solution of the governing differential equation.

To derive the differential equation, again consider the small element of the string, shown in Fig. 1–2. In addition to the tension at either end, two extra forces act on the element. As in Section 1.7, air resistance produces a force on the element in the y-direction equal to $-\kappa\rho\,\delta x(\partial y/\partial t)$. The external force is a given function of time and space, let it be $F(x, t)$ per unit mass and act in the positive y-direction. Then the equation of motion of the element in the y-direction is

$$\rho\,\delta x\frac{\partial^2 y}{\partial t^2}=T\sin(\Psi+\delta\Psi)-T\sin\Psi-\kappa\rho\,\delta x\frac{\partial y}{\partial t}+F\rho\,\delta x,$$

which by the same argument as in Chapter 1 leads to

$$\frac{\partial^2 y}{\partial t^2}+\kappa\frac{\partial y}{\partial t}-c^2\frac{\partial^2 y}{\partial x^2}=F(x,t). \tag{2.52}$$

To solve the differential equation we use the fact that, at any one time, the displacement y can be expanded in a Fourier sine series in the interval $0\leqslant x\leqslant l$ and that this series converges to the value of the displacement at all points within and bounding the interval. That is, at any fixed time,

$$y(x)=\sum_{n=1}^{\infty}D_n\sin\frac{n\pi x}{l}. \tag{2.53}$$

The values of the constants D_n will depend on the time chosen. But the expansion can be carried out at any time, therefore

$$y(x, t) = \sum_{n=1}^{\infty} \theta_n(t) \sin \frac{n\pi x}{l}, \tag{2.54}$$

where the θ_n are no longer constants but are functions of time. Equation (2.54) reduces to eqn (2.53) at any one fixed time. Similarly, the external force per unit mass can be expanded as a Fourier sine series at any fixed time and, therefore, at any time

$$F(x, t) = \sum_{n=1}^{\infty} \Phi_n(t) \sin \frac{n\pi x}{l}, \tag{2.55}$$

where the $\Phi_n(t)$ are functions of the time. Since $F(x, t)$ is given, the $\Phi_n(t)$ are immediately determinable from

$$\Phi_n(t) = \frac{2}{l} \int_0^l F(x, t) \sin \frac{n\pi x}{l} \, dx; \tag{2.56}$$

the $\theta_n(t)$ have to be determined from the governing differential equation (2.52) and the initial conditions. Substitution back into eqn (2.54) for the $\theta_n(t)$ then gives the displacement of the string. The $\theta_n(t)$ are called 'normal coordinates'.

Substitution from eqns (2.54) and (2.55) into eqn (2.52):

$$\sum_{n=1}^{\infty} \left[\frac{d^2\theta_n}{dt^2} + \kappa \frac{d\theta_n}{dt} + c^2 \left(\frac{n\pi}{l} \right)^2 \theta_n - \Phi_n \right] \sin \frac{n\pi x}{l} = 0.$$

Since Fourier series expansions are unique, the coefficient of each term must be zero at all times, i.e.

$$\frac{d^2\theta_n}{dt^2} + \frac{d\theta_n}{dt} + \left(\frac{n\pi c}{l} \right)^2 \theta_n = \Phi_n. \tag{2.57}$$

This ordinary differential equation has its constants of integration determined by the given initial values of displacement and velocity, $u(x)$ and $v(x)$ respectively. From eqn (2.54),

$$u(x) = y(x, 0) = \sum_{n=1}^{\infty} \theta_n(0) \sin \frac{n\pi x}{l}$$

and

$$v(x) = \left(\frac{\partial y}{\partial t} \right)_{t=0} = \sum_{n=1}^{\infty} \left(\frac{d\theta_n}{dt} \right)_{t=0} \sin \frac{n\pi x}{l}.$$

Using eqns (2.30), we see that

$$\theta_n = a_n \quad \text{and} \quad \frac{d\theta_n}{dt} = b_n \quad \text{at } t=0, \tag{2.58}$$

where a_n and b_n are the coefficients of the Fourier sine series expansions of $u(x)$ and $v(x)$ respectively in the interval $0 \leqslant x \leqslant l$.

Equation (2.57) with the initial condition of eqn (2.58) can be solved for θ_n. The Laplace transform of eqns (2.57) and (2.58) is

$$\left[p^2 + \kappa p + \left(\frac{n\pi c}{l}\right)^2\right]\bar{\theta}_n = pa_n + b_n + \kappa a_n + \bar{\Phi}_n,$$

where $\bar{f}(p) = \int_0^\infty e^{-pt} f(t)\, dt$. Let α_n and β_n be the roots of $p^2 + \kappa p + (n\pi c/l)^2 = 0$. Then,

$$\bar{\theta}_n = \frac{pa_n + b_n + \kappa a_n}{(p-\alpha_n)(p-\beta_n)} + \frac{\bar{\Phi}_n}{(p-\alpha_n)(p-\beta_n)}. \tag{2.59}$$

Separating into partial fractions,

$$\bar{\theta}_n = \frac{\alpha_n a_n + b_n + \kappa a_n}{\alpha_n - \beta_n} \frac{1}{p-\alpha_n} + \frac{\beta_n a_n + b_n + \kappa a_n}{\beta_n - \alpha_n} \frac{1}{p-\beta_n}$$
$$+ \left(\frac{1}{p-\alpha_n} - \frac{1}{p-\beta_n}\right) \frac{\bar{\Phi}_n}{\alpha_n - \beta_n}.$$

Therefore,

$$\theta_n = \frac{\alpha_n a_n + b_n + \kappa a_n}{\alpha_n - \beta_n} e^{\alpha_n t} + \frac{\beta_n a_n + b_n + \kappa a_n}{\beta_n - \alpha_n} e^{\beta_n t}$$
$$+ \frac{1}{\alpha_n - \beta_n} \int_0^t (e^{\alpha_n(t-\tau)} - e^{\beta_n(t-\tau)}) \Phi_n(\tau)\, d\tau. \tag{2.60}$$

Substituting back into eqn (2.54) for θ_n gives the displacement of the string. We shall now consider particular cases using the method of solution just developed.

First consider the case dealt with in previous sections, namely no air resistance and no external force, i.e. $\kappa = 0$, $F = 0$ hence $\Phi_n = 0$. Equation (2.59) gives

$$\alpha_n = \frac{n\pi c}{l} i \quad \text{and} \quad \beta_n = -\frac{n\pi c}{l} i,$$

where $\mathrm{i}=\sqrt{-1}$. Substituting in eqn (2.60),

$$\theta_n=\frac{(n\pi c/l)a_n\mathrm{i}+b_n}{(2n\pi c/l)\mathrm{i}}\exp\left(\mathrm{i}\frac{n\pi c}{l}t\right)-\frac{-(n\pi c/l)a_n\mathrm{i}+b_n}{(2n\pi c/l)\mathrm{i}}\exp\left(-\mathrm{i}\frac{n\pi c}{l}t\right)$$

$$=a_n\cos\left(\frac{n\pi ct}{l}\right)+\frac{l}{n\pi c}b_n\sin\left(\frac{n\pi ct}{l}\right).$$

Substituting back into eqn (2.54),

$$y=\sum_{n=1}^{\infty}\sin\frac{n\pi x}{l}\left(a_n\cos\frac{n\pi ct}{l}+\frac{l}{n\pi c}b_n\sin\frac{n\pi ct}{l}\right). \tag{2.61}$$

Comparison shows that eqn (2.61) is identical with eqn (2.11) when the initial conditions, eqns (2.30), have been used to determine the arbitrary constants A_n and B_n, eqns (2.31).

Since any physically possible transverse displacement of the string can be expressed, and expressed uniquely, in the form of equation (2.54) and since the solution of eqn (2.57) with the boundary conditions of eqn (2.58), given by eqn (2.60), is unique, it follows that the solution obtained by substituting back from eqns (2.60) into eqn (2.54) is both general and unique. In particular the solution given by eqn (2.61) is both general and unique whenever air resistance and external force are absent.

Secondly consider the case of air resistance but no external force. The air resistance will be assumed sufficiently small for κ to be less than $2\pi c/l$. Then, for all n, the discriminant of eqn (2.59), $[\kappa^2-(4n^2\pi^2c^2/l^2)]$, is negative and the α_n and β_n are given by

$$\alpha_n=-\frac{\kappa}{2}+\gamma_n\mathrm{i}\quad\text{and}\quad\beta_n=-\frac{\kappa}{2}-\gamma_n\mathrm{i},\quad\text{where}\quad\gamma_n=\left(\frac{n^2\pi^2c^2}{l^2}-\frac{\kappa^2}{4}\right)^{\frac{1}{2}}, \tag{2.62}$$

the positive sign of the square root being taken. Substituting in eqn (2.60),

$$\theta_n=\frac{(-\kappa/2+\gamma_n\mathrm{i})a_n+b_n+\kappa a_n}{2\gamma_n\mathrm{i}}\mathrm{e}^{(-\kappa/2+\gamma_n\mathrm{i})t}$$

$$-\frac{(-\kappa/2-\gamma_n\mathrm{i})a_n+b_n+\kappa a_n}{2\gamma_n\mathrm{i}}\mathrm{e}^{(-\kappa/2-\gamma_n\mathrm{i})t}$$

$$=\mathrm{e}^{-\kappa t/2}\left(a_n\cos\gamma_n t+\frac{b_n+(\kappa/2)a_n}{\gamma_n}\sin\gamma_n t\right).$$

Substituting back for θ_n and γ_n into eqn (2.54),

$$y(x,t)=\mathrm{e}^{-\kappa t/2}\sum_{n=1}^{\infty}\sin\frac{n\pi x}{l}\left\{a_n\cos\left[\left(\frac{n^2\pi^2c^2}{l^2}-\frac{\kappa^2}{4}\right)^{\frac{1}{2}}t\right]\right.$$

$$\left.\left[b_n+\frac{\kappa}{2}a_n\left(\frac{n^2\pi^2c^2}{l^2}-\frac{\kappa^2}{4}\right)^{-\frac{1}{2}}\right]\sin\left[\left(\frac{n^2\pi^2c^2}{l^2}-\frac{\kappa^2}{4}\right)^{\frac{1}{2}}t\right]\right\}. \quad (2.63)$$

Compare eqn (2.63) with eqn (2.61). It is seen that air resistance causes the amplitude to decay exponentially with time at a rate which is the same for all modes and that the frequency of the nth normal mode is no longer proportional to n. However, if the air resistance is so small that κ^2 can be neglected compared to $(4\pi^2c^2/l^2)$, then the frequencies of the modes are nearly proportional to n; but there is still a small exponential decay factor which will eventually damp out the vibration.

2.6. Point force, delta functions and resonance

The consideration of solutions obtained from eqn (2.60) continues in this section with the case of a sinusoidally oscillatory force of amplitude μ and radian frequency q applied at time $t=0$ to the string at a fixed point $x=X$. The string is assumed to be initially undisplaced and at rest and to vibrate against small air resistance. The subsequent motion is known as a forced oscillation.

What is the force distribution $F(x,t)$ in this case? We first treat a constant point force of unit magnitude. The force distribution per unit length is ρF. Consider a force distribution that is zero both for $x<X-\varepsilon$ and for $x>X+\varepsilon$ and equal to $1/(2\varepsilon)$ for $X-\varepsilon<x<X+\varepsilon$; Fig. 2.9 illustrates the distributions obtained by successive halving of the value of ε.

The total force applied to the string is the integral with respect to length of the force per unit lenth, i.e. $\int_0^l \rho F\,\mathrm{d}x$. In this case

$$\int_0^l \rho F\,\mathrm{d}x=\int_{x-\varepsilon}^{x+\varepsilon}\frac{1}{2\varepsilon}\,\mathrm{d}x=1.$$

The total force applied with the given force distribution is therefore unity. Now decrease ε. It is seen from the figure that the length over which the force is applied decreases, but the magnitude of the force per unit length at each point in the interval of application increases

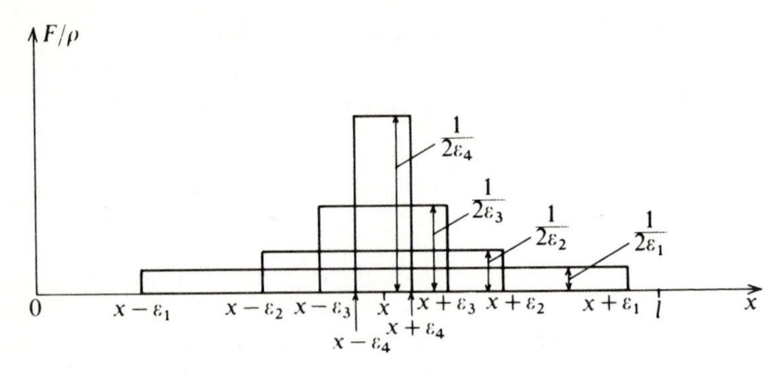

Fig. 2.9

and in such a way that the total area under the force distribution curve is constant and equal to unity. In the limit, as $\varepsilon \to 0$, the force distribution is zero at all points except X, where it is infinite. Since the total force remains unity, the limit is equivalent to a point force of unit magnitude applied at $x = X$.

It was the occurrence in applied mathematics of functions with the property of the force distribution we are considering that led P.A.M. Dirac in 1930 to define the delta function $\delta(x)$ by

$$\delta(x) = 0 \quad \text{for} \quad x \neq 0 \tag{2.64}$$

and

$$\int_{-\infty}^{\infty} \delta(x)\,\mathrm{d}x = 1. \tag{2.65}$$

No ordinary function of classical analysis satisfies eqns (2.64) and (2.65). Delta functions are treated rigorously in the theory of distributions. Below are developed the few properties we require. Since $\delta(x) = 0$ for $x \neq 0$, eqn (2.65) can be replaced by

$$\int_{-l}^{m} \delta(x)\,\mathrm{d}x = 1, \tag{2.66}$$

where l and m are any two positive numbers. In particular,

$$\lim_{\varepsilon \to 0} \int_{-\varepsilon}^{\varepsilon} \delta(x)\,\mathrm{d}x = 1. \tag{2.67}$$

Note that ε can be made arbitrarily small but not equal to zero. If $f(x)$ is any ordinary function continuous at $x = 0$,

$$\int_{-l}^{m} \delta(x) f(x)\,\mathrm{d}x = \int_{-l}^{m} \delta(x) f(0)\,\mathrm{d}x = f(0) \int_{-l}^{m} \delta(x)\,\mathrm{d}x = f(0), \quad (2.68)$$

since $\delta(x)=0$ except at $x=0$. An immediate corollary is

$$\int_{a-l}^{a+m} \delta(x-a) f(x)\mathrm{d}x \underset{x-a=y}{=} \int_{-l}^{m} \delta(y) f(y+a)\mathrm{d}y = f(a), \quad (2.69)$$

provided $f(x)$ is continuous at $x=a$.

The constant point force of unit magnitude applied at $x=X$, $0<X<l$, has force distribution

$$\rho F(x) = \delta(x-X). \quad (2.70)$$

Proof.

(i) If $x \neq X$, $F(x)=0$ and $\delta(x-X)=0$ by eqn (2.64).

(ii) $\int_0^l \rho F(x)\,\mathrm{d}x = \int_0^l \delta(x-X)\,\mathrm{d}x \underset{x-X=y}{=} \int_{-X}^{l-X} \delta(y)\,\mathrm{d}y = 1$ as required, since $l-X>0$ and $-X<0$.

Note that $\int_A^B \delta(x-a)\,\mathrm{d}x$ equals unity only if $A<a<B$. If $a>B$ or $a<A$, the integral is zero.

The point force $\mu \sin qt$ applied at $x=X$ has force distribution

$$\rho F(x, t) = \mu \sin qt\, \delta(x-X). \quad (2.71)$$

The corresponding values of $\Phi_n(t)$ are given by eqn (2.56) as

$$\Phi_n(t) = \frac{2}{l} \int_0^l \frac{\mu}{\rho} \sin qt\, \delta(x-X) \sin \frac{n\pi x}{l}\,\mathrm{d}x$$

$$= \frac{2\mu}{\rho l} \sin qt \int_0^l \delta(x-X) \sin \frac{n\pi x}{l}\,\mathrm{d}x$$

$$= \frac{2\mu}{\rho l} \sin qt \sin \frac{n\pi X}{l}, \quad (2.72)$$

by eqn (2.69).

It is convenient to work with exponential functions of time rather than with sines and cosines. If $\mathrm{R}[z]$ denotes the real part of z, then the equation for $\Phi_n(t)$ can be rewritten as

$$\Phi_n(t) = \mathrm{R}\left[-\frac{2\mu \mathrm{i}}{\rho l} \mathrm{e}^{\mathrm{i}qt} \sin \frac{n\pi X}{l} \right]. \quad (2.73)$$

Since the string is initially undisplaced and at rest, $a_n = b_n = 0$.

Substituting for Φ_n, a_n and b_n in eqn (2.60),

$$\theta_n(t)=\frac{1}{\rho}\,\mathrm{R}\left[\frac{1}{\alpha_n-\beta_n}\int_0^t\left(\mathrm{e}^{\alpha_n(t-\tau)}-\mathrm{e}^{\beta_n(t-\tau)}\right)\right.$$
$$\left.\times\left(-\frac{2\mu\mathrm{i}}{l}\,\mathrm{e}^{\mathrm{i}q\tau}\sin\frac{n\pi X}{l}\right)\mathrm{d}\tau\right].$$

On integration,

$$\theta_n(t)=\frac{1}{\rho}\,\mathrm{R}\left[\frac{1}{\alpha_n-\beta_n}\left(-\frac{2\mu\mathrm{i}}{l}\sin\frac{n\pi X}{l}\right)\right.$$
$$\left.\times\left(\mathrm{e}^{\alpha_n t}\frac{\mathrm{e}^{(\mathrm{i}q-\alpha_n)\tau}\big|_0^t}{\mathrm{i}q-\alpha_n}-\mathrm{e}^{\beta_n t}\frac{\mathrm{e}^{(\mathrm{i}q-\beta_n)\tau}\big|_0^t}{\mathrm{i}q-\beta_n}\right)\right].\tag{2.74}$$

Inserting the limits of integration and using eqns (2.62),

$$\theta_n(t)=-\frac{\mu}{\rho\gamma_n l}\sin\frac{n\pi X}{l}\,\mathrm{R}\left[\frac{\mathrm{e}^{\mathrm{i}qt}-\mathrm{e}^{\alpha_n t}}{\mathrm{i}q+(\kappa/2)-\gamma_n\mathrm{i}}-\frac{\mathrm{e}^{\mathrm{i}qt}-\mathrm{e}^{\beta_n t}}{\mathrm{i}q+(\kappa/2)+\gamma_n\mathrm{i}}\right]\tag{2.75}$$

$$=-\frac{\mu}{\rho\gamma_n l}\sin\frac{n\pi X}{l}\,\mathrm{R}\left[\frac{(\mathrm{e}^{\mathrm{i}qt}-\mathrm{e}^{-\kappa t/2}\mathrm{e}^{\gamma_n\mathrm{i}t})((\kappa/2)-\mathrm{i}q+\gamma_n\mathrm{i})}{(\kappa^2/4)+(q-\gamma_n)^2}\right.$$
$$\left.-\frac{(\mathrm{e}^{\mathrm{i}qt}-\mathrm{e}^{-\kappa t/2}\mathrm{e}^{-\gamma_n\mathrm{i}t})((\kappa/2)-\mathrm{i}q-\gamma_n\mathrm{i})}{(\kappa^2/4)+(q+\gamma_n)^2}\right]$$

$$=-\frac{\mu}{\rho\gamma_n l}\sin\frac{n\pi X}{l}$$
$$\times\left\{\frac{(\kappa/2)\cos qt-(\kappa/2)\mathrm{e}^{-\kappa t/2}\cos\gamma_n t+(q-\gamma_n)\sin qt-(q-\gamma_n)\mathrm{e}^{-\kappa t/2}\sin}{(\kappa^2/4)+(q-\gamma_n)^2}\right.$$

$$\left.-\frac{(\kappa/2)\cos qt-(\kappa/2)\mathrm{e}^{-\kappa t/2}\cos\gamma_n t+(q+\gamma_n)\sin qt+(q+\gamma_n)\mathrm{e}^{-\kappa t/2}\sin\gamma_n t}{(\kappa^2/4)+(q+\gamma_n)^2}\right\}$$

$$=2\frac{\dfrac{\mu}{\rho l}\sin\dfrac{n\pi X}{l}}{\left(\dfrac{n^2\pi^2c^2}{l^2}-q^2\right)^2+q^2\kappa^2}\left\{-q\kappa\cos qt+\left(\frac{n^2\pi^2c^2}{l^2}-q^2\right)\sin qt\right.$$

$$\left.+\mathrm{e}^{-\kappa t/2}\left[q\kappa\cos\gamma_n t+\frac{q}{2\gamma_n}\left(\kappa^2+2q^2-\frac{2_n^2\pi^2c^2}{l^2}\right)\sin\gamma_n t\right]\right\},\tag{2.76}$$

on substituting $\gamma_n^2+(\kappa^2/4)=n^2\pi^2c^2/l^2$ from eqn (2.62).

Substituting for θ_n in eqn (2.54) gives the displacement as

$$y = 2\frac{\mu}{\rho l}\sum_{n=1}^{\infty}\frac{\sin\dfrac{n\pi X}{l}\sin\dfrac{n\pi x}{l}}{\left(\dfrac{n^2\pi^2c^2}{l^2}-q^2\right)^2+q^2\kappa^2}$$

$$\times\left\{-q\kappa\cos qt+\left(\frac{n^2\pi^2c^2}{l^2}-q^2\right)\sin qt\right.$$

$$\left.+\mathrm{e}^{-\kappa t/2}\left[q\kappa\cos\gamma_n t+\frac{q}{2\gamma_n}\left(\kappa^2+2q^2-\frac{2n^2\pi^2c^2}{l^2}\right)\sin\gamma_n t\right]\right\}. \quad (2.77)$$

For each value of n, the last two terms in eqn (2.77) are multiplied by the exponentially decreasing time factor $\mathrm{e}^{-\kappa t/2}$ and therefore, by allowing sufficient time to elapse, the ratio of these two terms to the first two terms can be made arbitrarily small. The first two terms have the same frequency q and can be combined. Neglecting the last two terms, each value of n represents an oscillation in a normal mode with frequency q and normal amplitude

$$C_n = 2\frac{\mu}{\rho l}\sin\frac{n\pi X}{l}\left[\left(\frac{n^2\pi^2c^2}{l^2}-q^2\right)^2+q^2\kappa^2\right]^{-\frac{1}{2}}, \quad (2.78)$$

i.e. after a sufficient time has elapsed for the effect of the initial conditions to be negligible, the frequency of each mode and therefore the frequency of the whole vibration is the same as that of the applied force.

A normal amplitude will be zero if $\sin(n\pi X/l)=0$, i.e. if the mode has a node at the point of application of the force. For example, if the force is applied at the mid-point of the string, no even-order normal modes will be generated, only odd-order ones. Rayleigh in his theory of sound points out that if, after a string is struck at a point, that point is brought to rest at a later time, then the whole string will come to rest at that time because all modes which do not have nodes at this point will be brought to rest and the modes which have nodes at this point will not have been generated in the first place.

Assuming that $\kappa^2 \ll \pi^2c^2/l^2$, as usually occurs in practice, consider how the normal amplitudes vary with the frequency q of the imposed force. It is convenient to rewrite eqn (2.78) in the form

$$C_n = 2\frac{\mu}{\rho l}\sin\frac{n\pi X}{l}\left[\left(\frac{n^2\pi^2c^2}{l^2}-\frac{\kappa^2}{2}-q^2\right]^2+\kappa^2\left(\frac{n^2\pi^2c^2}{l^2}-\frac{\kappa^2}{4}\right)\right]^{-\frac{1}{2}} \quad (2.79)$$

C_n is a maximum when the first term in the square brackets is zero, i.e. when $q = q_n$, where

$$q_n = \left(\frac{n^2\pi^2c^2}{l^2} - \frac{\kappa^2}{2}\right)^{\frac{1}{2}}; \tag{2.80}$$

$C_n \simeq (2\mu/\rho n\pi c\kappa)\sin(n\pi X/l)$ at the maximum. When q differs appreciably from q_n, the first term in the square brackets dominates the second and C_n is of the order of $(2\mu/\rho n\pi c)\sin(n\pi X/l)$, which is of the order of κ times the value of C_n at the maximum. It follows that the amplitude response of the displacement at any point of the string plotted as a function of the forcing frequency will show a series of maxima, each one close to one of the natural frequencies, $n\pi c/l$, of the string (see Fig. 2.10).

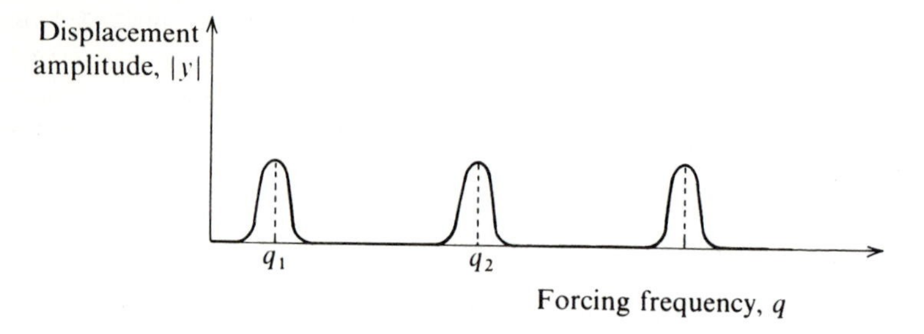

Fig. 2.10

This phenomenon is known as 'resonance'. However, no maximum will appear for a particular value of n if either the point at which the force is applied or the point at which the displacement is measured coincide with a node of the nth mode. In the absence of air resistance, $\kappa = 0$, the maxima occur at the natural frequencies and the corresponding amplitudes are infinite.

Another interesting result can be deduced from equation (2.77) because it is completely symmetrical in x and X; namely, the displacement produced at the point x due to the action of a force at the point X is equal to the displacement that would be produced at the point X if the same force acted at the point x.

2.7. Finite-duration point force

The last case we shall consider is that of the string acted on by a point force for a short period of time only. This is perhaps the most important practical application of this theory because it applies to the vibration of a wire in a piano. The hammer that strikes the piano wire is covered by a piece of felt and this causes the impact to be spread over a short period of time. We shall use the approximation, introduced by Helmholtz, that the force acting on the wire is of the form $\mu \sin qt$ while it acts and that it acts over a half-cycle only, i.e. for a time π/q. The value of q depends upon the mss and elasticity of the hammer and felt but not appreciably upon the velocity with which it strikes the string.

If we again take the point of impact to be $x = X$ then the problem is identical to the previous case for all times[†] $t \leqslant \pi/q$ and the solution for $t \leqslant \pi/q$ is therefore given by eqn (2.77). $\Phi_n(t)$ is given by eqn (2.72) for $t \leqslant \pi/q$ and $\Phi_n(t) = 0$ for $t \geqslant \pi/q$ by eqn (2.56). Therefore, on substituting for $\Phi_n(t)$ into eqn (2.60) for $t \geqslant \pi/q$, we can still use eqn (2.73) for $\Phi_n(t)$ provided we replace the upper limit of integration in eqn (2.60) by π/q because the integrand is zero for $t > \pi/q$. Equation (2.74) gives $\theta_n(t)$ provided that the upper limit of integration is replaced by π/q, i.e.

$$\theta_n(t) = \frac{1}{\rho} \mathrm{R}\left[\frac{1}{\alpha_n - \beta_n} \left(-\frac{2\mu \mathrm{i}}{l} \sin \frac{n\pi X}{l} \right) \times \left(\mathrm{e}^{\alpha_n t} \frac{\mathrm{e}^{(\mathrm{i}q - \alpha_n)\tau} \big|_0^{\pi/q}}{\mathrm{i}q - \alpha_n} - \mathrm{e}^{\beta_n t} \frac{\mathrm{e}^{(\mathrm{i}q - \beta_n)\tau} \big|_0^{\pi/q}}{\mathrm{i}q - \beta_n} \right) \right], \quad t \geqslant \frac{\pi}{q}.$$

Inserting the limits of integration, using eqns (2.62) and $\mathrm{e}^{\mathrm{i}\pi} = -1$,

$$\theta_n(t) = \frac{\mu}{\rho \gamma_n l} \sin \frac{n\pi X}{l} \mathrm{R}\left[\frac{\mathrm{e}^{\alpha_n (t - \pi/q)} + \mathrm{e}^{\alpha_n t}}{\mathrm{i}q - \alpha_n} - \frac{\mathrm{e}^{\beta_n (t - \pi/q)} + \mathrm{e}^{\beta_n t}}{\mathrm{i}q - \beta_n} \right], \quad t \geqslant \frac{\pi}{q}. \tag{2.81}$$

The terms in $\mathrm{e}^{\alpha_n t}$ and $\mathrm{e}^{\beta_n t}$ in eqn (2.81) are identical to those appearing in eqn (2.75). They give rise to the terms in $\cos \gamma_n t$ and $\sin \gamma_n t$ in eqn (2.76). The terms in $\mathrm{e}^{\alpha_n (t - \pi/q)}$ and $\mathrm{e}^{\beta_n (t - \pi/q)}$ give rise to the same terms but in $\cos \gamma_n (t - \pi/q)$ and $\sin \gamma_n (t - \pi/q)$. Hence, from eqn (2.81), by comparison with eqn (2.76),

† The string is at rest until $t = 0$ and therefore times $t < 0$ are not included in the analysis. To make this explicit, insert $H(t)$ on the r.h.s. of the equation for $\theta_n(t)$.

$$\theta_n(t)=\frac{2\dfrac{\mu}{\rho l}\sin\dfrac{n\pi X}{l}}{\left(\dfrac{n^2\pi^2c^2}{l^2}-q^2\right)^2+q^2\kappa^2}$$

$$\times\left\{e^{-\kappa t/2}\left[q\kappa\cos\gamma_n t+\frac{q}{2\gamma_n}\left(\kappa^2+2q^2-\frac{2n^2\pi^2c^2}{l^2}\right)\sin\gamma_n t\right]\right.$$

$$+e^{-\kappa(t-\pi/q)/2}\left[q\kappa\cos\gamma_n(t-\pi/q)\right.$$

$$\left.\left.+\frac{q}{2\gamma_n}\left(\kappa^2+2q^2-\frac{2n^2\pi^2c^2}{l^2}\right)\sin\gamma_n(t-\pi/q)\right]\right\},\quad t\geqslant\frac{\pi}{q}. \tag{2.82}$$

Substitution for the θ_n from eqn (2.82) into eqn (2.54) gives the displacement at later times.

For simplicity, we shall discuss the solution in the absence of air resistance, i.e. $\kappa=0$. In this case, eqn (2.82) becomes

$$\theta_n(t)=\frac{-2\mu q\sin\dfrac{n\pi X}{l}}{\rho l\gamma_n\left(\dfrac{n^2\pi^2c^2}{l^2}-q^2\right)}(\sin\gamma_n t+\sin\gamma_n(t-\pi/q))$$

and, from eqn (2.62), $\gamma_n=n\pi c/l$. Therefore

$$\theta_n(t)=-\frac{4\mu q\sin\dfrac{n\pi X}{l}}{\rho n\pi c\left(\dfrac{n^2\pi^2c^2}{l^2}-q^2\right)}\sin\left(\frac{n\pi c}{l}\left(t-\frac{\pi}{2q}\right)\right)\cos\frac{n\pi^2c}{2lq},\quad t\geqslant\frac{\pi}{q}. \tag{2.83}$$

Substituting in eqn (2.54),

$$y(x,t)=-\frac{4\mu q l^2}{\rho\pi c}\sum_{n=1}^{\infty}\frac{\sin\dfrac{n\pi X}{l}\sin\dfrac{n\pi x}{l}}{n(n^2\pi^2c^2-q^2l^2)}$$

$$\times\sin\frac{n\pi c}{l}\left(t-\frac{\pi}{q}\right)\cos\frac{n\pi^2c}{2lq},\quad t\geqslant\frac{\pi}{q}. \tag{2.84}$$

Equation (2.84) shows that the normal modes, which are absent after a piano wire has been struck by its hammer, include not only those that have a node at the point of impact ($\sin(n\pi X/l)=0$) but also those for which $n\pi^2 c/2lq$ is an odd multiple of $\pi/2$, i.e. those for which $n=(2N+1)\,lq/\pi c$, where N is an integer.

Now π/q is the duration of contact of the hammer on the wire, which depends on the mass and elasticity of hammer and felt, and $2l/c$ is the fundamental period of the string. Therefore, a particular mode, say the nth mode, will not be present if n is any odd integral multiple of the ratio of half the fundamental period to the duration of contact of the hammer. It is part of the art of the piano designer so to choose the mass and elasticity of the hammer to ensure the presence or the absence, as required, or particular harmonics. Note that, as well as the factors $\sin(n\pi X/l)$ and $\cos(n\pi^2 c/l)$, each term in eqn (2.84) is multiplied by a coefficient or order $1/n^3$ and therefore the amplitude of the higher harmonics decreases rapidly with n. If the air resistance and internal friction that occur in nature were taken into account, the above remarks would require but little modification. Instead of certain modes vanishing altogether, they become very small compared to those which do not vanish in the absence of the dissipative forces.

Exercise 2.8. Repeat the problem in Exercise 2.3, but include air resistance.

Exercise 2.9. Repeat the problem in Exercise 2.5, but include air resistance.

Exercise 2.10. To a string undisplaced and at rest, a transverse force distribution $\rho F(x,t)=\mu x(l-x)\sin qt$ is applied at time $t=0$. Including air resistance, find the displacement when $t>0$.

As stated at the beginning of Section 1.7, the effect of resistance in reality is more complicated than the simple proportionality to velocity which has been assumed. In fact, air resistance increases more rapidly than the velocity in a rather complicated manner and there are other resistive forces, such as internal friction and damping at the supports, which act on a vibrating string. However, the linear damping used in this section reproduces theoretically most of the important effects of the actual resistive forces, without requiring the introduction of more advanced mathematical techniques.

Answers to exercises

2.1. $y=\dfrac{\eta l}{\pi c}\sin\dfrac{\pi x}{l}\sin\dfrac{\pi ct}{l}+\varepsilon\sin\dfrac{2\pi x}{l}\cos\dfrac{2\pi ct}{l}.$

2.2. $y=\varepsilon_1 \sin\frac{\pi x}{l}\cos\frac{\pi ct}{l}+\varepsilon_3 \sin\frac{3\pi x}{l}\cos\frac{3\pi ct}{l}.$

2.3. $$y=\frac{16h}{\sqrt{2}\pi^2}\left(\sin\frac{\pi x}{l}\cos\frac{\pi ct}{l}+\frac{1}{3^2}\sin\frac{3\pi x}{l}\cos\frac{3\pi ct}{l}-\frac{1}{5^2}\sin\frac{5\pi x}{l}\cos\frac{5\pi ct}{l}-\frac{1}{7^2}\sin\frac{7\pi x}{l}\cos\frac{7\pi ct}{l}+\ldots\right).$$

2.4. $$y=\frac{6\sqrt{3}h}{\pi^2}\left(\sin\frac{\pi x}{l}\cos\frac{\pi ct}{l}-\frac{1}{5^2}\sin\frac{5\pi x}{l}\cos\frac{5\pi ct}{l}+\frac{1}{7^2}\sin\frac{7\pi x}{l}\cos\frac{7\pi ct}{l}-\frac{1}{11^2}\sin\frac{11\pi x}{l}\cos\frac{11\pi ct}{l}+\ldots\right).$$

2.5. $$y=\frac{8\alpha}{\pi^2}\left(\sin\frac{\pi x}{l}\cos\frac{\pi ct}{l}+\frac{1}{3^3}\sin\frac{3\pi x}{l}\cos\frac{3\pi ct}{l}+\frac{1}{5^3}\sin\frac{5\pi x}{l}\cos\frac{5\pi ct}{l}+\ldots\right)$$
$$+\frac{4l\beta}{\pi^3 c}\left(\sin\frac{\pi x}{l}\sin\frac{\pi ct}{l}-\frac{1}{3^3}\sin\frac{3\pi x}{l}\sin\frac{3\pi ct}{l}+\frac{1}{5^2}\sin\frac{5\pi x}{l}\sin\frac{5\pi ct}{l}-\ldots\right).$$

2.6. $$K=\frac{32\rho c^2h^2}{l\pi^2}\left(\sin\frac{2\pi ct}{l}+\frac{1}{3^2}\sin^2\frac{3\pi ct}{l}+\frac{1}{5^2}\sin^2\frac{5\pi ct}{l}+\ldots\right),$$
$$V=\frac{32Th^2}{l\pi^2}\left(\cos^2\frac{\pi ct}{l}+\frac{1}{3^2}\cos^2\frac{3\pi ct}{l}+\frac{1}{5^2}\cos^2\frac{5\pi ct}{l}+\ldots\right),$$
$$K+V=\frac{4Th^2}{l},\ \frac{8}{\pi^2}\text{ or about 81 per cent.}$$

2.7. $$K=\frac{32\rho c^2\alpha^2}{l\pi^4}\left[\sin^2\left(\frac{\pi ct}{l}-\frac{\pi}{4}\right)+\frac{1}{3^4}\sin^2\left(\frac{3\pi ct}{l}+\frac{\pi}{4}\right)+\frac{1}{5^4}\sin^2\left(\frac{5\pi ct}{l}-\frac{\pi}{4}\right)+\ldots\right],$$
$$V=\frac{32T^2\alpha^2}{l\pi^4}\left[\cos^2\left(\frac{\pi ct}{l}-\frac{\pi}{4}\right)+\frac{1}{3^4}\cos^2\left(\frac{3\pi ct}{l}+\frac{\pi}{4}\right)+\frac{1}{5^4}\cos^2\left(\frac{5\pi ct}{l}-\frac{\pi}{4}\right)+\ldots\right],$$

$$K+V=\frac{T\alpha^2}{3l}, \frac{96}{\pi^4} \text{ or about 98.5 per cent.}$$

2.8. Equation (63) with $a_n=\dfrac{16h}{n^2\pi^2}\sin\dfrac{n\pi}{2}\cos\dfrac{n\pi}{4}$ and $b_n=0$.

2.9. Equation (63) with $a_n=\dfrac{4\alpha}{n^3\pi^3}(1-(-)^n)$ and $b_n=\dfrac{4\beta}{n^2\pi^2}\sin\dfrac{n\pi}{2}$.

2.10.
$$y=\frac{4\mu l^2}{\rho\pi^3}\sum_{n=1}^{\infty}\frac{(1-(-)^n)\sin\dfrac{n\pi x}{l}}{n^2\left[\left(\dfrac{n^2\pi^2c^2}{l^2}-q^2\right)^2+q^2\kappa^2\right]}$$
$$\times\left\{-q\kappa\cos qt+\left(\frac{n^2\pi^2c^2}{l^2}-q^2\right)\sin qt\right.$$
$$\left.+\,\mathrm{e}^{-\kappa t/2}\left[q\kappa\cos\gamma_n t+\frac{q}{2\gamma_n}\left(\kappa^2+2q^2-\frac{2n^2\pi^2c^2}{l^2}\right)\sin\gamma_n t\right]\right\}$$

where

$$\gamma_n=\sqrt{\frac{n^2\pi^2c^2}{l^2}-\frac{\kappa^2}{4}}.$$

3 Long waves on canals

3.1. Governing equations; free and forced oscillations

A typical canal has still water of depth h of about 1 metre. Waves observed travelling along canals have wavelengths λ much larger than 1 metre. The height of the water surface above its rest position is known as the elevation η. $|\eta|$ rarely exceeds 10 centimetres. Consequently

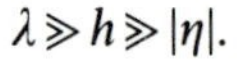

$$\lambda \gg h \gg |\eta|.$$

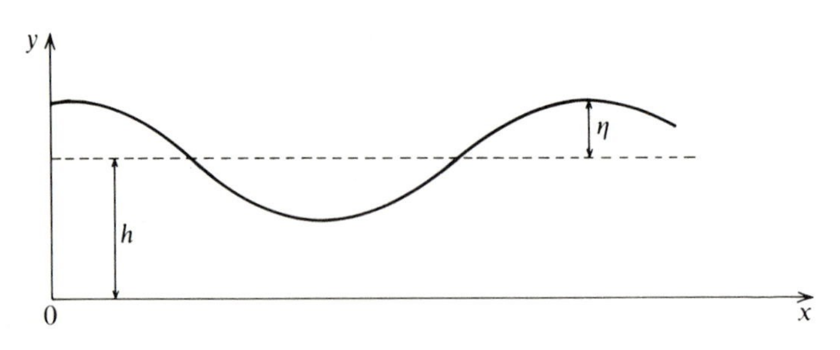

Fig. 3.1

It is consistent with the inequality $|\eta|/h \ll 1$ to assume changes in variables from their rest values are sufficiently small to enable squares and products of the changes to be neglected compared to terms linear in the changes. This is the linearization hypothesis common to many branches of applied mathematics. When the variable is zero in the rest state, then the change in the variable is equal to the value of the variable itself.

The relative smallness of the elevation η suggests that the vertical acceleration of the water particles is small. The theory of long waves on canals is based on the hypothesis that vertical acceleration is negligible. The consistency of this hypothesis is treated at the end of this section; the condition for its validity is found in the next chapter. The compressibility of water is sufficiently small to enable its density to be treated as constant.

Consider a straight canal with horizontal bed and parallel vertical sides. Let the axis of x be parallel to the length of the canal, the axis of y vertically upwards with $y=0$ as the plane of the bed. We shall look at motions which depend on x and y only. Viscous effects will not be included. The ordinate of the free surface at abscissa x and time t is $h+\eta(x, t)$. Let the components of the velocity of the water at any point be (u, v), where $u = u(x, y, t)$ and $v=v(x, y, t)$.

Since vertical acceleration is neglected, the pressure p at any point (x, y) exceeds the atmospheric pressure p_0 by an amount equal to ρg times the distance the point is below the surface, i.e.

$$p-p_0=\rho g(h+\eta-y), \tag{3.1}$$

where ρ is the (constant) density of water and g the acceleration due to gravity. Differentiating with respect to x,

$$\frac{\partial p}{\partial x}=\rho g\frac{\partial \eta}{\partial x}. \tag{3.2}$$

Since $\eta=\eta(x, t)$, both sides of the above equation are independent of y.

The equation of motion in the x-direction is

$$\rho\frac{\mathrm{D}u}{\mathrm{D}t}=-\frac{\partial p}{\partial x}, \tag{3.3}$$

where $\mathrm{D}/\mathrm{D}t$ is the material time derivative. It follows, since $\partial p/\partial x$ is independent of y, that the horizontal acceleration $\mathrm{D}u/\mathrm{D}t$ is the same for all particles in a plane $x=$constant. If the water was once at rest, then u is the same for all particles in this plane, which remains a plane with its x-coordinate constant, i.e. $u=u(x, t)$.

Now

$$\frac{\mathrm{D}u}{\mathrm{D}t}=\frac{\partial u}{\partial t}+u\frac{\partial u}{\partial x},$$

since $u=u(x, t)$, and by the linearity assumption the product $u(\partial u/\partial x)$ is neglected compared to $\partial u/\partial t$. Equation (2.3) becomes

$$\rho\frac{\partial u}{\partial t}=-\frac{\partial p}{\partial x}. \tag{3.4}$$

Let ξ be the displacement of a particle in the x-direction. Then

$$u=\frac{D\xi}{Dt}=\frac{\partial\xi}{\partial t} \tag{3.5}$$

by the linearity assumption. Elimination of u between eqns (3.4) and (3.5) gives

$$\rho\frac{\partial^2\xi}{\partial t^2}=-\frac{\partial p}{\partial x}, \tag{3.6}$$

and then elimination of p between eqns (3.2) and (3.6) gives

$$\frac{\partial^2\xi}{\partial t^2}=-g\frac{\partial\eta}{\partial x}. \tag{3.7}$$

Integrating eqn (5) with respect to t from τ to t,

$$\xi(x,t)=\int_\tau^t u(x,\theta)\mathrm{d}\theta+\xi(x,\tau).$$

It follows that if ξ is independent of y at some time τ, then it remains independent of y at later times t. This we take to be the case.

An equation of continuity can be found by integrating $\partial u/\partial x+\partial v/\partial y=0$ with respect to y and using the fact that $v=0$ on $y=0$:

$$v=-\int_0^y\frac{\partial u}{\partial x}\mathrm{d}y=-y\frac{\partial u}{\partial x} \tag{3.8}$$

since $\partial u/\partial x$ is independent of y. From eqn (3.8) it is seen that the vertical velocity of any particle is proportional to its height above the bottom. At the free surface $v=\partial\eta/\partial t$ and $y=h+\eta$.

Hence

$$\frac{\partial\eta}{\partial t}=-(h+\eta)\frac{\partial u}{\partial x}=-h\frac{\partial u}{\partial x}=-h\frac{\partial^2\xi}{\partial x\partial t} \tag{3.9}$$

on neglecting the product $\eta(\partial u/\partial x)$ compared to $h(\partial u/\partial x)$. Therefore,

$$\eta=-h\frac{\partial\xi}{\partial x}, \tag{3.10}$$

where the arbitrary function of integration $f(x)$ is zero provided there exists a time at which the water is at rest and $\eta=\xi\equiv 0$. Elimination of η from eqns (3.7) and (3.10) gives

$$\frac{\partial^2\xi}{\partial t^2}=gh\frac{\partial^2\xi}{\partial x^2}, \tag{3.11}$$

the classical wave equation with velocity $\sqrt{gh}$. Differentiation of eqn (3.11) with respect to x and substitution of $\eta=-h(\partial\xi/\partial x)$ gives

$$\frac{\partial^2\eta}{\partial t^2}=gh\frac{\partial^2\eta}{\partial x^2}. \tag{3.12}$$

The general solution of eqn (3.11) is D'Alemberts' solution

$$\xi=f(x-ct)+g(x+ct), \quad c=\sqrt{gh}, \tag{3.13}$$

and from eqn (3.10),

$$\eta=-h\{f'(x-ct)+g'(x+ct)\}. \tag{3.14}$$

The component of velocity in the x-direction is

$$u=\frac{\partial\xi}{\partial t}=-cf'(x-ct)+cg'(x+ct). \tag{3.15}$$

We can now investigate the circumstances in which the solution represented by eqn (3.13) will be consistent with the approximations used in deriving the solution.

The exact equation for the vertical component of momentum balance (Newton's 2nd law of motion) is

$$\rho\frac{\mathrm{D}v}{\mathrm{D}t}=-\frac{\partial p}{\partial y}-g\rho,$$

which on integration from y to $h+\eta$ gives

$$p-p_0=-g\rho(h+\eta-y)-\int_y^{h+\eta}\rho\frac{\mathrm{D}v}{\mathrm{D}t}\,\mathrm{d}y.$$

The neglect of vertical acceleration is justified provided

$$\beta(h+\eta)\ll g|\eta|, \tag{3.16}$$

where β is maximum vertical acceleration; $g\eta$ represents the increase in pressure due to gravitational force above the static value. Now if L is the distance between successive nodes of a wave, i.e. points at which $\eta=0$, the time elapsed between these nodes passing a fixed point is $L/\sqrt{gh}$. Therefore the vertical velocity is of the order of $(|\eta|\sqrt{gh})/L$ and the vertical acceleration is of order $(|\eta|gh)/L^2$. Putting $\beta=(|\eta|gh)/L^2$ in the inequality (3.16), the neglect of vertical acceleration is consistent if

$$h(h+\eta)\ll L^2. \tag{3.17}$$

Since $|\eta| \ll h$, eqn (3.17) can be replaced by

$$h \ll L. \tag{3.18}$$

For sinusoidal waves, $L = \frac{1}{2}\lambda$, where λ is the wavelength, and the criterion becomes

$$\frac{\lambda}{h} \gg 1. \tag{3.19}$$

This is the usual form of the criterion; the wavelength is large compared with the depth. Hence the nomenclature 'long waves on canals'. The 'canal' can be replaced by any body of incompressible liquid whose depth is small compared to the wavelength of the disturbance under consideration.

The second approximation was to use the linearity assumption to neglect the operator $u(\partial/\partial x)$ compared to $\partial/\partial t$. When the dependent variables depend on x and t through† $x - \sqrt{gh}\,t = \phi$, say,

$$u\frac{\partial}{\partial x} \equiv u\frac{\mathrm{d}}{\mathrm{d}\phi} \quad \text{and} \quad \frac{\partial}{\partial t} \equiv -\sqrt{gh}\frac{\mathrm{d}}{\mathrm{d}\phi}.$$

Hence the approximation is justified provided $|u| \ll \sqrt{gh} = c$, i.e. the particle velocity in the x-direction is small compared to the wave velocity. From eqns (3.14) and (3.15), when $g(x+ct) \equiv 0$, $|u|/c = |\eta|/h$ and therefore the second approximation is equivalent to

$$|\eta|/h \ll 1, \tag{3.20}$$

i.e. the surface elevation is small compared to the depth. Note that inequality (3.20) is quite distinct from inequality (3.18). It is possible that one inequality may remain valid when the other does not. The inequalities (3.18) and (3.20) will be satisfied for the general solution, equation (3.13), provided they are satisfied for the two travelling waves into which the general solution can be resolved. To sum up, the conditions for a consistent approximate theory of long waves on canals are $|\eta| \ll h \ll \lambda$, the observed inequalities of the first paragraph of this section.

The potential energy per unit breadth of a wave, or system of waves, due to the elevation or depression of the fluid above or below the mean level‡, is

---

† The positive moving wave.

‡ The potential energy of a mass M at height H is MgH.

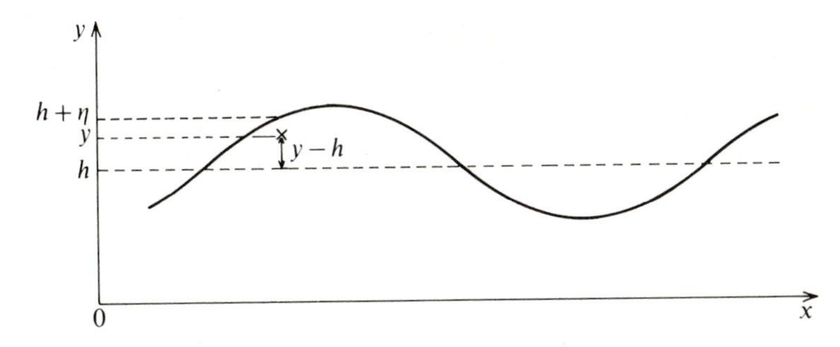

Fig. 3.2

$$V=\rho g\int\left(\int_{h}^{\eta+h}(y-h)\,\mathrm{d}y\right)\mathrm{d}x=\tfrac{1}{2}\rho g\int\eta^{2}\,\mathrm{d}x. \tag{3.21}$$

The fluid above the mean level has not only to fall back to that level but has also to fall further to occupy the 'hollow' left where the surface is below the mean level. Hence the integral is taken not only over positive values of η but also over negative ones.

The order of the ratio of the vertical velocity v of the fluid to its horizontal velocity u is

$$\frac{|\eta|\sqrt{gh}}{L}\bigg/\frac{c|\eta|}{h},$$

which equals h/L since $c=\sqrt{gh}$. The kinetic energy of the fluid per unit volume is $\frac{1}{2}\rho(u^2+v^2)=\frac{1}{2}\rho u^2(1+v^2/u^2)=\frac{1}{2}\rho u^2$, on neglecting h^2/L^2 compared to 1. If T is the kinetic energy per unit breadth, then

$$T=\tfrac{1}{2}\rho\int\left(\int_{0}^{h+\eta}u^{2}\,\mathrm{d}y\right)\mathrm{d}x=\tfrac{1}{2}\rho(h+\eta)\int u^{2}\,\mathrm{d}x \tag{3.22}$$

$$=\tfrac{1}{2}\rho h\int u^{2}\,\mathrm{d}x,$$

since u is independent of y and η is neglected compared to h. If the displacement of the surface represents a wave travelling in one direction only, from equations (3.13) and (3.14)

$$u=\pm\frac{c}{h}\eta, \tag{3.23}$$

and, therefore $V=T$ for a single travelling wave.

The above result $V=T$ has been proved by Rayleigh in the following manner. Any travelling wave $\xi=f(x-ct)$ can be considered as being set up by a stationary displacement $\xi=2f(x)$ released at time $t=0$. Two waves $\xi=f(x-ct)+f(x+ct)$ are released at this time. Each wave has one-quarter the potential energy of the stationary displacement because the amplitude of each wave is one half of that of the stationary wave. The kinetic energies of the two waves are equal by symmetry and, since total energy is conserved, each wave has kinetic energy equal to one-quarter of the stationary potential energy. The kinetic and potential energies of each wave are therefore identical.

Let us now look at free oscillations in a canal of length l with fixed vertical ends. The mathematical formulation of this problem is:

Solve

$$\frac{\partial^2 \xi}{\partial x^2}=\frac{1}{c^2}\frac{\partial^2 \xi}{\partial t^2}, \quad c^2=gh,$$

with

$$\xi=0 \quad \text{at} \quad x=0, l.$$

This is exactly the same problem mathematically as that for the transverse vibrations of a finite string with fixed ends, except that $c^2=gh$ instead of $c^2=T/\rho$. The solution is

$$\xi=\sum_{n=1}^{\infty} \sin\frac{n\pi x}{l}\left(A_n \cos\frac{n\pi ct}{l}+B_n \sin\frac{n\pi ct}{l}\right);$$

hence

$$\eta=-h\frac{\partial \xi}{\partial x}=-\frac{h\pi}{l}\sum_{n=1}^{\infty} n\cos\frac{n\pi x}{l}\left(A_n \cos\frac{n\pi ct}{l}+B_n \sin\frac{n\pi ct}{l}\right).$$

In each mode the water rises and falls up and down the vertical face at each end. The constants A_n and B_n are determined from the initial conditions; two are required for all x, $0\leqslant x\leqslant l$, either two on ξ or two on η or one each on ξ and η. The base cyclic frequency is

$$\frac{c}{2l}=\frac{\sqrt{gh}}{2l}.$$

As a second example consider a canal opening into a tidal estuary. The canal has length l, $x=l$ is a vertical fixed end and at the end $x=0$, which opens into the tidal estuary, the elevation η varies with the tidal frequency. Therefore the boundary conditions are $\xi=0$ at $x=l$ and $\eta=A\cos\omega t$ at $x=0$. Since this is a forced oscillation problem, look for a solution of the form $\xi=X(x)\cos\omega t$, the steady-state solution. Substitute for ξ into the governing partial differential equation (3.11) to give

$$X''(x)+\frac{\omega^2}{c^2}X(x)=0,\quad X(x)=A'\cos\left(\frac{\omega}{c}x\right)+B\sin\left(\frac{\omega}{c}x\right),$$

where A' and B are constants.

For $X(l)=0$, $A'=-B\tan(\omega l/c)$,

whence

$$X(x)=B\left(\sin\left(\frac{\omega}{c}x\right)-\tan\left(\frac{\omega}{c}l\right)\cos\left(\frac{\omega}{c}x\right)\right),$$

$$\xi=B\left(\sin\left(\frac{\omega}{c}x\right)-\tan\left(\frac{\omega}{c}l\right)\cos\left(\frac{\omega}{c}x\right)\right)\cos\omega t, \tag{3.24}$$

$$\eta=-h\frac{\partial\xi}{\partial x}=-\frac{B\omega h}{c}\left(\cos\left(\frac{\omega}{c}x\right)+\tan\left(\frac{\omega}{c}l\right)\sin\left(\frac{\omega}{c}x\right)\right)\cos\omega t. \tag{3.25}$$

For $\eta=A\cos\omega t$ at $x=0$,

$$-\frac{B\omega h}{c}=A,\quad B=-\frac{c}{\omega h}A.$$

Substitute back for B into eqns (3.24) and (3.25):

$$\xi=\frac{c}{\omega h}A\frac{\sin\left(\frac{\omega}{c}(l-x)\right)}{\cos\left(\frac{\omega}{c}l\right)}\cos\omega t\ \text{ and }\ \eta=A\frac{\cos\left(\frac{\omega}{c}(l-x)\right)}{\cos\left(\frac{\omega}{c}l\right)}\cos\omega t \tag{3.26}$$

Note that ξ and η become infinite if $\cos(\omega l/c)=0$ or $\omega l/c=(n+\frac{1}{2})\pi$; the least value of cyclic frequency for this resonance phenomenon is given by $n=0$ and

$$\frac{\omega}{2\pi}=\frac{1}{4}\frac{c}{l}=\frac{\sqrt{gh}}{4l}.$$

Typical values are

$$\tau=\frac{2\pi}{\omega}=\frac{1}{2}\text{day}=12\times 60^2\ \text{s},\quad \omega=\frac{2\pi}{\tau}=\frac{2\pi}{12\times 60^2}\ \text{s}^{-1}$$

$$=1.454\times 10^{-4}\ \text{s}^{-1},$$

$$l=100\ \text{km}=10^7\ \text{cm},\quad h=1\ \text{m}=10^2\ \text{cm},$$

$$g=980\ \text{cm s}^{-2},\quad c=\sqrt{gh}=313.0\ \text{cm s}^{-1}\ (\simeq 10\ \text{ft s}^{-1}),$$

$$\frac{\omega}{c}l=4.645=1.479\pi,\quad \cos 4.645=-0.067.$$

As x increases from 0 to l, $\omega x/c$ increases from 0 to 1.479π. In fact 1.479π is not far from the second resonance value 1.5π or $3\pi/2$. For some values of $l-x$, η/A is considerably greater than 1. Hence the amplitude of the oscillation is considerably greater at some points up the canal than at its mouth. The theory only holds for $\eta \ll h$ and so the tidal effects must be very small at the mouth of the canal for the theory to hold. The theory ignores viscosity and friction between water and canal walls, which would decrease the amplitude of oscillation. However, as Lamb points out, the theory does illustrate the exaggeration of oceanic tides which occur in shallow seas and estuaries. Incidentally, the time taken for a pulse to travel along the canal is $10^7/313\ \text{s}=8.875\ \text{hours}\simeq 9\ \text{hours}$.

3.2. Change of canal section and junctions

So far we have treated waves travelling along a canal of constant rectangular section, i.e. of constant depth and implicitly of constant breadth. We now consider what happens when a wave (the incident wave) reaches a change of section, which change includes a fixed vertical end or a junction with one or more other canals.

First consider the reflection of a wave at a fixed vertical end. At such an end $\xi=0$. The general solution is

$$\xi=f(x-ct)+g(x+ct)=F\left(t-\frac{x}{c}\right)+G\left(t+\frac{x}{c}\right).$$

If the origin is taken at the end, $\xi=0$ at $x=0$; i.e.

$$0=F(t)+G(t).$$

Hence

$$G(t) = -F(t) \quad \text{and} \quad \xi(x, t) = F\left(t - \frac{x}{c}\right) - F\left(t + \frac{x}{c}\right),$$

i.e. the wave $F(t - x/c)$ travelling on the canal, $-\infty < x \leqslant 0$, is incident on the end $x = 0$ and the wave reflected back is $-F(t + x/c)$. The incident wave is reflected back unaltered except for a change of sign. Using eqn (3.10)

$$\eta(x, t) = \frac{h}{c}\left(F'\left(t - \frac{x}{c}\right) + F'\left(t + \frac{x}{c}\right)\right).$$

The water moves up and down the fixed end $x = 0$ at double the elevation of the incident wave.

Now consider a canal which changes section at $x = 0$. The canal on both sides of $x = 0$ is rectangular, but for $x < 0$ the breadth and depth for still water are b_0 and h_0 respectively and for $x > 0$ they are b_1 and h_1. It is convenient to work in terms of the elevation η. Let a wave $\eta = F(t - x/c_0)$ travel on the section of the canal $x < 0$ and be incident on the section change at $x = 0$. In general there will be a reflected wave $\eta = G(t + x/c_0)$ on $x < 0$ and a transmitted wave $\eta = J(t - x/c_1)$ on $x > 0$ (Fig. 3.3).

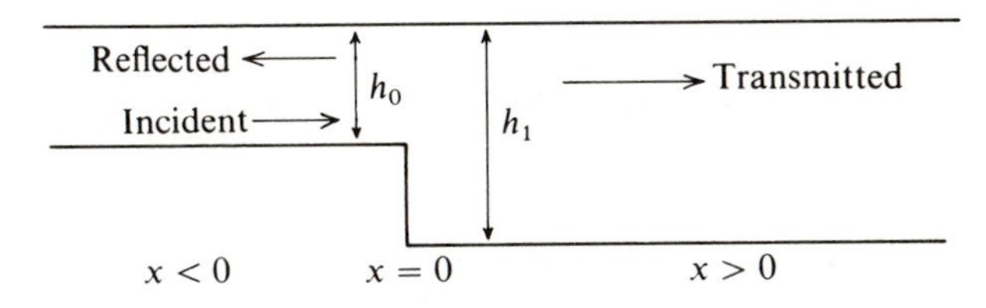

Fig. 3.3

$$c_0 = \sqrt{gh_0} \tag{3.27}$$

and

$$c_1 = \sqrt{gh_1} \tag{3.28}$$

The problem is to determine G and J in terms of F. The total values of η either side of $x = 0$ are

$$\eta = F\left(t - \frac{x}{c_0}\right) + G\left(t + \frac{x}{c_0}\right) \quad \text{for } x < 0, \tag{3.29}$$

$$\eta = J\left(t - \frac{x}{c_1}\right) \quad \text{for } x > 0. \tag{3.30}$$

For continuity of pressure $\lim_{x\to-0} \eta = \lim_{x\to+0} \eta$, or

$$F(t)+G(t)=J(t). \tag{3.31}$$

The mass flow across $x=0$ must be continuous. Hence, since $u=u(x, t)$,

$$b_0 h_0 u_0 = b_1 h_1 u_1, \quad \text{where} \quad u_0 = \lim_{x\to-0} u \quad \text{and} \quad u_1 = \lim_{x\to+0} u.$$

Since $u=\partial\xi/\partial t$ and $\eta=-h(\partial\xi/\partial x)$,

$$u=\frac{c}{h}\eta \quad \text{when} \quad \xi=f\left(t-\frac{x}{c}\right)$$

and

$$u=-\frac{c}{h}\eta \quad \text{when} \quad \xi=f\left(t+\frac{x}{c}\right).$$

Therefore,

$$u_0=\frac{c_0}{h_0}F(t)-\frac{c_0}{h_0}G(t) \quad \text{and} \quad u_1=\frac{c_1}{h_1}J(t);$$

hence, substituting in $b_0 h_0 u_0 = b_1 h_1 u_1$,

$$b_0 c_0 F(t) - b_0 c_0 G(t) = b_1 c_1 J(t)$$

or

$$F(t)-G(t)=\frac{b_1 c_1}{b_0 c_0} J(t). \tag{3.32}$$

Solving eqns (3.31) and (3.32) for $G(t)$ and $J(t)$,

$$J(t)=\frac{2b_0 c_0}{b_0 c_0 + b_1 c_1} F(t) \tag{3.33}$$

and

$$G(t)=\frac{b_0 c_0 - b_1 c_1}{b_0 c_0 + b_1 c_1} F(t). \tag{3.34}$$

Therefore the reflected wave is

$$\eta=G\left(t+\frac{x}{c_0}\right)=\frac{b_0 c_0 - b_1 c_1}{b_0 c_0 + b_1 c_1} F\left(t+\frac{x}{c_0}\right) \tag{3.35}$$

and the transmitted wave is

$$\eta = J\left(t - \frac{x}{c_1}\right) = \frac{2b_0c_0}{b_0c_0 + b_1c_1} F\left(t - \frac{x}{c_1}\right). \tag{3.36}$$

What happens when a wave is incident on a canal junction? Pressure must be continuous at the junction and hence η must be continuous as the junction is approached from each branch. If the incident wave travels on canal 0, there will be a reflected wave on canal 0 and transmitted waves on canals 1 and 2 (Fig. 3.4). The *net* mass flow into the junction must be zero. All the canals have constant, but in general different, values of depth and breadth. Let s measure distance along each canal. The junction will be taken as $s=0$ on all canals; s will increase on canals 1 and 2, so that $s \geqslant 0$ on these canals; s will decrease on canal 0, so that $s \leqslant 0$ on this canal. With these conventions the results below will correspond to those already obtained for a change of section. In fact a change of section is just a special case of a junction with only one canal leading out of as well as into the junction.

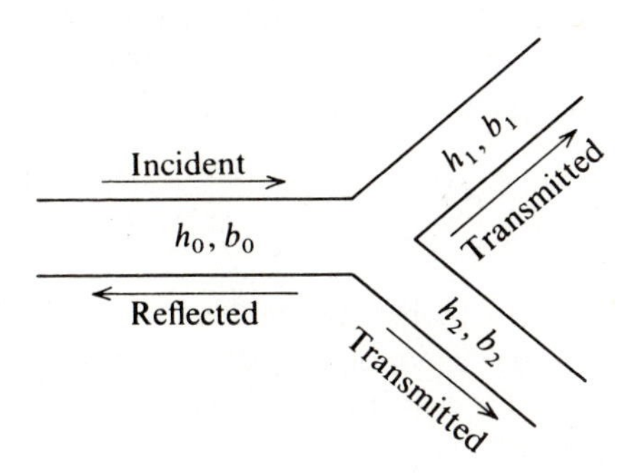

Fig. 3.4

Canal 0: $\eta = F\left(t - \dfrac{s}{c_0}\right) + G\left(t + \dfrac{s}{c_0}\right), \quad c_0 = \sqrt{gh_0}, \quad s < 0.$

Canal 1: $\eta = J_1\left(t - \dfrac{s}{c_1}\right), \quad c_1 = \sqrt{gh_1}, \quad s > 0.$

Canal 2: $\eta = J_2\left(t - \dfrac{s}{c_2}\right), \quad c_2 = \sqrt{gh_2}, \quad s > 0.$

As before $u = \pm(c\eta/h)$ when $\xi = f(t \mp x/c)$. Hence,

$$\lim_{s\to 0} u_0 = \frac{c_0}{h_0} F(t) - \frac{c_0}{h_0} G(t),$$

$$\lim_{s\to 0} u_1 = \frac{c_1}{h_1} J_1(t),$$

$$\lim_{s\to 0} u_2 = \frac{c_2}{h_2} J_2(t).$$

For continuity of pressure η must be continuous:

$$F(t) + G(t) = J_1(t) = J_2(t).$$

For continuity of mass flow:

$$b_0c_0 F(t) - b_0c_0 G(t) = b_1c_1 J_1(t) + b_2c_2 J_2(t).$$

Eliminating $J_2(t)$ from the second equation,

$$b_0c_0 F(t) - b_0c_0 G(t) = (b_1c_1 + b_2c_2) J_1(t).$$

The equations for $G(t)$ and $J_1(t)$ are the same as those previously found in the change of section analysis for $G(t)$ and $J(t)$, eqns (3.31) and (3.32), except that $b_1c_1 + b_2c_2$ replaces b_1c_1. Hence,

$$J_1(t) = J_2(t) = \frac{2b_0c_0}{b_0c_0 + b_1c_1 + b_2c_2} F(t), \quad G(t) = \frac{b_0c_0 - b_1c_1 - b_2c_2}{b_0c_0 + b_1c_1 + b_2c_2} F(t).$$

Clearly, this result can be extended to the case of a junction of $n+1$ canals, where the incident and reflected waves travels on canal 0 and the transmitted waves travel on canals 1 to n. With an obvious extension of notation,

$$J_i(t) = \frac{2b_0c_0}{\Sigma} F(t) \quad \text{and} \quad G(t) = \frac{2b_0c_0 - \Sigma}{\Sigma} F(t), \tag{3.37}$$

where $i = 1, 2, 3, \ldots, n$ and

$$\Sigma = b_0c_0 + b_1c_1 + b_2c_2 + \ldots + b_nc_n. \tag{3.38}$$

3.3. A brief introduction to Bessel functions

Before treating canals of varying section, the solutions of eqn (3.39) below are required. When solutions of a partial differential equation can be found by the method of separation of variables, ordinary differential equations are frequently derived which cannot be solved in terms of elementary functions or in terms of finite combinations of elementary functions.† Probably the most frequently occurring of these equations is Bessel's equation of order n,

$$\frac{d^2R}{d\rho^2}+\frac{1}{\rho}\frac{dR}{d\rho}+\left(1-\frac{n^2}{\rho^2}\right)R=0, \tag{3.39}$$

for R as a function of ρ where n is a non-negative constant, frequently but not always an integer. The solutions of eqn (3.39) are known as Bessel functions. By the method of solution in series, it can be shown that the two independent solutions of eqn (3.39) are $J_n(\rho)$ and $J_{-n}(\rho)$, where

$$J_n(\rho)=\sum_{s=0}^{\infty}\frac{(-)^s}{s!(n+s)!}\left(\frac{\rho}{2}\right)^{n+2s}, \tag{3.40}$$

provided n is not an integer. When n is an integer,

$$J_{-n}(\rho)=(-)^n J_n(\rho)$$

and the second solution of Bessel's equation, independent of $J_n(\rho)$, is

$$Y_n(\rho)=\frac{2}{\pi}J_n(\rho)\ln\frac{\rho}{2}-\frac{1}{\pi}\sum_{s=0}^{\infty}(-)^s\frac{(\rho/2)^{n+2s}}{s!(n+s)!}\{F(s)+F(n+s)\}$$
$$-\frac{1}{\pi}\sum_{s=0}^{n-1}\frac{(n-s-1)!}{s!}\left(\frac{\rho}{2}\right)^{-n+2s}, \tag{3.41}$$

where the digamma function $F(s)$ is given by

$$F(s)=\frac{d}{ds}(\ln s!). \tag{3.42}$$

Tables of $J_n(\rho)$ and $Y_n(\rho)$ have been constructed for the commonly occurring values of n and ρ and computational algorithms exist for their evaluation.

As $\rho\to 0$,

$$J_n(\rho)\to\frac{1}{n!}\left(\frac{\rho}{2}\right)^n, \tag{3.43}$$

so that

$$J_0(\rho)\to 1 \quad\text{and}\quad J_n(\rho)\to 0 \quad\text{for } n>0 \text{ as } \rho\to 0. \tag{3.44}$$

† Sums, differences, products and quotients of rational, exponential and trigonometric functions.

Also, as $\rho \to 0$,

$$Y_n(\rho) \to \frac{2}{\pi} J_n(\rho) \ln \rho - \frac{1}{\pi}(n-1)! \left(\frac{\rho}{2}\right)^{-n}, \tag{3.45}$$

so that

$$Y_0(\rho) \to \frac{2}{\pi} \ln \rho \to -\infty \quad \text{as } \rho \to +0 \tag{3.46}$$

and, for $n > 0$,

$$Y_n(\rho) \to -\frac{1}{\pi}(n-1)! \left(\frac{\rho}{2}\right)^{-n} \to -\infty \quad \text{as } \rho \to +0. \tag{3.47}$$

The behaviour of the solutions of eqn (3.39) as $\rho \to \infty$ can be found by substituting $R(\rho) = \rho^{-\frac{1}{2}} S(\rho)$ to give

$$\frac{d^2 S}{d\rho^2} + \left(1 + \frac{\frac{1}{4} - n^2}{\rho^2}\right) S = 0.$$

For ρ sufficiently large, the term in $1/\rho^2$ can be neglected compared to 1 and $S(\rho)$ then behaves like $\sin \rho$ or $\cos \rho$. Hence the solutions of eqn (3.39) behave like $\rho^{-\frac{1}{2}} \sin \rho$ and $\rho^{-\frac{1}{2}} \cos \rho$ as $\rho \to \infty$. Both $J_n(\rho)$ and either $J_{-n}(\rho)$ or $Y_n(\rho)$ behave like linear combinations of $\rho^{-\frac{1}{2}} \cos \rho$ and $\rho^{-\frac{1}{2}} \sin \rho$ as $\rho \to \infty$; hence $J_n(\rho)$ and $J_{-n}(\rho)$ or $Y_n(\rho)$ are oscillatory with period 2π and with amplitude decreasing as $\rho^{-\frac{1}{2}}$ as $\rho \to \infty$.

The two most commonly occurring Bessel functions are $J_0(\rho)$ and $J_1(\rho)$; these are sketched in Fig. 3.5. Note that they oscillate with decreasing amplitude as ρ increases.

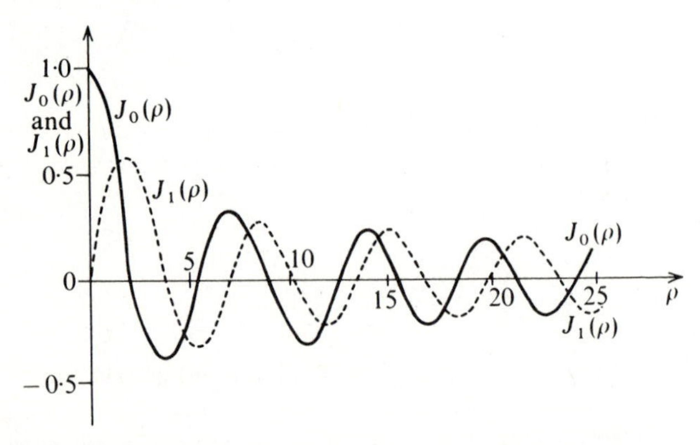

Fig. 3.5

Differentiation of eqn (3.40) with respect to ρ gives

$$\rho J_n'(\rho) = n J_n(\rho) - \rho J_{n+1}(\rho) \tag{3.48}$$

and

$$\rho J_n'(\rho) = \rho J_{n-1}(\rho) - n J_n(\rho). \tag{3.49}$$

Special cases are

$$J_0'(\rho) = -J_1(\rho) \tag{3.50}$$

and

$$J_1'(\rho) = J_0(\rho) - \frac{1}{\rho} J_1(\rho). \tag{3.51}$$

Subtraction of eqns (3.48) and (3.49) gives the recurrence relation

$$2\pi J_n(\rho) = \rho(J_{n-1}(\rho) + J_{n+1}(\rho)). \tag{3.52}$$

3.4. Rectangular canals of varying section

For a canal of arbitrary section and breadth b, the mean depth h is defined by

$$h = \frac{1}{b} \int_0^b h(z)\, \mathrm{d}z,$$

where $h(z)$ is the depth a distance z from one side (Fig. 3.6). It is assumed that such a canal can be treated as a canal of rectangular section of breadth b and depth h. This section is concerned with canals for which b and h are functions of x.

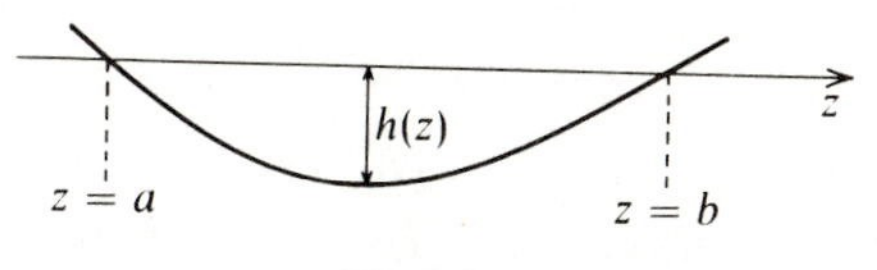

Fig. 3.6

We now choose $y = 0$ as the free surface so that the bottom of the canal is $y = -h(x)$ (Fig. 3.7). Then, neglecting vertical acceleration,

$$p - p_0 = \rho g(\eta - y), \quad \frac{\partial p}{\partial x} = \rho g \frac{\partial \eta}{\partial x}, \quad \text{independent of } y.$$

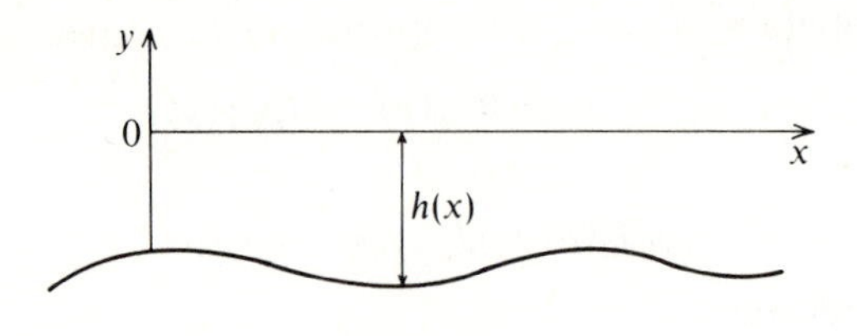

Fig. 3.7

The equation of motion in the horizontal direction is

$$-\frac{1}{\rho}\frac{\partial p}{\partial x}=\frac{\mathrm{D}u}{\mathrm{D}t}=\frac{\partial u}{\partial t};\quad \text{neglecting } u\frac{\partial u}{\partial x}\quad \text{compared to } \frac{\partial u}{\partial t}.$$

Hence

$$-g\frac{\partial \eta}{\partial x}=-\frac{1}{\rho}\frac{\mathrm{d}p}{\mathrm{d}x}=\frac{\partial u}{\partial t}=\frac{\partial^2 \xi}{\partial t^2},\quad \text{as before.} \tag{3.7}$$

An alternative method to that used previously will be used to derive the continuity equation. The volume of fluid δV entering the planes bounded by x and $x+\mathrm{d}x$ (Fig. 3.8) since the fluid was at rest ($\xi=\partial\xi/\partial t=0$) is

$$\delta V=(\xi S)_x-(\xi S)_{x+\mathrm{d}x}=-\frac{\partial}{\partial x}(\xi S)\,\mathrm{d}x,$$

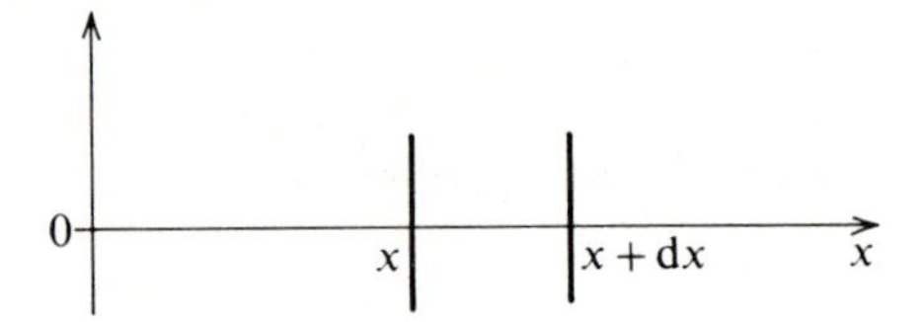

Fig. 3.8

where $S(x, t)$ is the cross-sectional area of the canal at abscissa x and time t. Since $S(x, t)=(h(x)+\eta(x, t))b(x)$, $\xi S=\xi h(x)b(x)$ to first order when $\eta \ll h$. δV must equal the increase in volume of water between x and $x+\mathrm{d}x$, i.e. $\delta V=\eta b\,\mathrm{d}x$. Therefore,

$$-\frac{\partial}{\partial x}(\xi S)=\eta b$$

or

$$\eta=-\frac{1}{b}\frac{\partial}{\partial x}(hb\xi). \tag{3.53}$$

Elimination of η between eqns (3.7) and (3.53) gives the equation for ξ:

$$g\frac{\partial}{\partial x}\left(\frac{1}{b}\frac{\partial}{\partial x}(hb\xi)\right)=\frac{\partial^2\xi}{\partial t^2}. \tag{3.54}$$

To obtain the equation for η, multiply through eqn (3.54) by $-hb$ and differentiate with respect to x:

$$\frac{\partial}{\partial x}\left(hbg\frac{\partial\eta}{\partial x}\right)=-\frac{\partial}{\partial x}\left(\frac{\partial^2(hb\xi)}{\partial t^2}\right)=\frac{\partial^2}{\partial t^2}\left(-\frac{\partial}{\partial x}(hb\xi)\right)=\frac{\partial^2}{\partial t^2}(b\eta)$$
$$=b\frac{\partial^2\eta}{\partial t^2},$$

where eqn (3.53) has been used twice, along with the fact that h and b are independent of t. Hence,

$$\frac{\partial^2\eta}{\partial t^2}=\frac{g}{b}\frac{\partial}{\partial x}\left(hb\frac{\partial\eta}{\partial x}\right). \tag{3.55}$$

The equations for ξ and η are identical only when h and b are constants.

As an example, consider a canal of constant depth whose breadth varies as distance from the end $x=0$. Then $h=$constant and $b=\beta x$ where β is a constant. The equations for ξ and η are

$$gh\frac{\partial}{\partial x}\left(\frac{1}{x}\frac{\partial}{\partial x}(x\xi)\right)=\frac{\partial^2\xi}{\partial t^2} \tag{3.56}$$

and

$$\frac{\partial^2\eta}{\partial t^2}=gh\frac{1}{x}\frac{\partial}{\partial x}\left(x\frac{\partial\eta}{\partial x}\right). \tag{3.57}$$

Suppose that the end $x=a$ opens into a tidal estuary at a point for which

$$\eta_{x=a}=A\cos\omega t.$$

As a second boundary condition it is sufficient to require that η is finite at $x=0$.

As this is a forced oscillation problem, look for a solution of the form $\eta=f(x)\cos\omega t$. Substitute in eqn (3.57):

$$-\frac{\omega^2}{gh}f=\frac{\mathrm{d}^2f}{\mathrm{d}x^2}+\frac{1}{x}\frac{\mathrm{d}f}{\mathrm{d}x} \tag{3.58}$$

or

$$\frac{d^2f}{dx^2}+\frac{1}{x}\frac{df}{dx}+k^2f=0, \tag{3.59}$$

where

$$k=\frac{\omega}{\sqrt{gh}}. \tag{3.60}$$

Put $kx=\xi$, then

$$\frac{df}{dx}=\frac{df}{d\xi}\frac{d\xi}{dx}=k\frac{df}{d\xi}, \quad \frac{d^2f}{dx^2}=k^2\frac{d^2f}{d\xi^2}$$

and eqn (3.59) becomes

$$\frac{d^2f}{d\xi^2}+\frac{1}{\xi}\frac{df}{d\xi}+f=0. \tag{3.61}$$

This is Bessel's equation of order zero with solution, since n equals zero an integer,

$$f(\xi)=CJ_0(\xi)+BY_0(\xi); \quad f(x)=CJ_0(kx)+BY_0(kx),$$

where C and B are arbitrary constants.

For $\eta_{x=0}=f(0)\cos\omega t$ to be finite, $B=0$. Hence

$$\eta(x, t)=CJ_0(kx)\cos\omega t. \tag{3.62}$$

To satisfy the condition at $x=a$,

$$A\cos\omega t=\eta_{x=a}=CJ_0(ka)\cos\omega t \quad \text{or} \quad C=\frac{A}{J_0(ka)}.$$

Substitute back for C into eqn (3.62):

$$\eta(x, t)=\frac{A}{J_0(ka)}J_0(kx)\cos\omega t, \tag{3.63}$$

where

$$k=\frac{\omega}{\sqrt{gh}}. \tag{3.60}$$

Equation (3.63) is the solution of this example. Resonance occurs when $J_0(ka)=0$, i.e. for radian frequencies ω such that $\omega a/\sqrt{gh}$ is a zero of $J_0(\xi)$. There are an infinite number of such zeros, see Fig. 3.5. The least positive zero occurs at $\xi=2.405$ and hence the least resonant radian frequency is $2.405\sqrt{gh}/a$.

3.5. Vibrations of a sheet of water; membrane analogy

Disturbances will be considered on a plane sheet of water of uniform depth h. The governing partial differential equation will be derived and the solution representing vibrations will be found. The bottom will be taken as the plane $z=0$ with the z-axis vertically upwards. The elevation ζ is now a function of x, y and t (Fig. 3.9). Again neglecting vertical acceleration,

$$p-p_0=\rho g(h+\zeta-z).$$

Hence,

$$\frac{\partial p}{\partial x}=\rho g\frac{\partial \zeta}{\partial x} \quad \text{and} \quad \frac{\partial p}{\partial y}=\rho g\frac{\partial \zeta}{\partial y}.$$

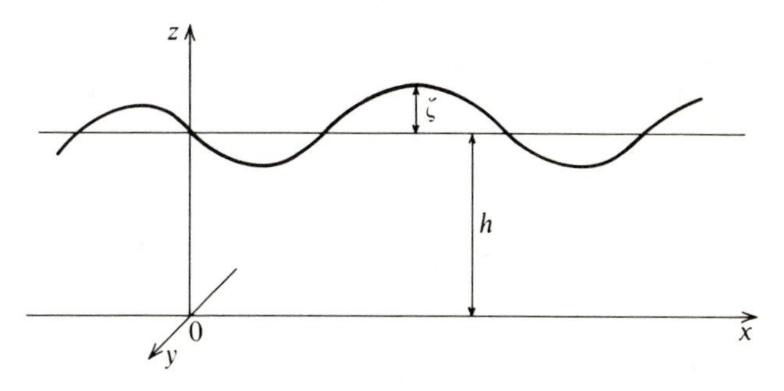

Fig. 3.9

Therefore $\zeta=\zeta(x, y, t)$; both sides of the above two equations are independent of z.

The equations of motion in the x and y directions, after linearization, are

$$-\frac{\partial p}{\partial x}=\rho\frac{\partial u}{\partial t} \quad \text{and} \quad -\frac{\partial p}{\partial y}=\rho\frac{\partial v}{\partial t}.$$

If ξ and η are the displacements of a particle in the x and y directions respectively,

$$u=\frac{\partial \xi}{\partial t} \quad \text{and} \quad v=\frac{\partial \eta}{\partial t}.$$

On elimination of p, u and v:

$$g\frac{\partial \zeta}{\partial x}=\frac{1}{\rho}\frac{\partial p}{\partial x}=-\frac{\partial u}{\partial t}=-\frac{\partial^2 \xi}{\partial t^2} \tag{3.64}$$

and

$$g\frac{\partial \zeta}{\partial y}=\frac{1}{\rho}\frac{\partial p}{\partial y}=-\frac{\partial v}{\partial t}=-\frac{\partial^2 \eta}{\partial t^2}. \tag{3.65}$$

Consider the rate of flow of water into the cylinder standing on the rectangular element with sides δx and δy, Fig. 3.10. This rate of flow must equal the rate of increase of volume of water within the cylinder. Since u and v are independent of z,

$$(uh\,\delta y)_x-(uh\,\delta y)_{x+\delta x}+(vh\,\delta x)_y-(vh\,\delta \zeta)_{y+\delta y}=\delta x\,\delta y\,\frac{\partial}{\partial t}(h+\zeta)$$

or

$$-h\frac{\partial u}{\partial x}-h\frac{\partial v}{\partial y}=\frac{\partial \zeta}{\partial t},$$

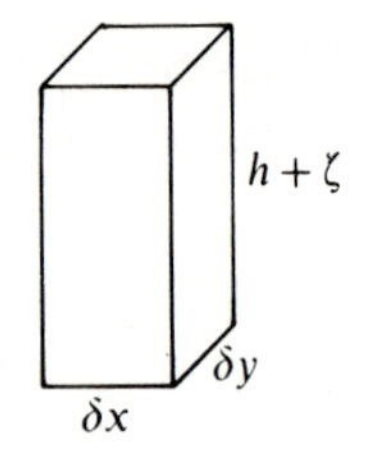

Fig. 3.10

on neglecting quantities of higher order. Substituting $u=\partial\xi/\partial t$ and $v=\partial\eta/\partial t$,

$$-h\left(\frac{\partial^2 \xi}{\partial x \partial t}+\frac{\partial^2 \eta}{\partial y \partial t}\right)=\frac{\partial \zeta}{\partial t}$$

or

$$\zeta=-h\left(\frac{\partial \xi}{\partial x}+\frac{\partial \eta}{\partial y}\right), \tag{3.66}$$

provided there existed a time at which the water was at rest with all particles undisplaced.

Differentiate eqn (3.66) twice with respect to t and substitute for $\partial^2\xi/\partial t^2$ and $\partial^2\eta/\partial t^2$ from eqns (3.64) and (3.65):

$$\frac{\partial^2 \zeta}{\partial t^2}=c^2\left(\frac{\partial^2 \zeta}{\partial x^2}+\frac{\partial^2 \zeta}{\partial y^2}\right) \tag{3.67}$$

where $c = \sqrt{gh}$, or

$$\frac{\partial^2 \zeta}{\partial t^2} = c^2 \nabla^2 \zeta,$$

where ∇^2 is the two-dimensional Laplacian operator.

The equations

$$g\frac{\partial \zeta}{\partial x} = -\frac{\partial^2 \xi}{\partial t^2} \quad \text{and} \quad g\frac{\partial \zeta}{\partial y} = -\frac{\partial^2 \eta}{\partial t^2}$$

may be written in two-dimensional vector form as

$$g \,\text{grad}\, \zeta = -\frac{\partial^2 \boldsymbol{\xi}}{\partial t^2} \tag{3.68}$$

where $\boldsymbol{\xi} \equiv (\xi, \eta)$ is the displacement vector. On a vertical fixed boundary, outward normal $\boldsymbol{n}$, $\boldsymbol{\xi} \cdot \boldsymbol{n} = 0$ for all t, hence $\dfrac{\mathrm{d}^2 \boldsymbol{\xi}}{\mathrm{d}t^2} \cdot \boldsymbol{n} = 0$ and the boundary condition on ζ is

$$0 = \text{grad}\, \zeta \cdot \boldsymbol{n} = \frac{\partial \zeta}{\partial n}. \tag{3.69}$$

In particular, if the water is bounded by a circularly cylindrical vertical wall, then, with the origin on the axis of the cylinder,

$$\frac{\partial \zeta}{\partial r} = 0 \quad \text{at} \quad r = a, \tag{3.70}$$

where a is the radius of the cylinder.

Since long waves on a canal satisfy the same partial differential equation as transverse waves on a string, one would suspect that two-dimensional long waves on water would satisfy the same partial differential equation as transverse waves on a membrane, the two-dimensional analogue of a string. This is now shown to be the case.

The membrane in its rest configuration occupies the plane $z = 0$ and is stretched to a tension T per unit length, uniform in all directions and the same at all points. In motion the transverse displacement of the membrane is

$$z = z(x, y, t)$$

Consider any surface area S of the membrane bounded by the closed curve C with projection S' bounded by C' on the x–y plane (Fig. 3.11a). The two-dimensional analogues of the approximations used in the transverse vibrations of strings are that the displacement

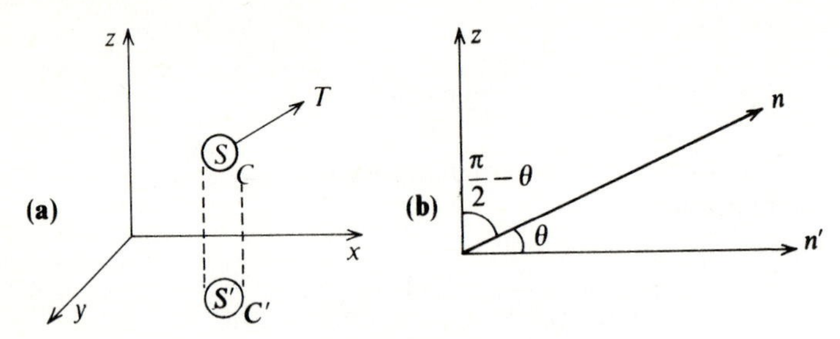

Fig. 3.11

of the membrane in the $x-y$ plane can be neglected and that, if $\pi/2-\theta$ is the angle which the principal normal $\boldsymbol{n}$ to C makes with the z-axis, θ^2 can be neglected compared to 1. If ρ is the rest areal density of the membrane, the equation of motion of S in the z-direction is

$$\int_{S'} \rho \frac{\partial^2 z}{\partial t^2} \, \mathrm{d}S' = \int_C T \sin\theta \, \mathrm{d}s = \int_C T \frac{\partial z}{\partial n} \, \mathrm{d}s.$$

Now (Fig. 3.11b) $\dfrac{\partial z}{\partial n} = \cos\theta \dfrac{\partial z}{\partial n'} = \dfrac{\partial z}{\partial n'}$ and $\mathrm{d}s = \sec\theta \, \mathrm{d}s' = \mathrm{d}s'$, twice neglecting θ^2 compared to 1.
Hence

$$\int_C T \frac{\partial z}{\partial n} \, \mathrm{d}s = \int_{C'} T \frac{\partial z}{\partial n'} \, \mathrm{d}s' = \int_{S'} \operatorname{div}(T \operatorname{grad} z) \, \mathrm{d}S' = \int_{S'} T\nabla^2 z \, \mathrm{d}S',$$

where the operators grad, div and $\nabla^2 (=\partial^2/\partial x^2 + \partial^2/\partial y^2)$ and the divergence theorem are two-dimensional.

Since this equation is true for any surface area S,

$$\rho \frac{\partial^2 z}{\partial t^2} = T\nabla^2 z \quad \text{or} \quad \frac{\partial^2 z}{\partial t^2} = c^2 \nabla^2 z, \tag{3.71}$$

where

$$c^2 = T/\rho. \tag{3.72}$$

The best known example of the transverse vibrations of a membrane is the sound made by a drum. For a drum, the boundary condition is $z=0$ at all points on the boundary. Equation (3.72) is only formally identical to eqn (1.11a). In eqn (3.72), T is force per unit length and ρ is mass per unit area; however the ratio T/ρ still has dimensions of velocity squared.

Consider the free oscillations of a sheet of water bounded by vertical walls. For the time dependence to be sinusoidal, substitute

$$\zeta=\bar{\zeta}\mathrm{e}^{-\mathrm{i}\omega t}, \quad \text{where } \bar{\zeta}=\bar{\zeta}(x, y);$$

so that

$$\nabla^2\bar{\zeta}+k^2\bar{\zeta}=0, \tag{3.73}$$

where

$$k^2=\frac{\omega^2}{c^2} \tag{3.74}$$

and $\partial\bar{\zeta}/\partial n=0$ on the boundary.

(i) *Rectangular boundary*: $\partial\bar{\zeta}/\partial x=0$ on $x=0$, a and $\partial\bar{\zeta}/\partial y=0$ on $y=0$, b. The most general solution of eqn (3.73) for $\bar{\zeta}$ under these boundary conditions is

$$\bar{\zeta}=\sum_{m=0}^{\infty}\sum_{n=0}^{\infty}A_{mn}\cos\frac{m\pi x}{a}\cos\frac{n\pi y}{b} \quad \text{with} \quad k^2=\pi^2\left(\frac{m^2}{a^2}+\frac{n^2}{b^2}\right),$$

where m and n are non-negative integers, not both zero. $m=n=0$ gives $\zeta=\text{constant}$, not a vibration.
The period of oscillation is

$$\frac{2\pi}{\omega}=\frac{2\pi}{ck}=\frac{2}{c}\left(\frac{m^2}{a^2}+\frac{n^2}{b^2}\right)^{-\frac{1}{2}}.$$

This is a maximum for $a>b$ when $m=1$ and $n=0$; the maximum period is $2a/c$.

(ii) *Circular boundary*: $\partial\bar{\zeta}/\partial r=0$ on $r=a$. In cylindrical polar co-ordinates, eqn (3.73) becomes

$$\frac{\partial^2\bar{\zeta}}{\partial r^2}+\frac{1}{r}\frac{\partial\bar{\zeta}}{\partial r}+\frac{1}{r^2}\frac{\partial^2\bar{\zeta}}{\partial\theta^2}+k^2\bar{\zeta}=0. \tag{3.75}$$

For single-valuedness the solution must be either periodic in θ with period $2\pi/n$ or independent of θ. Therefore look for a solution of the form

$$\bar{\zeta}=f(r)\begin{matrix}\cos\\ \sin\end{matrix}\, n\theta, \quad n+\text{ve integer or zero.} \tag{3.76}$$

Hence

$$\frac{\mathrm{d}^2 f}{\mathrm{d}r^2}+\frac{1}{r}\frac{\mathrm{d}f}{\mathrm{d}r}+\left(k^2-\frac{n^2}{r^2}\right)f=0.$$

Put $kr=\rho$, so that the equation for f becomes

$$\frac{d^2f}{d\rho^2}+\frac{1}{\rho}\frac{df}{d\rho}+\left(1-\frac{n^2}{\rho^2}\right)f=0, \qquad (3.77)$$

which is identical to eqn (3.39), Bessel's equation of order n. The solution which is finite at the origin is $f(\rho)=A_nJ_n(\rho)$, or

$$f(r)=A_nJ_n(kr).$$

Hence

$$\zeta=A_nJ_n(kr)\begin{matrix}\cos\\ \sin\end{matrix}\, n\theta\, e^{-i\omega t}. \qquad (3.78)$$

The boundary condition on $r=a$ is $\partial\zeta/\partial r=0$ or $(d/dr)J_n(kr)=0$. Now,

$$\frac{d}{dr}J_n(kr)=k\frac{\partial}{\partial(kr)}J_n(kr)=kJ_n'(kr),$$

where a prime denotes a derivative of a function with respect to its argument. Hence the boundary condition can be written $(J_n'(kr))_{r=a}=0$ or, for short,

$$J_n'(ka)=0. \qquad (3.79)$$

The corresponding normal radian frequencies ω for free oscillations, found by substituting $k=\omega/c=\omega/\sqrt{gh}$ in eqn (3.79), are

$$J_n'\left(\frac{\omega a}{\sqrt{gh}}\right)=0.$$

For $n=0$, all modes are axisymmetric, so that the waves have circular ridges and furrows. The least positive root of $0=J_0'(ka)=J_1(ka)$ is $ka=3.8317$, so that the lowest normal frequency is given by $\omega a/\sqrt{gh}=3.8317$ or $\omega=3.8317\sqrt{gh}/a$.

For $n=1$, using the 'cos θ' solution,

$$\zeta=A_1J_1(kr)\cos\theta\, e^{-i\omega t}, \quad \text{where } J_1'(ka)=0.$$

Note that the line $\theta=\pm\pi/2$ is a node. For modes higher than the lowest, there are nodal circles on which $J_1(kr)=0$.

Exercises

3.1. A travelling wave $\eta=\varepsilon e^{i\omega(t-x/c)}$ in an open water channel of depth h and breadth b is incident upon a junction with a second open channel, of depth αh and breadth βb; the reflected wave

$$\eta = R\varepsilon e^{i\omega(t+x/c)}$$

is produced, together with a transmitted wave. Find the value of R. Find the ratio of the height of the transmitted wave to that of the incident wave, and show that it approaches 2 when α and β are small.

3.2. Show that the elevation η in a canal of constant breadth b but varying depth $h(x)$ satisfies the equation

$$\frac{\partial^2 \eta}{\partial t^2} = g\frac{\partial}{\partial x}\left(h\frac{\partial \eta}{\partial x}\right).$$

If the depth of a canal with ends $x=0$, a is $h(x)=\alpha x$, if η is finite at $x=0$ and if the end $x=a$ opens into a tidal estuary for which $\eta_{x=a}=A\cos\omega t$, show that

$$\eta = A\frac{J_0(2\kappa^{\frac{1}{2}}x^{\frac{1}{2}})}{J_0(2\kappa^{\frac{1}{2}}a^{\frac{1}{2}})}\cos\omega t.$$

3.3. A rectangular membrane is rigidly fixed along its boundaries, $x=0$, $x=a$, $y=0$ and $y=b$. By the method of separation of variables, derive the solution

$$z(x,y,t) = \sum_{m,n=1}^{\infty} A_{mn}\cos(\omega_{mn}t+\varepsilon_{mn})\sin\frac{m\pi x}{a}\sin\frac{n\pi y}{b},$$

where m and n are positive integers and A_{mn} and ε_{mn} are constants. Determine ω_{mn} and verify that the frequencies of the higher harmonics are not in general integral multiples of the fundamental.

Answers to exercises

3.1. $R = \dfrac{(\alpha^{\frac{1}{2}}\beta-1)}{(\alpha^{\frac{1}{2}}\beta+1)}.$

Ratio of heights $= \dfrac{2}{(\alpha^{\frac{1}{2}}\beta+1)} \to 2$ for small α and β.

3.3. $\omega_{mn} = c\pi\sqrt{\dfrac{m^2}{a^2}+\dfrac{n^2}{b^2}}.$

4 Surface waves on relatively deep water

4.1. Governing equations; dispersion equation

The theory of long waves on canals was based on the approximation that the vertical acceleration was negligible and it was shown that the theory was consistent with this approximation provided that the depth h was small compared to the wavelength λ. We now consider waves where h may not be small compared to λ and where, therefore, vertical acceleration must be included.

Attention will be confined to plane motion, the x-axis is horizontal, the y-axis is vertically upwards and there is no dependence on z (Fig. 4.1). The undisturbed level of the water is $y=0$. Water of constant depth h will be treated, so that the bottom is $y=-h$. The components of velocity of the water are $u(x, y)$ and $v(x, y)$ in the x- and y-directions respectively and there is no motion in the z-direction.

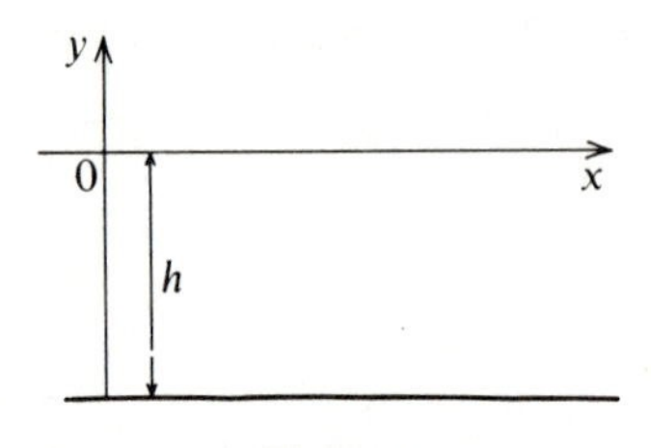

Fig. 4.1

The motion is both irrotational and equivoluminal, so that respectively

$$\frac{\partial u}{\partial y}=\frac{\partial v}{\partial x} \tag{4.1}$$

and

$$\frac{\partial u}{\partial x}+\frac{\partial v}{\partial y}=0. \tag{4.2}$$

Equation (4.1) is the necessary and sufficient condition for the existence of a velocity potential ϕ such that

$$u=-\frac{\partial\phi}{\partial x} \tag{4.3}$$

and

$$v=-\frac{\partial\phi}{\partial y}. \tag{4.4}$$

Substitution from eqns (4.3) and (4.4) into eqn (4.2) gives

$$\frac{\partial^2\phi}{\partial x^2}+\frac{\partial^2\phi}{\partial y^2}=0. \tag{4.5}$$

In vector form equations (4.3) and (4.4) can be written

$$\boldsymbol{v}=-\operatorname{grad}\phi \tag{4.6}$$

where $\boldsymbol{v}\equiv(u, v)$. At a fixed boundary with unit normal $\boldsymbol{n}$ at any point, $\boldsymbol{v}\cdot\boldsymbol{n}=0$ on the boundary. Using eqn (4.6),

$$\frac{\partial\phi}{\partial n}=0 \quad \text{on a fixed boundary.} \tag{4.7}$$

In particular, on the bottom,

$$\frac{\partial\phi}{\partial y}=0 \quad \text{on } y=-h. \tag{4.8}$$

The equation for the pressure p and the boundary condition on the free surface will now be found. The equations of motion are

$$\frac{\mathrm{D}\boldsymbol{v}}{\mathrm{D}t}=\boldsymbol{X}-\frac{1}{\rho}\operatorname{grad}p, \tag{4.9}$$

where $\boldsymbol{X}\equiv(X, Y)$ is the body force per unit mass, ρ is the constant density and p is the pressure. Take x-components and expand $\mathrm{D}/\mathrm{D}t$ $(=\partial/\partial t+\boldsymbol{v}\cdot\boldsymbol{\nabla})$:

$$\frac{\partial u}{\partial t}+u\frac{\partial u}{\partial x}+v\frac{\partial u}{\partial y}=X-\frac{1}{\rho}\frac{\partial p}{\partial x}.$$

If the body force has potential Ω, $\boldsymbol{X}=-\operatorname{grad}\Omega$. Substitute for X, use eqn (4.1) for $\partial u/\partial y$ and eqn (4.3) for u in the term $\partial u/\partial t$:

$$-\frac{\partial^2\phi}{\partial x\partial t}+u\frac{\partial u}{\partial x}+v\frac{\partial v}{\partial x}=-\frac{\partial\Omega}{\partial x}-\frac{1}{\rho}\frac{\partial p}{\partial x}.$$

On integration,

$$\frac{p}{\rho}=\frac{\partial\phi}{\partial t}-\Omega-\tfrac{1}{2}q^2+F_1(t,y),$$

where $F_1(t, y)$ is an arbitrary function of t and y. By similarity, the same argument applied to the y-component of the equation of motion, eqn (4.9), gives

$$\frac{p}{\rho}=\frac{\partial\phi}{\partial t}-\Omega-\tfrac{1}{2}q^2+F_2(t,x).$$

The last two equations are only consistent if

$$\frac{p}{\rho}=\frac{\partial\phi}{\partial t}-\Omega-\tfrac{1}{2}q^2+F(t),$$

which is Bernoulli's equation for incompressible fluids. Provided the motion is sufficiently small to neglect the squares of small quantities,

$$\frac{p}{\rho}=\frac{\partial\phi}{\partial t}-\Omega+F(t).$$

For surface waves on water, $\Omega=gy$, so that

$$\frac{p}{\rho}=\frac{\partial\phi}{\partial t}-gy+F(t). \tag{4.10}$$

On the free surface $y=\eta$ the elevation and $p=p_0$ the atmospheric pressure, which will be treated as a constant. Substitution in eqn (4.10) gives

$$\frac{p_0}{\rho}=\left(\frac{\partial\phi}{\partial t}\right)_{y=\eta}-g\eta+F(t).$$

Now

$$\begin{aligned}\left(\frac{\partial\phi}{\partial t}\right)_{y=\eta}&=\left(\frac{\partial\phi}{\partial t}\right)_{y=0}+\eta\left(\frac{\partial^2\phi}{\partial\eta\partial t}\right)_{y=0}+\text{higher-order terms}\\&=\left(\frac{\partial\phi}{\partial t}\right)_{y=0},\end{aligned}$$

to first order in small quantities. Hence,

$$\frac{p_0}{\rho}=\left(\frac{\partial\phi}{\partial t}\right)_{y=0}-g\eta+F(t). \tag{4.11}$$

Subtract eqn (4.11) from eqn (4.10):

$$\frac{p}{\rho}=\frac{p_0}{\rho}+\frac{\partial\phi}{\partial t}-\left(\frac{\partial\phi}{\partial t}\right)_{y=0}-gy+g\eta. \tag{4.12}$$

Since it is only the derivatives $\partial\phi/\partial x$ and $\partial\phi/\partial y$ that have physical significance, an arbitrary function of time may be included in ϕ. In particular $\int^t (p_0/\rho - F(\tau))\,d\tau$ may be included in ϕ so that eqn (4.11) becomes

$$g\eta=\left(\frac{\partial\phi}{\partial t}\right)_{y=0}, \tag{4.13}$$

and substitution of eqn (4.13) into eqn (4.12) gives

$$\frac{p}{\rho}=\frac{p_0}{\rho}+\frac{\partial\phi}{\partial t}-gy. \tag{4.14}$$

If the elevation is sufficiently small for the normal to the free surface to make a small angle with the vertical, i.e. $|\partial\eta/\partial x|\ll 1$, the condition that the normal component of the fluid velocity must equal the normal velocity of the surface itself can be replaced by the equality of the y-components of the two velocities, i.e.

$$-\left(\frac{\partial\phi}{\partial y}\right)_{y=0}=\frac{\partial\eta}{\partial t}. \tag{4.15}$$

Elimination of η between eqns (4.13) and (4.15) gives

$$\frac{\partial^2\phi}{\partial t^2}+g\frac{\partial\phi}{\partial y}=0, \quad \text{on } y=0. \tag{4.16}$$

This is the required boundary condition on $y=0$. The pressure p is given by eqn (4.14).

Surface waves on water are governed by Laplace's equation. The time variation comes in through the boundary condition on the free surface. Section 4.5 will shed further light on this phenomenon.

Let us investigate whether solutions exist to

$$\frac{\partial^2\phi}{\partial x^2}+\frac{\partial^2\phi}{\partial y^2}=0, \quad \text{with } \frac{\partial\phi}{\partial y}=0 \quad \text{on } y=-h,$$

$$\text{and } \frac{\partial^2\phi}{\partial t^2}+g\frac{\partial\phi}{\partial y}=0 \quad \text{on } y=0,$$

which represents harmonic waves travelling in the positive x-direction, i.e. waves of the form

$$\phi = Y(y)\,e^{-i(\omega t - kx)}, \tag{4.17}$$

where $Y(y)$ is real and ω and k are positive. Substitution in Laplace's equation gives

$$-k^2 Y + Y'' = 0$$

or

$$Y(y) = A\,e^{ky} + Be^{-ky}, \tag{4.18}$$

where A and B are real.

The condition on $y = -h$ is $Y'(-h) = 0$, or

$$kAe^{-kh} - kBe^{kh} = 0, \quad B = Ae^{-2kh}. \tag{4.19}$$

Substituting back for B into eqn (4.18) and then for $Y(y)$ into eqn (4.17):

$$\phi = 2Ae^{-kh}\cosh k(y+h)\,e^{-i(\omega t - kx)}. \tag{4.20}$$

The condition on $y = 0$ must now be satisfied. From eqn (4.20),

$$\frac{\partial \phi}{\partial y} = 2Ae^{-kh} k \sinh k(y+h) e^{-i(\omega t - kx)}$$

and

$$\frac{\partial^2 \phi}{\partial t^2} = -\omega^2 2Ae^{-kh}\cosh k(y+h) e^{-i(\omega t - kx)}.$$

Put $y = 0$ and substitute into eqn (4.16):

$$\omega^2 \cosh kh = gk \sinh kh. \tag{4.21}$$

The free surface condition gives the relationship between the radian frequency ω and the wavenumber k.

The wave velocity c is given by:

$$c = \frac{\omega}{k} = \left(\frac{g}{k}\tanh kh\right)^{\frac{1}{2}}. \tag{4.22}$$

In terms of the wavelength λ ($k = 2\pi/\lambda$),

$$c = \left(\frac{g\lambda}{2\pi}\tanh\frac{2\pi h}{\lambda}\right)^{\frac{1}{2}}. \tag{4.23}$$

The wave velocity depends on wavelength. Surface waves are dispersive. Equation (4.22) or (4.23) is known as the dispersion equation.

For $h/\lambda \gg 1$,

$$\tanh\frac{2\pi h}{\lambda} \simeq 1, \quad \text{so that } c \simeq \left(\frac{g\lambda}{2\pi}\right)^{\frac{1}{2}}. \tag{4.24}$$

For $h/\lambda \ll 1$,

$$\tanh\frac{2\pi h}{\lambda} \simeq \frac{2\pi h}{\lambda} \quad \text{and} \quad c \simeq \sqrt{gh}. \tag{4.25}$$

The latter result is in accord with the theory of long waves on canals.

The elevation is given by

$$\eta = \frac{1}{g}\left(\frac{\partial \phi}{\partial t}\right)_{y=0} = -\frac{2A}{g}\mathrm{i}\omega \mathrm{e}^{-kh}\cosh kh\,\mathrm{e}^{-\mathrm{i}(\omega t - kx)}.$$

Taking the real part,

$$\eta = \frac{2A\omega}{g}\mathrm{e}^{-kh}\cosh kh\,\sin(kx - \omega t).$$

It is the amplitude of the elevation which is observed. Let it equal a. Then

$$\eta = a\sin(kx - \omega t), \quad a = \frac{2A\omega}{g}\mathrm{e}^{-kh}\cosh kh$$

and

$$\phi = \frac{ga}{\omega}\frac{\cosh k(y+h)}{\cosh kh}\cos(kx - \omega t) = \frac{a\omega}{k}\frac{\cosh k(y+h)}{\sinh kh}\cos(kx - \omega t), \tag{4.26}$$

on using eqn (4.21). The components of particle velocity are

$$u = -\frac{\partial \phi}{\partial x} = a\omega\frac{\cosh k(y+h)}{\sinh kh}\sin(kx - \omega t) \tag{4.27}$$

and

$$v = -\frac{\partial \phi}{\partial y} = -a\omega\frac{\sinh k(y+h)}{\sinh kh}\cos(kx - \omega t) \tag{4.28}$$

In terms of particle displacement (U, V), $u = \partial U/\partial t$ and $v = \partial V/\partial t$; hence,

$$U = a\frac{\cosh k(y+h)}{\sinh kh}\cos(kx - \omega t) \tag{4.29}$$

and

$$V = a\frac{\sinh k(y+h)}{\sinh kh}\sin(kx - \omega t). \tag{4.30}$$

The equation of a particle path is given by eliminating $kx-\omega t$:

$$\frac{U^2}{\left(a\,\dfrac{\cosh k(y+h)}{\sinh kh}\right)^2}+\frac{V^2}{\left(a\,\dfrac{\sinh k(y+h)}{\sinh kh}\right)^2}=1. \tag{4.31}$$

This is the equation of an ellipse with semi-major axis in the x-direction of magnitude

$$a\,\frac{\cosh k(y+h)}{\sinh kh}$$

and with semi-minor axis in the y-direction of magnitude

$$a\,\frac{\sinh k(y+h)}{\sinh kh}.$$

The lengths of both semi-axes decrease with depth. When $y=-h$, $V=0$, $U\neq 0$ and particles oscillate along the bottom; for a real liquid, viscosity would prevent this oscillation.

We are now in a position to investigate when the neglect of vertical acceleration is justified and hence the range of validity of the previous theory of long waves on canals. We have already shown that the required condition eqn (3.16) is

$$\beta(h+\eta)\ll g|\eta|,$$

where β is the maximum vertical acceleration on a vertical line. On linearization, this condition is

$$\beta h\ll g|\eta|.$$

From surface wave theory, the vertical acceleration is

$$\frac{\partial v}{\partial t}=-gak\,\frac{\sinh k(y+h)}{\cosh kh}\sin(kx-\omega t).$$

The acceleration is a maximum with respect to y at $y=0$, and so

$$\beta=-gak\tanh kh\sin(kx-\omega t).$$

Now, $g\eta=ag\sin(kx-\omega t)$, hence,

$$\left|\frac{\beta h}{g\eta}\right|=kh\tanh kh.$$

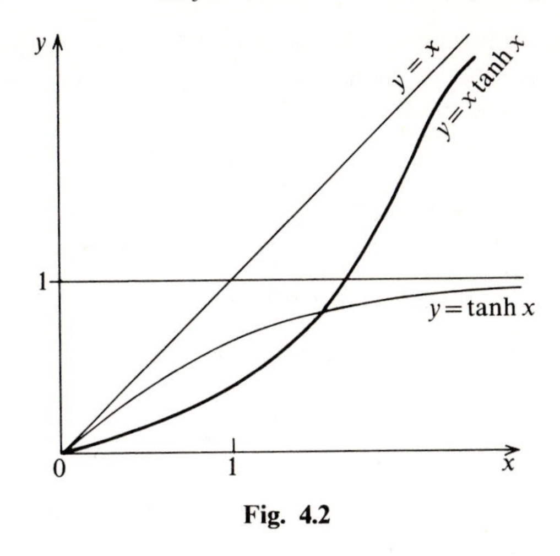

Fig. 4.2

$x \tanh x$ is plotted against x in Fig. 4.2.

$$\left|\frac{\beta h}{g\eta}\right| \ll 1, \quad \text{if } kh \tanh kh \ll 1 \text{ or } kh \ll 1.$$

Since $k = 2\pi/\lambda$, the condition is satisfied if $h \ll \lambda$, i.e. if depth is small compared to wavelength. We have now proved that $h \ll \lambda$ is not just the condition that the long-wave theory is consistent but it is also the condition for its validity, always within the limitations of linear theory.

The reader should note that the validity of an approximation in an applied mathematical theory cannot be proved from the theory itself, only the consistency of the approximation within the theory can be shown. To prove validity it is necessary to construct a theory not using the approximation and then to determine in what circumstances the approximation is valid. This latter process is far more difficult. For a new theory many years may elapse before the validity of the approximations used are established. Only the consistency, not the validity, of the assumption $u = 0$ in the theory of transverse waves on strings was proved at the end of section 1.1.

Frequently, to find the validity of an approximate theory it is necessary to resort to experiment. The criterion for validity is now practical. Does the theory predict the experimental results to within a pre-stated degree of accuracy? It it does, theory is then accepted within the range of successful prediction. The range of validity of the linearization hypothesis is determined by separate experiments for each particular theory. The range is often different for different materials for the same theory.

When $kh \ll 1$, $\cosh kh \simeq 1$, $\sinh k(h+y) \simeq k(h+y)$ and $\sin kh \simeq kh$. Substitution into eqns (4.27) to (4.30) reproduces the velocities and displacements of shallow water theory.

4.2. Potential and kinetic energy

The potential energy of a travelling wave per unit area of surface is V, where

$$V = \frac{\rho g}{2\lambda} \int_0^\lambda \eta^2 \, \mathrm{d}x,$$

by the same argument used in deriving eqn (3.21). Hence,

$$V = \tfrac{1}{4} \rho g a^2, \tag{4.32}$$

since $\eta = a \sin(kx - \omega t)$ and the integral of $\sin^2(kx - \omega t)$ over a wavelength is $\lambda/2$. The corresponding kinetic energy is T, where

$$\begin{aligned} T &= \frac{\rho}{2\lambda} \int_0^\lambda \left(\int_{-h}^0 (u^2 + v^2) \mathrm{d}y \right) \mathrm{d}x \\ &= \frac{\rho}{2\lambda} \int_0^\lambda \int_{-h}^0 \left(\left(\frac{\partial \phi}{\partial x} \right)^2 + \left(\frac{\partial \phi}{\partial y} \right)^2 \right) \mathrm{d}x \, \mathrm{d}y \\ &= \frac{\rho}{2\lambda} \int_S \operatorname{grad} \phi \, . \operatorname{grad} \phi \, \mathrm{d}S \\ &= \frac{\rho}{2\lambda} \int_S \operatorname{div}(\phi \operatorname{grad} \phi) \, \mathrm{d}S, \end{aligned}$$

since $\nabla^2 \phi = 0$. In Fig. 4.3 the interior S of the rectangle is bounded by the closed curve C.

$$T = \frac{\rho}{2\lambda} \int_C \phi \operatorname{grad} \phi \cdot \boldsymbol{n} \, \mathrm{d}s$$

by the divergence theorem.

Now $\phi \operatorname{grad} \phi$ has the same values on $x = 0$ and $x = \lambda$ and $\boldsymbol{n}$ is in opposite senses on these two lines. The contributions from these two lines to the integral cancel,

$$\operatorname{grad} \phi \cdot \boldsymbol{n} = -\frac{\partial \phi}{\partial y} \quad \text{on } y = -h \quad \text{and} \quad \frac{\partial \phi}{\partial y} = 0 \quad \text{on } y = -h.$$

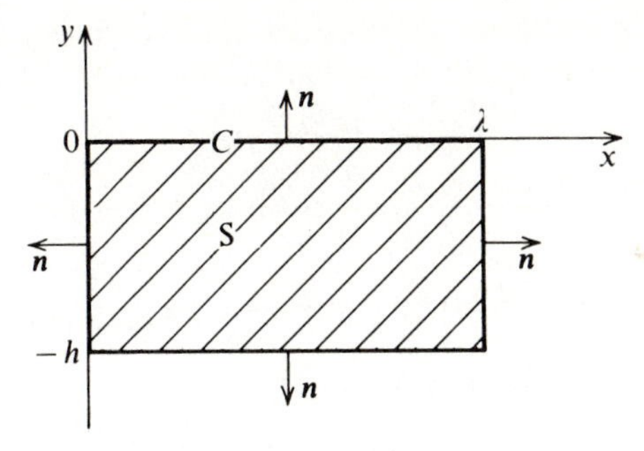

Fig. 4.3

Hence there remains

$$T = \frac{\rho}{2\lambda}\int_0^{\lambda}\left(\phi\frac{\partial\phi}{\partial y}\right)_{y=0}\mathrm{d}x$$

$$= \frac{\rho}{2\lambda}\int_0^{\lambda}\frac{ga}{\omega}\cos(kx-\omega t)\frac{gak}{\omega}\tanh kh\cos(kx-\omega t)\mathrm{d}x$$

$$= \frac{\rho}{4}\frac{g^2a^2k}{\omega^2}\tanh kh$$

$$= \tfrac{1}{4}\rho g a^2, \qquad (4.33)$$

on using the dispersion equation (4.21).

Equations (4.32) and (4.33) show that the potential and kinetic energies per unit surface area are equal for a travelling sinusoidal wave and are equal to $\frac{1}{4}\rho g a^2$. The total energy per unit area is $\frac{1}{2}\rho g a^2$.

4.3. Oscillations of water in a tank

The oscillations of water in a tank of constant depth h and length l can be treated by looking for solutions of the form

$$\phi = X(x)\,Y(y)\cos\omega t. \qquad (4.34)$$

Substituting in $\partial^2\phi/\partial x^2 + \partial^2\phi/\partial y^2 = 0$,

$$\frac{X''}{X} = -\frac{Y''}{Y} = -\mu^2,$$

where the sign of the separation constant is chosen to give exponential solutions in y and sinusoidal solutions in x. Hence,

$$Y(y)=A'e^{\mu y}+Be^{-\mu y}.$$

For $\partial\phi/\partial y=0$ at $y=-h$, $Y'(-h)=0$ and $B=A'e^{-2\mu h}$ as before. Therefore,

$$Y(y)=A'e^{-\mu h}(e^{\mu(y+h)}+e^{-\mu(y+h)})=A\cosh\mu(y+h), \qquad (4.35)$$

where $A=2A'e^{-\mu h}$. Solving $X''+\mu^2X=0$,

$$X(x)=C'\cos\mu x+D\sin\mu x. \qquad (4.36)$$

If the ends of the tank are $x=0$ and $x=l$, then the condition $\partial\phi/\partial x=0$ at the ends requires $X'(0)=X'(l)=0$. Since

$$X'(x)=-\mu C'\sin\mu x+\mu D\cos\mu x,$$

$D=0$ and $\mu C'\sin\mu l=0$. If $C'=0$, $\phi=0$; if $\mu=0$, $\phi=C\cos\omega t$ and $C=0$ from eqn (4.16). In both cases the water is at rest. Hence, for oscillations, $\sin\mu l=0$ or $\mu l=n\pi$.

If $n=0$, $\mu=0$, which has already been excluded. Therefore n is a positive integer and

$$\mu=\frac{n\pi}{l}, \quad n=1, 2, 3 \ldots. \qquad (4.37)$$

Substituting back for D and μ into eqn (4.36) and (4.35),

$$X(x)=C'\cos\frac{n\pi x}{l}, \quad Y(y)=A\cosh\frac{n\pi}{l}(y+h).$$

Then from eqn (4.34),

$$\phi=C_n\cos\frac{n\pi x}{l}\cosh\frac{n\pi}{l}(y+h)\cos\omega_n t, \quad n=1, 2, 3, \ldots \qquad (4.38)$$

and C_n is constant.

Since a different value of ω is to be expected for each value of n, a suffix n has been added to ω in eqn (4.38). The relation between ω_n and n is given by substitution for ϕ from eqn (4.39) into the free surface condition, eqn (4.16):

$$-\omega_n^2\cosh\frac{n\pi h}{l}+g\frac{n\pi}{l}\sinh\frac{n\pi h}{l}=0,$$

or

$$\omega_n = \left(\frac{n\pi g}{l} \tanh \frac{n\pi h}{l} \right)^{\frac{1}{2}}. \tag{4.39}$$

This is the equation for the normal frequencies of oscillation. The normal modes are given by substituting back for ω_n from eqn (4.39) into eqn (4.38).

If the tank is deep compared to its length so that $h/l \gg 1$, then $\tanh(n\pi h/l) \simeq 1$, and the normal frequencies are

$$\omega_n \simeq \left(\frac{n\pi g}{l} \right)^{\frac{1}{2}}. \tag{4.40}$$

The fundamental has radian frequency

$$\omega_1 = \left(\frac{\pi g}{l} \tanh \frac{\pi h}{l} \right)^{\frac{1}{2}} \simeq \left(\frac{\pi g}{l} \right)^{\frac{1}{2}}, \quad \text{when } \frac{h}{l} \gg 1.$$

When $(n\pi h/l) \ll 1$, $\tanh(n\pi h/l) \simeq n\pi h/l$ and $\omega_n \simeq n\pi \sqrt{gh}/l$. This result is only valid for the lower modes—eventually n becomes large enough to violate the inequality. For the fundamental mode, $n=1$, the cyclic frequency is

$$\left(\frac{\omega}{2\pi} \right)_{n=1} = \frac{\sqrt{gh}}{2l},$$

which agrees with the result previously obtained for oscillations on canals.

The analysis can be simplified a little when the depth is very much greater than all the other lengths: h is effectively treated as infinite. In the solution for $Y(y)$,

$$Y(y) = A\mathrm{e}^{\mu y} + B\mathrm{e}^{-\mu y},$$

one requires that $Y(y)$ remains finite as $y \to -\infty$, hence $B=0$ and

$$Y(y) = A\mathrm{e}^{\mu y}.$$

The free surface condition then gives $-\omega^2 + g\mu = 0$ and μ is given as before by $\mu = n\pi/l$. Hence, the normal frequencies are

$$\omega = \left(\frac{n\pi g}{l} \right)^{\frac{1}{2}}$$

as above. Treating h as infinite is often called the 'deep' water approximation.

The analysis for surface waves can be extended to the case where there is dependence on the third coordinate, z. The conditions for irrotational motion are $\partial u/\partial y = \partial v/\partial x$, $\partial v/\partial z = \partial w/\partial y$ and $\partial w/\partial x = \partial u/\partial z$, which are the necessary and sufficient conditions for a velocity potential ϕ to exist such that

$$u = -\frac{\partial \phi}{\partial x}, \quad v = -\frac{\partial \phi}{\partial y} \quad \text{and} \quad w = -\frac{\partial \phi}{\partial z};$$

u, v and w are the components of the velocity vector. The condition of incompressibility is

$$\frac{\partial u}{\partial x} + \frac{\partial v}{\partial y} + \frac{\partial w}{\partial z} = 0.$$

Hence

$$\nabla^2 \phi = 0, \quad \text{where } \nabla^2 \equiv \frac{\partial^2}{\partial x^2} + \frac{\partial^2}{\partial y^2} + \frac{\partial^2}{\partial z^2}.$$

The free surface condition as before is

$$\frac{\partial^2 \phi}{\partial t^2} + g\frac{\partial \phi}{\partial y} = 0, \quad \text{on } y = 0.$$

The condition at a fixed vertical boundary remains $\partial \phi/\partial n = 0$.

Exercise 4.1. A rectangular basin has length l and breadth b, and its depth is much greater than either. Determine the normal modes of oscillation of its water surface, by seeking solutions of Laplace's equation exponential in y which satisfy the boundary condition at the surface, and then picking combinations of these which satisfy the boundary conditions at the walls. (Take the walls as $x = 0$, $x = l$, $z = 0$ and $z = b$ and the y-axis vertically upwards.)

4.4. Group velocity

The equation for the velocity of sinusoidal surface waves is

$$c = \frac{\omega}{k} = \left(\frac{g}{k} \tanh kh\right)^{\frac{1}{2}}. \tag{4.22}$$

The waves are dispersive, i.e. waves of different frequency or wavelength travel at different speeds. It follows that if a wave whose shape is not harmonic is resolved into its harmonic components by Fourier analysis, and if at a later time the components are added together again, then the components will have travelled different distances and the final shape will be different from the initial one.

Consider two harmonic waves of equal amplitude and of slightly different wavelengths, and hence of velocity and frequency, travelling in the same sense, i.e.

$$\eta = a\sin(kx-\omega t) + a\sin((k+\delta k)x-(\omega+\delta\omega)t),$$

and ask at what speed do the maxima of the total disturbancc travel. $\delta\omega$ and δk are not independent. They are related by eqn (4.22). Using the formula for the sum of two sines,

$$\eta = 2a\sin((k+\tfrac{1}{2}\delta k)x-(\omega+\tfrac{1}{2}\delta\omega)t)\cos(\tfrac{1}{2}\delta k\,x-\tfrac{1}{2}\delta\omega\,t).$$

At a fixed time t the sine term has wavelength $2\pi/(k+\frac{1}{2}\delta k)$ and the cosine term $4\pi/\delta k$. For $\delta k \ll k$, the wavelength of the latter is far greater. The net result is the oscillation of the sine term with its amplitude slowly varying between 0 and 1 (Fig. 4.4). As $\delta k \to 0$, the peaks of the cosine term become further and further apart. The velocity of each peak is $\delta\omega/\delta k$. In the limit as $\delta k \to 0$, each peak is separated by an infinite distance from its neighbours, and the velocity becomes $\mathrm{d}\omega/\mathrm{d}k$. It is called the group velocity and is denoted by v_g; c is called the 'phase' or 'wave' velocity. Since $c=\omega/k$,

$$v_g = \frac{\mathrm{d}\omega}{\mathrm{d}k} = \frac{\mathrm{d}(kc)}{\mathrm{d}k} = c + k\frac{\mathrm{d}c}{\mathrm{d}k}. \tag{4.41}$$

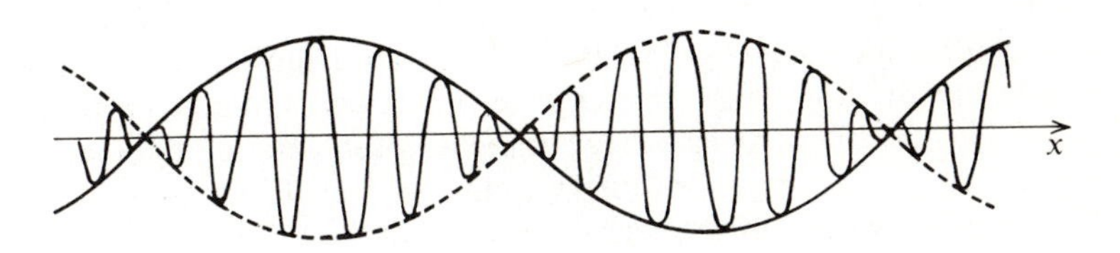

Fig. 4.4

If c is independent of k, i.e. there is no dispersion, then $\mathrm{d}c/\mathrm{d}k=0$ and $v_g=c$. The wave and group velocities are identical (and constant) in the absence of dispersion. In terms of c and $\lambda=2\pi/k$,

$$v_g = c + k\frac{\mathrm{d}c}{\mathrm{d}\lambda}\frac{\mathrm{d}\lambda}{\mathrm{d}k} = c - k\frac{2\pi}{k^2}\frac{\mathrm{d}c}{\mathrm{d}\lambda} = c - \lambda\frac{\mathrm{d}c}{\mathrm{d}\lambda}. \tag{4.42}$$

The above equations hold for any dispersive medium. For water waves, when

$$c=\left(\frac{g}{k}\tanh kh\right)^{\frac{1}{2}},$$

$$v_g=\frac{\mathrm{d}(kc)}{\mathrm{d}k}=\frac{\mathrm{d}}{\mathrm{d}k}(gk\tanh kh)^{\frac{1}{2}}$$

$$=\frac{\frac{1}{2}g}{(gk\tanh kh)^{\frac{1}{2}}}(\tanh kh+kh\,\mathrm{sech}^2\,kh)$$

or

$$v_g=\frac{1}{2}c\left(1+\frac{2kh}{\sinh 2kh}\right). \tag{4.43}$$

$$\frac{v_g}{c}\to\frac{1}{2}\text{ as } kh\to\infty \quad\text{and}\quad \frac{v_g}{c}\to 1 \text{ as } kh\to 0. \tag{4.44}$$

The former limit is $h \gg \lambda$, very deep water, and the latter $h \ll \lambda$. In the latter limit, $c\to\sqrt{gh}$, there is no dispersion and $v_g\to c$.

Can these results be extended to a small packet or group of waves whose wavenumbers are in the range $-\varepsilon+k_0<k<\varepsilon+k_0$? That is,

$$\eta=\int_{-\varepsilon+k_0}^{\varepsilon+k_0} a(k)\,\mathrm{e}^{\mathrm{i}(kx-\omega t)}\,\mathrm{d}k, \tag{4.45}$$

where $\omega=\omega(k)$ is given by the dispersion equation.

First consider $\int_{-\infty}^{\infty}\phi(x)\mathrm{e}^{\mathrm{i}f(x)}\mathrm{d}x$, where the real function $f(x)$ in general varies much more rapidly than the real function $\phi(x)$. In these circumstances $\phi(x)$ varies very little as the argument of the exponential increases by $2\pi\mathrm{i}$ and the net contribution of the integrand to the integral is practially zero over this increase of argument. This result ceases to be true in the neighbourhood of points at which $f'(x)=0$, i.e. near stationary points of $f(x)$. The main contribution to the value of the integral comes from the neighbourhood of such points. This argument is known as the principle of 'stationary phase' and is due to Kelvin.

Next consider

$$\int_{-\infty}^{\infty} a(k)\,\mathrm{e}^{\mathrm{i}(kx-\omega t)}\,\mathrm{d}k,\quad \omega=\omega(k).$$

The integral can be written

$$\int_{-\infty}^{\infty} a(k)\mathrm{e}^{\mathrm{i}t(kx/t-\omega)}\,\mathrm{d}k. \tag{4.46}$$

By choosing t sufficiently large, the argument of the exponential can be made to vary arbitrarily more rapidly with k than $a(k)$ because ω is a function of k. Applying the principle of stationary phase to this integral, the main contribution will come from those values of k for which

$$0=\frac{\mathrm{d}}{\mathrm{d}k}\left(t\left(\frac{k}{t}x-\omega\right)\right)=t\left(\frac{x}{t}-\frac{\mathrm{d}\omega}{\mathrm{d}k}\right)$$

or

$$\frac{\mathrm{d}\omega}{\mathrm{d}k}=\frac{x}{t}. \tag{4.47}$$

Now suppose $a(k)$ is zero outside the range $-\varepsilon+k_0<k<\varepsilon+k_0$. The integral (4.46) becomes equal to the integral (4.45) and only for those values of x/t in the neighbourhood of $(\mathrm{d}\omega/\mathrm{d}k)_{k=k_0}$ will the value of η be appreciable. But $x/t=(\mathrm{d}\omega/\mathrm{d}k)_{k=k_0}$ represents a disturbance travelling with velocity $(\mathrm{d}\omega/\mathrm{d}k)_{k=k_0}$, i.e. with the group velocity evaluated at $k=k_0$.

It has been shown that the main part of the disturbance representing the superposition of a small group of waves with wavenumbers in the neighbourhood of k_0 travels with the group velocity v_g evaluated at k_0. The greater the time t, the more accurate is the use of the principle of stationary phase used in deriving this result. Note that

$$\eta(x,t)=\int_{-\infty}^{\infty}a(k)\,\mathrm{e}^{\mathrm{i}(kx-\omega t)}\,\mathrm{d}k \quad \text{at } t=0$$

becomes

$$\eta(x,0)=\int_{-\infty}^{\infty}a(k)\,\mathrm{e}^{\mathrm{i}kx}\,\mathrm{d}k.$$

Hence $a(k)$ is the Fourier transform of $\eta(x,0)$ apart from a constant factor.

We shall next treat the energy velocity of a wave on water. The energy velocity is defined as the time average of the rate of transmission of energy per unit width divided by the energy per unit surface area of the wave. The latter has already been evaluated (eqns (4.32) and (4.33)); it is

$$T+V=\tfrac{1}{2}\rho g a^2. \tag{4.48}$$

The rate of transmission of energy across unit width of wave, E, is

$$E=\int_{-h}^{0}pu\,\mathrm{d}y. \tag{4.49}$$

Substitute for p and u from eqns (4.14) and (4.3):

$$E = -\int_{-h}^{0} \left(p_0 + \rho \frac{\partial \phi}{\partial t} - \rho g y \right) \frac{\partial \phi}{\partial x} \, \mathrm{d}y. \tag{4.50}$$

Now

$$\phi = \frac{ga}{\omega} \frac{\cosh k(y+h)}{\cosh kh} \cos(kx - \omega t).$$

The time average of both $\cos(kx-\omega t)$ and $\sin(kx-\omega t)$ over a period is zero, whereas the time average of $\sin^2(kx-\omega t)$ is $\frac{1}{2}$. If $\bar{E}$ denotes the time average of E over a period, then

$$\begin{aligned}
\bar{E} &= \frac{1}{2} \rho \frac{g^2 a^2 k}{\omega \cosh^2 kh} \int_{-h}^{0} \cosh^2 k(y+h) \, \mathrm{d}y \\
&= \frac{\rho g^2 a^2 k}{4\omega \cosh^2 kh} \left[y + \frac{1}{2k} \sinh 2k(y+h) \right]_{-h}^{0} \\
&= \frac{\rho g^2 a^2 k}{4\omega \cosh^2 kh} \left(h + \frac{1}{2k} \sinh 2kh \right).
\end{aligned} \tag{4.51}$$

The energy velocity, v_E, is given by substitution from eqns (4.48) and (4.51) in

$$v_E = \bar{E}/(T+V) \tag{4.52}$$

as

$$v_E = \frac{gkh}{2\omega} \operatorname{sech}^2 kh \left(1 + \frac{1}{2kh} \sinh 2kh \right).$$

Hence,

$$\begin{aligned}
v_E &= \frac{q}{2\omega} \tanh kh \left(\frac{2kh}{\sinh 2kh} + 1 \right) \\
&= \frac{g}{2\omega} \frac{\omega^2}{gk} \left(\frac{2kh}{\sinh 2kh} + 1 \right) \\
&= \frac{1}{2} c \left(1 + \frac{2kh}{\sinh 2kh} \right).
\end{aligned} \tag{4.53}$$

Comparison of eqns (4.53) and (4.43) shows that

$$v_E = v_g, \tag{4.54}$$

i.e. the energy velocity is equal to the group velocity.

We have now studied three ways in which the group velocity arises. Which treatment is the most satisfactory is a matter of individual preference, for example some people prefer to define group velocity as the energy velocity. The important fact is that a velocity, denoted by v_g and related to the wave velocity c by the equation

$$v_g = c + k\frac{dc}{dk} \tag{4.41}$$

arises in many contexts in mathematical physics and is as important as the wave velocity in any dispersive system. In a non-dispersive system, the two velocities are equal.

Exercise 4.2. The velocity potential for surface waves in a liquid possessing surface tension T satisfies

$$\frac{\partial^2\phi}{\partial x^2} + \frac{\partial^2\phi}{\partial y^2} = 0 \quad \text{in the interior}$$

and

$$\frac{\partial^2\phi}{\partial t^2} + g\frac{\partial\phi}{\partial y} - \frac{T}{\rho}\frac{\partial^3\phi}{\partial x^2\partial y} = 0 \quad \text{at } y = 0.$$

Find the group velocity v_g in a liquid of infinite depth of waves with surface displacement $\eta = a\sin k(x - ct)$. Show that $v_g/c \to \frac{1}{2}$ as $k \to 0$ and $v_g/c \to \frac{3}{2}$ as $k \to \infty$.

Exercise 4.3. The equation, satisfied by the displacement $y(x, t)$ in the flexural vibration of bars, is

$$\frac{\partial^2 y}{\partial t^2} + \alpha^2\frac{\partial^4 y}{\partial x^4} = 0, \quad \alpha \text{ constant.}$$

Find the phase and group velocities of harmonic waves, travelling in the positive sense along an infinite bar, as a function of the wavelength.

4.5. Evaluation of wave packets

The shape of the wave packet

$$\eta = \int_{-\varepsilon + k_0}^{\varepsilon + k_0} a(k)e^{i(kx - \omega t)}\,dk \tag{4.45}$$

can be found for large t. A more detailed treatment of the matter of this paragraph can be found in any text on asymptotics which includes the method of stationary phase. Consider

$$u = \int_a^b \phi(x)e^{if(x)}\,dx, \tag{4.55}$$

where $f'(x)$ has only one zero, at $x=\alpha$ say, and $a<\alpha<b$. As already stated, the main contribution to the integral will come from the neighbourhood of $x=\alpha$. In this neighbourhood, $\phi(x)$ is approximately constant and equal to $\phi(\alpha)$ and $f(x)$ can be expanded as

$$f(x)=f(\alpha)+\tfrac{1}{2}(x-\alpha)^2 f''(\alpha), \tag{4.56}$$

provided higher-order terms are sufficiently small. This condition will be treated later. Little error will be introduced by replacing a and b by $-\infty$ and $+\infty$, because the contributions from the integrand cancel except near $x=\alpha$. Therefore,

$$u=\phi(\alpha)\mathrm{e}^{\mathrm{i}f(\alpha)}\int_{-\infty}^{\infty}\mathrm{e}^{\frac{1}{2}\mathrm{i}\xi^2 f''(\alpha)}\,\mathrm{d}\xi$$

on putting $x-\alpha=\xi$. Since

$$\int_{-\infty}^{\infty}\mathrm{e}^{\pm \mathrm{i}m^2\delta^2}\,\mathrm{d}\delta=\frac{\sqrt{\pi}}{m}\mathrm{e}^{\pm \mathrm{i}\pi/4},$$

$$u=\frac{\sqrt{\pi}\,\phi(\alpha)}{\sqrt{|\frac{1}{2}f''(\alpha)|}}\,\mathrm{e}^{\mathrm{i}(f(\alpha)\pm\pi/4)}, \tag{4.57}$$

taking the upper sign if $f''(\alpha)>0$, and the lower sign if $f''(\alpha)<0$.

In the case of the wave packet $f(k)=kx-\omega t$ and $\phi(k)=a(k)$, we have already seen that the integral (4.45) is only appreciable for those values of x and t in the neighbourhood of $x/t=(\mathrm{d}\omega/\mathrm{d}k)_{k=k_0}$, eqn (4.47). For these values of x and t,

$$f'(k_0)=x-t\left(\frac{\mathrm{d}\omega}{\mathrm{d}k}\right)_{k=k_0}=0, \quad f''(k_0)=-\left(\frac{\mathrm{d}^2\omega}{\mathrm{d}k^2}\right)_{k=k_0}t.$$

Hence, using eqn (4.57),

$$\eta=\frac{\sqrt{2\pi}\,a(k_0)}{\sqrt{t|\mathrm{d}^2\omega/\mathrm{d}k^2|_{k=k_0}}}\cos(k_0x-\omega_0t\pm\pi/4), \tag{4.58}$$

where the ambiguous sign is the opposite of that of $(\mathrm{d}^2\omega/\mathrm{d}k^2)_{k=k_0}$.

Equation (4.58) is only valid in the neighbourhood of $x/t=(\mathrm{d}\omega/\mathrm{d}k)_{k=k_0}$. Away from this neighbourhood, η is negligible. Equations (4.47) and (4.58) give only a limited amount of information. Equation (4.47) states that, for t sufficiently large, the main part of the disturbance travels with the group velocity $(\mathrm{d}\omega/\mathrm{d}k)_{k=k_0}$ and eqn (4.58) states the main part is oscillatory with known amplitude decreasing with time as $t^{-\frac{1}{2}}$ and with phase differing by $\pi/4$ from the component $k=k_0$ of the original wave packet. The equations (4.47) and (4.58) do not tell the size of the wave packet, i.e. how far, at fixed t, on either side of $x=(\mathrm{d}\omega/\mathrm{d}k)_{k=k_0}\,t$ eqn (4.58) remains valid.

Equation (4.57) will next be applied to a δ-function elevation of deep water, for which $\omega^2=gk$, at the origin at time $t=0$. Since†

$$\delta(x)=\frac{1}{2\pi}\int_{-\infty}^{\infty} \mathrm{e}^{\mathrm{i}kx}\,\mathrm{d}k, \tag{4.59}$$

the elevation

$$\eta(x,t)=\frac{1}{2\pi}\int_{-\infty}^{\infty} \mathrm{e}^{\mathrm{i}(kx-\omega t)}\,\mathrm{d}k \tag{4.60}$$

represents a wave travelling in the positive sense; comparison of eqn (4.60) when $t=0$ with eqn (4.59) shows that

$$\eta(x,0)=\delta(x), \tag{4.61}$$

and represents that part of the initial elevation that travels in the positive sense of x. Assuming that the deep water extends from $x=-\infty$ to $x=+\infty$, by symmetry a wave identical to that of eqn (4.60) but with x replaced by $-x$ travels in the negative sense of x. The initial elevation to produce both waves is $2\delta(x)$. We are concerned here only with the wave in the positive sense.

What is the shape of the wave given by eqn (4.60) for large t? For a particular pair of values of x and t, only those values of k in the neighbourhood of $\mathrm{d}\omega/\mathrm{d}k=x/t$ will contribute to the shape. For all values of x/t for which there is an equal value of $\mathrm{d}\omega/\mathrm{d}k$ for some k, $-\infty<k<\infty$, there will be in general a non-negligible value of η.

Since $\omega^2=gk$,

$$\frac{x}{t}=\frac{\mathrm{d}\omega}{\mathrm{d}k}=\tfrac{1}{2}g^{\frac{1}{2}}k^{-\frac{1}{2}}$$

Therefore,

$$k=\frac{gt^2}{4x^2} \tag{4.62}$$

and

$$\omega=g^{\frac{1}{2}}k^{\frac{1}{2}}=\frac{gt}{2x}. \tag{4.63}$$

† Equation (4.59) is discussed at the end of this section.

It is seen from eqn (4.62) that for all values of x/t there is a corresponding value of k. Also

$$\phi(k)=\frac{1}{2\pi} \tag{4.64}$$

and

$$f(k)=kx-\omega t, \tag{4.65}$$

so that

$$f''(k)=-t\,\frac{\mathrm{d}^2\omega}{\mathrm{d}k^2}. \tag{4.66}$$

Hence,

$$f(k)=\frac{gt^2}{4x^2}x-\frac{gt}{2x}t=-\frac{gt^2}{4x} \tag{4.67}$$

and

$$\begin{aligned} f''(k)&=-t(-\tfrac{1}{4}g^{\frac{1}{2}}k^{-\frac{3}{2}})=\tfrac{1}{4}\,tg^{\frac{1}{2}}\,\frac{8x^3}{g^{\frac{3}{2}}t^3}\\ &=\frac{2x^3}{gt^2}. \end{aligned} \tag{4.68}$$

Substituting from eqns (4.64), (4.67) and (4.68) into eqn (4.57) with u replaced by η:

$$\eta=\frac{\sqrt{\pi}\,(1/2\pi)}{\sqrt{(x^3/gt^2)}}\exp\mathrm{i}\left(-\frac{gt^2}{4x}+\frac{\pi}{4}\right).$$

On taking the real part of the exponential and using $\cos(-\theta)=\cos\theta$,

$$\eta=\frac{g^{\frac{1}{2}}t}{2\pi^{\frac{1}{2}}x^{\frac{3}{2}}}\cos\left(\frac{gt^2}{4x}-\frac{\pi}{4}\right), \tag{4.69}$$

which is the equation for the elevation for large values of t†.

We next investigate the validity of terminating the series for $f(x)$ in eqn (4.56) after the second term. The third term in the series is $(1/3!)(x-\alpha)^3 f'''(\alpha)$ and a criterion for the validity of the termination is

$$\left|\frac{(1/3!)(x-\alpha)^3 f'''(\alpha)}{(1/2!)(x-\alpha)^2 f''(\alpha)}\right|\ll 1,$$

† In the language of asymptotic analysis, the equality sign in eqn (4.68) is replaced by $\underset{t\to\infty}{\sim}$

or

$$\left|\frac{(x-\alpha)f'''(\alpha)}{3f''(\alpha)}\right| \ll 1. \tag{4.70}$$

The principle of stationary phase is based on the postulate that only a small region in the neighbourhood of the stationary point (for which $f'(x)=0$) contributes to the integral. We suppose that over this region $f(x)$ changes by a few multiples of 2π so that an estimate of $(1/2!)(x-\alpha)^2|f''(\alpha)|$, which approximately equals $f(x)-f(\alpha)$, is $c^2/2$, where c is about 6.

Hence,

$$\frac{1}{3}(x-\alpha)|f''(\alpha)|^{\frac{1}{2}}=\frac{c}{3}$$

and the inequality (4.70) becomes

$$\left|\frac{f'''(\alpha)}{(f''(\alpha))^{\frac{3}{2}}}\right| \ll \frac{1}{2}. \tag{4.71}$$

In the application to harmonic wave packets,

$$f''(k)=-\frac{\mathrm{d}^2\omega}{\mathrm{d}k^2}t, \quad f'''(k)=-\frac{\mathrm{d}^3\omega}{\mathrm{d}k^3}t$$

and the inequality becomes

$$\left|\frac{\mathrm{d}^3\omega/\mathrm{d}k^3}{t^{\frac{1}{2}}(\mathrm{d}^2\omega/\mathrm{d}k^2)^{\frac{3}{2}}}\right| \ll \frac{1}{2}. \tag{4.72}$$

For the δ-function initial elevation on deep water,

$$\frac{\mathrm{d}^2\omega}{\mathrm{d}k^2}=-\frac{2x^3}{gt^3}, \quad \frac{\mathrm{d}^3\omega}{\mathrm{d}k^3}=\tfrac{3}{8}g^{\frac{1}{2}}k^{-\frac{5}{2}}=12\frac{x^3}{g^2t^3}$$

and the inequality (4.72) becomes

$$\frac{6\sqrt{2}x^{\frac{1}{2}}}{g^{\frac{1}{2}}t} \ll 1. \tag{4.73}$$

Hence, the termination of the series used in evaluating the stationary phase integral is valid if

$$g^{\frac{1}{2}}t \gg 6\sqrt{2}x^{\frac{1}{2}}.$$

We now consider eqn (4.59). The argument below assumes that the value of the integral $\int_0^n \cos kx\,\mathrm{d}x$ $(=(1/k)\sin nx)$ as $n\to\infty$ can be taken as its average value as $n\to\infty$, i.e. zero. For $x\neq 0$,

$$\int_{-n}^{n} \mathrm{e}^{\mathrm{i}kx}\,\mathrm{d}k = \int_{-n}^{n} (\cos kx + \mathrm{i}\sin kx)\mathrm{d}k$$

$$= 2\int_{0}^{n} \cos kx\,\mathrm{d}k,$$

since $\cos x$ is even and $\sin x$ is odd. Using the assumption just stated, for $x \neq 0$,

$$\int_{-\infty}^{\infty} \mathrm{e}^{\mathrm{i}kx}\mathrm{d}x = \lim_{n\to\infty}\int_{-n}^{n} \mathrm{e}^{\mathrm{i}kx}\mathrm{d}x = 2\lim_{n\to\infty}\int_{0}^{n}\cos kx\,\mathrm{d}k = 0. \quad (4.74)$$

For $x=0$, $\mathrm{e}^{\mathrm{i}kx}=1$ and the integral diverges.

Integrate the r.h.s. of eqn (4.59) with respect to x from $-\varepsilon$ to ε:

$$\frac{1}{2\pi}\int_{-\varepsilon}^{\varepsilon}\mathrm{d}x\int_{-\infty}^{\infty}\mathrm{e}^{\mathrm{i}kx}\,\mathrm{d}k = \frac{1}{2\pi}\int_{\infty}^{\infty}\mathrm{d}k\int_{-\varepsilon}^{\varepsilon}\mathrm{e}^{\mathrm{i}kx}\,\mathrm{d}x$$

$$= \frac{1}{2\pi}\int_{\infty}^{\infty}\mathrm{d}k\,\frac{\mathrm{e}^{\mathrm{i}kx}}{\mathrm{i}k}\bigg|_{x=-\varepsilon}^{\varepsilon}$$

$$= \frac{1}{\pi}\int_{\infty}^{\infty}\frac{\sin k\varepsilon}{k}\,\mathrm{d}k$$

$$\underset{k\varepsilon=y}{=} \frac{1}{\pi}\int_{\infty}^{\infty}\frac{\sin y}{y}\,\mathrm{d}y$$

$$= \frac{2}{\pi}\int_{0}^{\infty}\frac{\sin y}{y}\,\mathrm{d}y = \frac{2}{\pi}\cdot\frac{\pi}{2}$$

Therefore,

$$\frac{1}{2\pi}\int_{-\varepsilon}^{\varepsilon}\mathrm{d}x\int_{\infty}^{\infty}\mathrm{e}^{\mathrm{i}kx}\,\mathrm{d}k = 1. \quad (4.75)$$

Equations (4.74) and (4.75) show that $(1/2\pi)\int_{\infty}^{\infty}\mathrm{e}^{\mathrm{i}kx}\,\mathrm{d}x$ satisfies the defining properties of the δ-function. A rigorous proof of eqn (4.59) requires the theory of distributions which was invented to overcome analytical problems arising in the theory of δ-functions†.

4.6. Complex variable approach

An alternative treatment of surface waves can be made using complex variables. The general solution to Laplace's equation (4.5),

† See, for example: Gel'fand, I. M. and Shilov, G. E. *Generalized functions*, Vol. 1, or Zemanian, A. H. *Distribution theory and transform analysis*.

where ϕ is a function of x, y and t, can be written

$$\phi = \tfrac{1}{2}(f(z, t) + \bar{g}(\bar{z}, t)), \tag{4.76}$$

where $z = x + \mathrm{i}y$, $\bar{z} = x - \mathrm{i}y$ and f and $\bar{g}$ are arbitrary functions of their arguments. Since ϕ is real,

$$\mathrm{I}[\bar{g}(\bar{z}, t)] = -\mathrm{I}[f(z, t)] = \mathrm{I}[\bar{f}(\bar{z}, t)]$$

on taking the complex conjugate of $f(z, t)$. Hence,

$$\mathrm{I}[\bar{g}(\bar{z}, t) - \bar{f}(\bar{z}, t)] = 0$$

and it follows from the Cauchy–Riemann equations that $\bar{g}(\bar{z}, t) - \bar{f}(\bar{z}, t)$ is at most a real function of t, $T(t)$ say. Substitute back for $\bar{g}(\bar{z}, t)$ into eqn (4.76):

$$\phi = \tfrac{1}{2}(f(z, t) + \bar{f}(\bar{z}, t) + T(t)),$$

or

$$\phi = \mathrm{R}[f(z, t)] \tag{4.77}$$

where $\frac{1}{2}T(t)$ has been included in each of $f(z, t)$ and $\bar{f}(\bar{z}, t)$. $f(z, t)$ is known as the complex potential, its real part being the velocity potential ϕ.

Since $f(z, t)$ is a function of the complex variable z (and of the real variable t), its derivative with respect to z at any point in the complex x, y-plane is the same in all directions, i.e.

$$\frac{\partial f}{\partial z} = \frac{\partial f}{\partial x} = \frac{1}{\mathrm{i}}\frac{\partial f}{\partial y}. \tag{4.78}$$

Using $\phi = \mathrm{R}[f]$ and eqns (4.78),

$$\frac{\partial \phi}{\partial y} = \frac{\partial}{\partial y}\mathrm{R}[f] = \mathrm{R}\left[\frac{\partial f}{\partial y}\right] = \mathrm{R}\left[\mathrm{i}\frac{\partial f}{\partial z}\right] = -\mathrm{I}\left[\frac{\partial f}{\partial z}\right],$$

eqn (4.8) becomes

$$\mathrm{I}\left[\frac{\partial f}{\partial z}\right] = 0, \quad \text{on } y = -h, \text{ i.e. on } \mathrm{I}[z] = -h, \tag{4.79}$$

and eqn (4.16) becomes

$$\mathrm{R}\left[\frac{\partial^2 f}{\partial t^2} + \mathrm{i}g\frac{\partial f}{\partial z}\right] = 0, \quad \text{on } y = 0, \text{ i.e. on } \mathrm{I}[z] = 0. \tag{4.80}$$

Since on $y=0$, $z=x$ and since $\partial f/\partial z=\partial f/\partial x$, eqn (4.80) can be written

$$\mathrm{R}\left[\frac{\partial^2 f_0}{\partial t^2}+\mathrm{i}g\frac{\partial f_0}{\partial x}\right]=0, \tag{4.81}$$

where

$$f_0(x,t)=(f(z,t))_{y=0}. \tag{4.82}$$

The problem of solving eqn (4.5) for $\phi(x, y, t)$ subject to eqns (4.8) and (4.16) has been replaced by finding a complex function $f(z, t)$ to satisfy eqn (4.79) and either eqn (4.80) or (4.81).

Consider $f=A\mathrm{e}^{\mathrm{i}(\omega t-kz)}$, A a real constant. Substitution of $f_0=A\mathrm{e}^{\mathrm{i}(\omega t-kx)}$ in eqn (4.81) gives

$$-\omega^2+gk=0. \tag{4.83}$$

$\mathrm{I}[\partial f/\partial z]=\mathrm{I}[-\mathrm{i}k\mathrm{e}^{\mathrm{i}(\omega t-kx)+ky}]=-k\mathrm{e}^{ky}\cos(\omega t-kx)$ and eqn (4.79) is satisfied for $y=-h$ when $h\to\infty$. $f=A\mathrm{e}^{\mathrm{i}(\omega t-kz)}$ is the complex potential for a sinusoidal wave travelling in the positive sense of x of angular frequency ω on water of infinite depth. The dispersion equation (4.83) is the same as the limit of eqn (4.22) as $h\to\infty$.

For sinusoidal waves on water of finite depth, consider

$$f=A\mathrm{e}^{\mathrm{i}(\omega t-kz)}+B\mathrm{e}^{-\mathrm{i}(\omega t-kz)},\quad A \text{ and } B \text{ real constants.}$$

Then

$$\frac{\partial^2 f_0}{\partial t^2}+\mathrm{i}g\frac{\partial f_0}{\partial x}=(-\omega^2+gk)A\mathrm{e}^{\mathrm{i}(\omega t-kx)}+(-\omega^2-gk)B\mathrm{e}^{-\mathrm{i}(\omega t-kx)}$$

and eqn (4.81) requires

$$A(-\omega^2+gk)-B(-\omega^2-gk)=0. \tag{4.84}$$

Now

$$\frac{\partial f}{\partial z}=-\mathrm{i}k\,A\mathrm{e}^{\mathrm{i}(\omega t-kx)+ky}+\mathrm{i}k\,B\mathrm{e}^{-\mathrm{i}(\omega t-kx)-ky}$$

$$I\left[\frac{\partial f}{\partial z}\right]=(-kA\mathrm{e}^{ky}+kB\mathrm{e}^{-ky})\cos(\omega t-kx),$$

and eqn (4.79) requires

$$kA\mathrm{e}^{-kh}-kB\mathrm{e}^{kh}=0. \tag{4.85}$$

Elimination of A/B from eqns (4.84) and (4.85) gives the dispersion equation (4.22).

A sufficient condition for eqn (4.80) to be satisfied is

$$\frac{\partial^2 f}{\partial t^2}+\mathrm{i}g\frac{\partial f}{\partial z}=0. \tag{4.86}$$

Equation (4.86) is formally the same as the heat conduction equation

$$\kappa\frac{\partial^2 T}{\partial x^2}=\frac{\partial T}{\partial t}$$

provided T is replaced by f, x by t, κ by i/g and t by z. It follows that any solution of the heat conduction equation becomes a solution of eqn (4.86) and hence of eqn (4.81) after the above changes of variable. It becomes a solution of a surface wave problem if it also satisfies eqn (4.79).

We shall consider the particular solution of the heat conduction equation, which after the above change of variables becomes

$$f(z,t)=Bz^{-\frac{1}{2}}\mathrm{e}^{\mathrm{i}gt^2/(4z)}\int_0^{\sqrt{-\mathrm{i}g/z}(t/2)}\mathrm{e}^{\Psi^2}\,\mathrm{d}\Psi, \tag{4.87}$$

where B is a complex constant at our disposal.

$$\frac{\partial f_0(x,t)}{\partial t}=B\left(\tfrac{1}{2}\mathrm{i}gtx^{-\frac{3}{2}}\int_0^{\sqrt{-\mathrm{i}g/x}(t/2)}\mathrm{e}^{\Psi^2+\mathrm{i}gt^2/4x}\,\mathrm{d}\Psi+\tfrac{1}{2}\sqrt{-\mathrm{i}g/x}\right).$$

Since

$$\eta=\frac{1}{g}\frac{\partial\phi_0}{\partial t}=\frac{1}{g}R\left[\frac{\partial f_0}{\partial t}\right],$$

the second term on the r.h.s. of the above equation will not appear in η if we choose

$$B=B_0\sqrt{-\mathrm{i}}=B_0\mathrm{e}^{-\mathrm{i}\pi/4},\quad B_0\text{ real},$$

to give

$$\eta(x,t)=B_0R\left[\tfrac{1}{2}\mathrm{e}^{\mathrm{i}\pi/4}\,tx^{-\frac{3}{2}}\int_0^{\sqrt{-\mathrm{i}g/x}(t/2)}\mathrm{e}^{\Psi^2+\mathrm{i}gt^2/4x}\,\mathrm{d}\Psi\right].$$

Change the variable of integration for Ψ to θ, where $\Psi=\mathrm{e}^{-\mathrm{i}\pi/4}\theta$

$$\eta(x,t)=B_0R\left[\tfrac{1}{2}tx^{-\frac{3}{2}}\int_0^{\sqrt{g/x}(t/2)}\mathrm{e}^{-\mathrm{i}(\theta^2-gt^2/4x)}\,\mathrm{d}\theta\right] \tag{4.88}$$

$$=\tfrac{1}{2}B_0tx^{-\frac{3}{2}}\int_0^{\sqrt{gt^2/4x}}\cos(\theta^2-gt^2/4x)\mathrm{d}\theta. \tag{4.89}$$

For $t=0$, $\eta=0$ for $x\neq 0$ and $\eta(0,0)$ is undetermined. Now,

$$\int_0^\infty \eta(x,t)\mathrm{d}x=\tfrac{1}{2}B_0 t\int_0^\infty \mathrm{d}x\, x^{-\frac{3}{2}}\int_0^{\sqrt{gt^2/4x}} \cos(\theta^2-gt^2/4x)\mathrm{d}\theta$$

and, on putting $\xi=\sqrt{(gt^2/4x)}$, where ξ is a new variable of integration replacing x,

$$\int_0^\infty \eta(x,t)\,\mathrm{d}x=\frac{2B_0}{\sqrt{g}}\int_0^\infty \mathrm{d}\xi\int_0^\xi \cos(\theta^2-\xi^2)\mathrm{d}\theta$$

$$=\frac{2B_0}{\sqrt{g}}\int_0^\infty\left\{\cos\xi^2\int_0^\xi\cos\theta^2\,\mathrm{d}\theta+\sin\xi^2\int_0^\xi\sin\theta^2\,\mathrm{d}\theta\right\}\mathrm{d}\xi.$$

On changing the order of integration (see Fig. 4.5),

$$\int_0^\infty \eta(x,t)\mathrm{d}x=\frac{2B_0}{\sqrt{g}}\int_0^\infty\left\{\cos\theta^2\int_\theta^\infty\cos\xi^2\,\mathrm{d}\xi+\sin\theta^2\int_\theta^\infty\sin\xi^2\,\mathrm{d}\xi\right\}\mathrm{d}\theta.$$

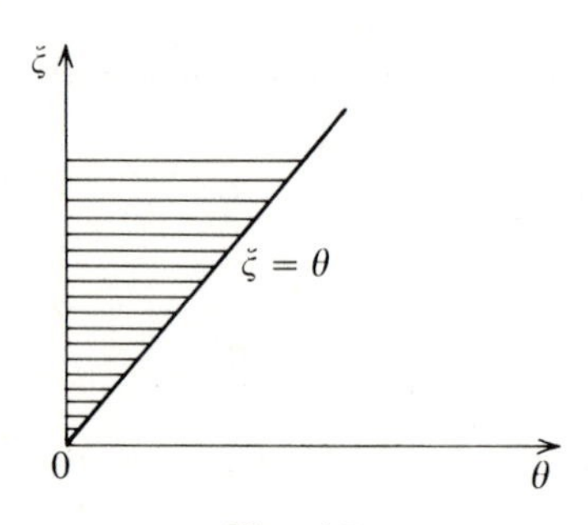

Fig. 4.5

Interchange the variables of integration θ and ξ in the last equation and add to the previous equation:

$$2\int_0^\infty \eta(x,t)\mathrm{d}x=\frac{2B_0}{\sqrt{g}}\left[\left(\int_0^\infty\cos\theta^2\,\mathrm{d}\theta\right)^2+\left(\int_0^\infty\sin\theta^2\,\mathrm{d}\theta\right)^2\right].$$

Since

$$\int_0^\infty\cos\theta^2\,\mathrm{d}\theta=\int_0^\infty\sin\theta^2\,\mathrm{d}\theta=\frac{\sqrt{\pi}}{2\sqrt{2}},$$

$$\int_0^\infty \eta(x,t)\mathrm{d}x=\frac{B_0\pi}{4\sqrt{g}}. \tag{4.90}$$

For $x \leqslant 0$ in eqn (4.89), put $x = \xi e^{i\pi}$,

$$\eta(x,t) = \tfrac{1}{2} B_0 t i \, \xi^{-\frac{3}{2}} \int_0^{-i\sqrt{g/4\xi}(t/2)} \cos(\theta^2 + gt^2/4\xi) \, d\theta.$$

Put $\theta = -i\phi$,

$$\eta(x,t) = +B_0 t \, \xi^{-\frac{3}{2}} \int_0^{\sqrt{g/\xi}(t/2)} \cos\left(\phi^2 - \frac{gt^2}{4\xi}\right) d\phi. \qquad (4.91)$$

Comparison of the right-hand sides of eqns (4.89) and (4.91) shows that

$$\eta(x,t) = \eta(\xi,t) = \eta(-x,t), \qquad (4.92)$$

and hence $\eta(x,t)$ is even in x. Then from eqn (4.90),

$$\int_{-\infty}^{\infty} \eta(x,t) \, dx = 2 \int_0^{\infty} \eta(x,t) \, dx = \tfrac{1}{2} B_0 \frac{\pi}{\sqrt{g}}.$$

Therefore, since $\eta(x,0) = 0$ for $x \neq 0$,

$$\eta(x,0) = \delta(x) \qquad (4.93)$$

if

$$B_0 = \frac{2\sqrt{g}}{\pi}. \qquad (4.94)$$

Substitute back for B_0 into eqn (4.87) to give

$$f(z,t) = \frac{2\sqrt{g}}{\pi} e^{-i\pi/4} z^{-\frac{1}{2}} e^{igt^2/4z} \int_0^{\sqrt{-ig/z}(t/2)} e^{\Psi^2} d\Psi. \qquad (4.95)$$

Equation (4.87) is the complex potential for an initial elevation in the form of a delta function, eqn (4.93), on water of infinite depth provided it can be shown that $f(z,t) = f(x+iy,t) \to 0$ as $y \to -\infty$.

Put $z = 1/\zeta$ in eqn (4.87), then

$$f\left(\frac{1}{\zeta}, t\right) = \frac{2\sqrt{g}}{\pi} e^{-i\pi/4} \zeta^{\frac{1}{2}} e^{igt^2\zeta/\Psi} \int_0^{\sqrt{-ig\zeta}(t/2)} e^{\Psi^2} d\Psi.$$

For $|\zeta| \ll 1$, expand the exponentials as power series:

$$\begin{aligned} f\left(\frac{1}{\zeta}, t\right) &= \frac{2\sqrt{g}}{\pi} e^{-i\pi/4} \zeta^{\frac{1}{2}} \left(1 + \frac{igt^2}{4}\zeta + \dots\right) \\ &\quad \times \left(\sqrt{-ig\zeta}\frac{t}{2} + \frac{1}{3}\sqrt{-ig\zeta}\frac{t}{2}\right)^3 + \dots \\ &= -\frac{igt}{\pi}\left(\zeta + \frac{1}{6} igt^2 \zeta^2 + \dots\right). \end{aligned} \qquad (4.96)$$

$f(z, t)$ is analytic in the ζ-plane about $\zeta=0$ and its Taylor series has infinite radius of convergence. It is therefore analytic everywhere in the z-plane except at the origin. $f(z, t)\to 0$ for all z as $|z|\to\infty$, in particular as $y\to-\infty$.

Equations (4.69) and (4.73) are easily found by the complex variable method. Substitution from eqn (4.94) into eqn (4.88) gives

$$\eta(x, t)=\frac{\sqrt{g}}{\pi}tx^{-\frac{3}{2}}\mathrm{R}\left[\mathrm{e}^{\mathrm{i}gt2/4x}\int_0^{\sqrt{gt^2/4x}}\mathrm{e}^{-\mathrm{i}\theta^2}\mathrm{d}\theta\right].$$

For $\beta\gg 1$,

$$\begin{aligned}\int_\beta^\infty \mathrm{e}^{-\mathrm{i}\phi^2}\mathrm{d}\phi&=\int_\beta^\infty -2\mathrm{i}\phi\,\mathrm{e}^{-2\mathrm{i}\phi^2}\frac{1}{-2\mathrm{i}\phi}\mathrm{d}\phi\\&=\left|\mathrm{e}^{-\mathrm{i}\phi^2}\frac{\mathrm{i}}{2\phi}\right|_\beta^\infty+\frac{1}{2}\int_\beta^\infty\frac{1}{\beta^2}\mathrm{e}^{-\mathrm{i}\phi^2}\mathrm{d}\phi\\&=-\frac{\mathrm{i}}{2\beta}\mathrm{e}^{-\mathrm{i}\beta^2}+\mathrm{O}\left(\frac{1}{\beta^2}\right).\end{aligned}$$

Also

$$\int_0^\infty \mathrm{e}^{-\mathrm{i}\phi^2}\mathrm{d}\phi=\int_0^\infty(\cos\phi^2-\mathrm{i}\sin\phi^2)\mathrm{d}\phi=\frac{\sqrt{\pi}}{2\sqrt{2}}(1-\mathrm{i})$$

Hence,

$$\eta(x, t)\underset{\frac{gt^2}{4x}\to\infty}{\sim}\frac{\sqrt{g}}{\pi}tx^{-\frac{3}{2}}\mathrm{R}\left[\mathrm{e}^{\mathrm{i}gt^2/4x}\left(\frac{\sqrt{\pi}}{2\sqrt{2}}(1-\mathrm{i})+\frac{\mathrm{i}}{2\beta}\mathrm{e}^{-\mathrm{i}\beta^2}+\mathrm{O}\left(\frac{1}{\beta^2}\right)\right)\right],$$

$$\beta=\sqrt{\frac{gt^2}{4x}}. \tag{4.97}$$

Therefore,

$$\eta(x, t)\underset{\frac{gt^2}{4x}\to\infty}{\sim}\frac{g^{\frac{1}{2}}t}{\pi^{\frac{1}{2}}x^{\frac{3}{2}}}\frac{1}{2\sqrt{2}}\left(\cos\frac{gt^2}{4x}+\sin\frac{gt^2}{4x}+\mathrm{O}\left(\frac{gt^2}{4x}\right)^{-\frac{1}{2}}\right)$$

or

$$\eta(x, t)\underset{\frac{gt^2}{4x}\to\infty}{\sim}\frac{g^{\frac{1}{2}}t}{2\pi^{\frac{1}{2}}x^{\frac{3}{2}}}\cos\left(\frac{gt^2}{4x}-\frac{\pi}{4}\right), \tag{4.69}$$

since $1/\sqrt{2}=\cos(\pi/4)=\sin(\pi/4)$. The retention of just the first term in the series on the r.h.s. of eqn (4.97) is justified if $1/\beta\ll\sqrt{\pi}$ or $g^{\frac{1}{2}}t\gg(2/\sqrt{\pi})x^{\frac{1}{2}}$, a criterion of the same order of magnitude as inequality (4.73).

The complex variable approach can be extended to include variable pressure on the free surface, capillary effects and Rayleigh's approximation for small dissipation. The analogues of the many solutions to the heat conduction equation obtained by Bluman and Cole† await investigation. The complex variable approach is discussed by J. V. Weyhausen and E. V. Lattone in Section 22 of their article in *Handbuch der Physik*, vol. 9, where further references are given.

Answers to exercises

4.1. $\phi = A_{mn} \cos\dfrac{n\pi x}{l} \cos\dfrac{m\pi z}{b} \exp\left(\sqrt{\dfrac{n^2}{l^2}+\dfrac{m^2}{b^2}}\,\pi y\right) \cos\left(g^{\frac{1}{2}}\pi^{\frac{1}{2}}\left(\dfrac{n^2}{l^2}+\dfrac{m^2}{b^2}\right)^{\frac{1}{4}} t\right).$

4.2. $\dfrac{v_g}{c} = 1 + \dfrac{1}{2}\dfrac{Tk^2-\rho g}{Tk^2+\rho g}.$

4.3. $c = 2\pi\alpha/\lambda,\ v_g = 2c.$

† Bluman, G. W. and Cole, J. D. (1969). The general similarity solution of the heat equation. *J. Math. Mech.*, **18**, 1025–42.

5 Characteristics and boundary conditions

5.1. Quasi-linear first-order partial differential equations in two independent variables

First the definition of quasi-linearity: an equation is said to be quasi-linear if it is linear in its highest-order terms. A linear equation is always quasilinear, but the converse is not true in general. For example,

$$f\frac{\partial f}{\partial x}+\frac{\partial f}{\partial y}=0 \quad \text{is non-linear and quasi-linear;}$$

$$\left(\frac{\partial f}{\partial x}\right)^2+\frac{\partial f}{\partial y}=0 \quad \text{is non-linear and } \textit{not} \text{ quasi-linear.}$$

The most general first-order quasi-linear partial differential equation (p.d.e.) for the dependent variable z as a function of the two independent variables x and y is

$$P\frac{\partial z}{\partial x}+Q\frac{\partial z}{\partial y}=R, \tag{5.1}$$

where P, Q and R are functions of x, y and z but not of the derivatives of z. The simplest example of eqn (5.1) is

$$\frac{\partial z}{\partial x}=0,$$

with solution $z=f(y)$, where f is an arbitrary function of y. To determine $f(y)$, we need to be given z for all y at one value of x for each value of y. The straight lines in Fig. 5.1 alongside are lines of constant y; z is constant on each line. At one point on each line, z must be given: z is then known at all points on each line. For example, if z is given on the curve C from point A to point B, then z is known on all the straight lines $y=$ constant passing through C; z is not known on any straight line above B or below A and z must only

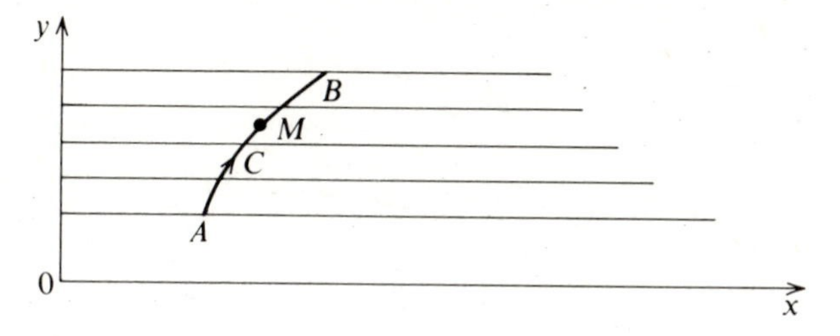

Fig. 5.1

be given at one point on each line y = constant. If z is given at two points, then if the values are unequal there is a contradiction and if the values are equal the information is redundant. Also note that if z is discontinuous on C at M, then z is discontinuous across the straight line y = constant passing through M.

We return to eqn (5.1). Consider the family of curves in the x, y-plane with slopes $\mathrm{d}x/P = \mathrm{d}y/Q$. If P, Q, R are single-valued functions of x, y, z, there is one curve through each point in the x, y-plane once z is known at that point. The variation of $z(= z(x, y))$ along this curve is

$$\mathrm{d}z = \frac{\partial z}{\partial x}\,\mathrm{d}x + \frac{\partial z}{\partial y}\,\mathrm{d}y \tag{5.2}$$

$$= \left(\frac{\partial z}{\partial x} + \frac{Q}{P}\frac{\partial z}{\partial y}\right)\mathrm{d}x \quad \text{since} \quad \mathrm{d}y = \frac{Q}{P}\,\mathrm{d}x$$

$$= \frac{R}{P}\,\mathrm{d}x \ \text{ by eqn (5.1).}$$

Hence

$$\frac{\mathrm{d}x}{P} = \frac{\mathrm{d}y}{Q} = \frac{\mathrm{d}z}{R} \tag{5.3}$$

on such a curve. Eqns (5.3) have been derived from eqn (5.1) and the general relation (5.2) for the differential of a function of two variables. From eqns (5.2) and (5.3), eqn (5.1) can be derived†. Hence eqns (5.1) and (5.3) are equivalent. Equation (5.1) can be solved by solving eqns (5.3) or vice-versa. Which way round we choose is a matter of

---

† Let $\mathrm{d}k$ equal each of the fractions in eqns (5.3). Substitute $\mathrm{d}x = P\mathrm{d}k$, $\mathrm{d}y = Q\mathrm{d}k$ and $\mathrm{d}z = R\mathrm{d}k$ into eqn (5.2) and divide through by $\mathrm{d}k$.

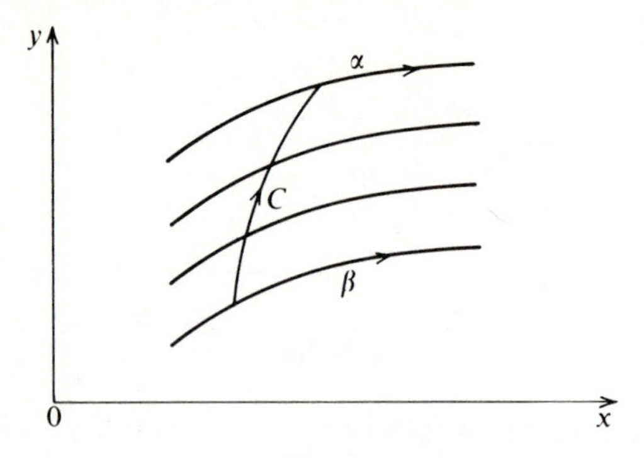

Fig. 5.2

algebraic convenience. In practice, eqns (5.3) are usually solved. Equations (5.3) are called the associated equations.

Equations (5.3) mean that, given z at some point $A(x, y)$, we can continue the solution along the curve through A for which $\mathrm{d}x/P = \mathrm{d}y/Q$ to a neighbouring point $B(x+\mathrm{d}x, y+\mathrm{d}y)$ and calculate z at B, i.e. $z_B = z_A + \mathrm{d}z$, where $\mathrm{d}z = (R/P)\mathrm{d}x$ or $(R/Q)\mathrm{d}y$. Then from B we proceed a further step, etc. We may also go from A in the opposite sense. We end up with a curve through A along which z is known. Now suppose z is given along some curve C in the x, y-plane and that $\mathrm{d}y/\mathrm{d}x$ on C is nowhere equal to Q/P. Then we can construct the curves through each point on C which satisfy $\mathrm{d}x/P = \mathrm{d}y/Q = \mathrm{d}z/R$. If we assume that the curves do not intersect or form an envelope, then we obtain a region covered by curves in which the solution is known, see Fig. 5.2. The region is bounded by the curves α and β through the endpoints of C. The solution is not known outside the region. Each of the curves on which $\mathrm{d}x/P = \mathrm{d}y/Q = \mathrm{d}z/R$ is known as a 'characteristic' of the p.d.e.

$$P\frac{\partial z}{\partial x} + Q\frac{\partial z}{\partial y} = R.$$

If Q/P is independent of z, then the characteristics (but not the values of z) are independent of the boundary conditions. In particular Q/P is independent of z for a linear p.d.e. If P and Q are constants, then the characteristics are parallel straight lines; e.g. the equation previously considered, $\partial z/\partial x = 0$, for which $\mathrm{d}y/\mathrm{d}x = 0/1 = 0$, or $y = \text{constant}$.

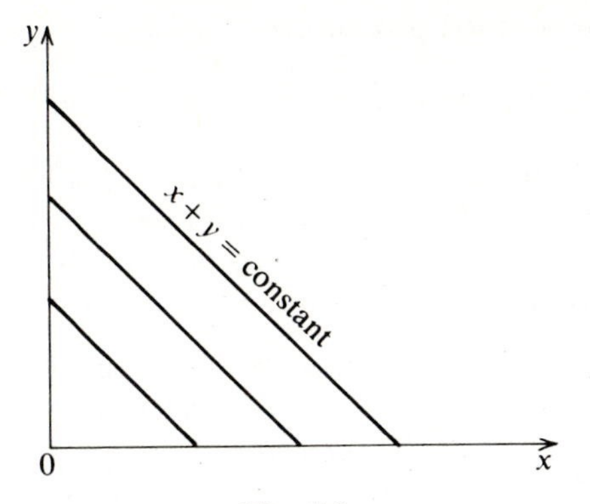

Fig. 5.3

Take as an example the particular p.d.e.:

$$\frac{\partial z}{\partial x}-\frac{\partial z}{\partial y}=1 \quad \text{with } z=x^2 \text{ on } y=0.$$

Since $P=1$, $Q=-1$, $R=1$, the associated equations are $dx/1 = dy/-1 = dz/1$. The characteristics are given by $dy/dx=-1$ or $x+y=$ constant, α say; α is different on each characteristic. On $x+y=\alpha$, $dz/dx=1$ or $z=x+\beta$, where β is a constant, in general different on each characteristic. The value of z must be given at one point on each characteristic. Since $y=0$ intersects each characteristic once and only once, this condition is satisfied, Fig. 5.3.

There are two alternative ways to proceed.

Method 1. When $y=0$, $x=\alpha$ from $x+y=\alpha$ and $z=\alpha^2$ from $z=x^2$ on $y=0$. Hence from $z=x+\beta$, $\beta=\alpha^2-\alpha$:

$$z=x+\alpha^2-\alpha \quad \text{on} \quad x+y=\alpha.$$

Eliminating α,

$$z=x+(x+y)^2-(x+y)=(x+y)^2-y.$$

Method 2. $z-x$ is constant when $x+y$ is constant. A functional relationship $\eta=f(\xi)$ implies that η is constant when ξ is constant. Hence

$$z-x=f(x+y).$$

Since $z=x^2$ when $y=0$, $f(x)=x^2-x$:

$$f(x+y)=(x+y)^2-(x+y) \quad \text{and} \quad z=x+(x+y)^2-(x+y)=(x+y)^2-y.$$

To sum up the solution procedure:

Integrate associated equations; this gives two arbitrary constants α and β.
Method 1: Boundary conditions give $\beta = \beta(\alpha)$. Eliminate α and β between the two integrated equations and $\beta = \beta(\alpha)$.
Method 2: Introduce arbitrary function which is determined by the boundary condition.

Whichever method is used, the solution should be checked. From $z = (x+y)^2 - y$, we have $z = x^2$ when $y = 0$, $\partial z/\partial x = 2(x+y)$ and $\partial z/\partial y = 2(x+y) - 1$ and the p.d.e. is satisfied.

The equations $\mathrm{d}x/P = \mathrm{d}y/Q = \mathrm{d}z/R$ can often be solved by the 'method of multipliers'. If we put each of these fractions equal to $\mathrm{d}k$, then $\mathrm{d}x = P\,\mathrm{d}k$, $\mathrm{d}y = Q\,\mathrm{d}k$, $\mathrm{d}z = R\,\mathrm{d}k$, so that $l\,\mathrm{d}x + m\,\mathrm{d}y + n\,\mathrm{d}z = (lP + mQ + nR)\,\mathrm{d}k$, where the multipliers l, m and n are arbitrary functions of x, y and z. There are two main cases in which this method is useful.

Case (i): $lP + mQ + nR = 0$ and $l\,\mathrm{d}x + m\,\mathrm{d}y + n\,\mathrm{d}z$ is a perfect differential.

Case (ii): $l\mathrm{d}x + m\,\mathrm{d}y + n\,\mathrm{d}z$ is the differential of $lP + mQ + nR$, possibly multiplied by a constant.

An example of case (i):

$$\frac{\mathrm{d}x}{z(x+y)} = \frac{\mathrm{d}y}{z(x-y)} = \frac{\mathrm{d}z}{x^2+y^2}.$$

$$\text{Each fraction} = \mathrm{d}k = \frac{x\,\mathrm{d}x - y\,\mathrm{d}y - z\,\mathrm{d}z}{xz(x+y) - yz(x-y) - z(x^2+y^2)}$$

$$= \frac{x\,\mathrm{d}x - y\,\mathrm{d}y - z\,\mathrm{d}z}{0} = \frac{\frac{1}{2}\mathrm{d}(x^2 - y^2 - z^2)}{0}.$$

Here we have chosen $l = x$, $m = -y$, $n = -z$. Since

$$\mathrm{d}k = \frac{l\,\mathrm{d}x + m\,\mathrm{d}y + n\,\mathrm{d}z}{lP + mQ + nR} \quad \text{and} \quad \mathrm{d}k = \frac{\mathrm{d}x}{P} = \frac{\mathrm{d}y}{Q} = \frac{\mathrm{d}z}{R}$$

and so is not zero or infinity, $\frac{1}{2}\mathrm{d}(x^2 - y^2 - z^2) = 0$. Hence, one integral is $x^2 - y^2 - z^2 = \lambda$, constant. We can also choose $l = y$, $m = x$, $n = -z$, so that each fraction equals

$$\frac{y\,dx + x\,dy - z\,dz}{yz(x+y) + xz(x-y) - z(x^2+y^2)} = \frac{d(xy - \frac{1}{2}z^2)}{0};$$

another integral is $xy - \frac{1}{2}z^2 = \mu$, constant. The choice of multipliers is on a trial-and-error basis and experience is useful. It is an art, not a science.

An example of case (ii):

$$\frac{dx}{y} = \frac{dy}{x} = \frac{dz}{z}.$$

The last term is already of the required form (dz is the differential of z). If we take $l = 1, m = 1, n = 0$ and then $l = 1, m = -1, n = 0$, we obtain

$$\frac{dz}{z} = \frac{dx + dy}{y + x} = \frac{dx - dy}{y - x},$$

whence $\ln z = \ln(x+y) + \ln\lambda = -\ln(x-y) + \ln\mu$, where $\ln\lambda$ and $\ln\mu$ are the constants of integration. Hence

$$z = \lambda(x+y) = \frac{\mu}{x-y}.$$

In other cases $dx/P = dy/Q = dz/R$ can be completely solved if one integral is easily found. This integral is then used to eliminate one variable from the remaining equation to produce an ordinary differential equation between the two remaining variables, e.g.

$$\frac{dx}{1} = \frac{dy}{-2} = \frac{dz}{3x^2 \sin(y+2x)}$$

Since $dy/dx = -2$, $y + 2x = \lambda$. Then

$$\frac{dz}{3x^2 \sin\lambda} = \frac{dx}{1}, \quad z = x^3 \sin\lambda + \mu.$$

We shall now solve the p.d.e. $y(\partial z/\partial x) + x(\partial z/\partial y) = z$ with $z = x^3$ on $y = 0$ and $z = y^3$ on $x = 0$ and investigate the continuity of z and its first derivatives across $x = \pm y$. The associated equations are those of the example of case (ii) above, namely $dx/y = dy/x = dz/z$ with integrals $z = \lambda(x+y) = \mu/(x-y)$. The characteristics are given by the equation between x and y as $x^2 - y^2 = \mu/\lambda = \alpha$, constant.

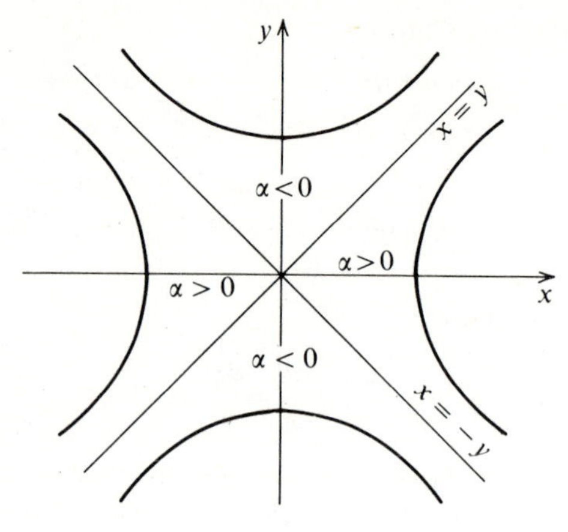

Fig. 5.4

The characteristics are rectangular hyperbolae, shown in Fig. 5.4. It is always advisable to draw a figure when solving p.d.e.'s from the associated equations. One can see at once whether the correct form of boundary condition is given, i.e. one value of z on each characteristic. In this problem, either the x-axis or the y-axis cuts each hyperbola once and only once. Both axes cut both asymptotes $y=\pm x$ at the origin, but $z=0$ on both axes at $x=y=0$ so there is no contradiction.

When $\alpha>0$, $x=\pm\sqrt{\alpha}$ when $y=0$. From the boundary condition, $z=\pm\alpha^{\frac{3}{2}}$ when $x=\pm\sqrt{\alpha}$ and $y=0$. Substituting in $z=\lambda(x+y)$, $\lambda=\alpha$. Hence,

$$z=\alpha(x+y) \quad \text{on} \quad x^2-y^2=\alpha.$$

Eliminating α, $z=(x^2-y^2)(x+y)$ when $x^2-y^2>0$.

When $\alpha<0$, $y=\pm\sqrt{-\alpha}$ when $x=0$. From the boundary condition, $z=\pm(-\alpha)^{\frac{3}{2}}$ when $y=\pm\sqrt{-\alpha}$ and $x=0$. Substituting in $z=\lambda(x+y)$, $\lambda=-\alpha$. Hence,

$$z=-\alpha(x+y) \quad \text{on} \quad y^2-x^2=-\alpha.$$

Eliminating α,

$$z=(y^2-x^2)(x+y) \quad \text{when} \quad y^2-x^2>0.$$

The reader should now check that the p.d.e. and boundary conditions are satisfied.

We next investigate the continuity of z, $\partial z/\partial x$ and $\partial z/\partial y$ across $x=y$ and $x=-y$, the boundaries of the regions within each of which z and its derivatives are continuous. Because of the factor x^2-y^2 (or y^2-x^2), $z\to 0$ in all regions as $x\to \pm y$. Hence z is continuous across $x=\pm y$.

Before calculating $\partial z/\partial x$ and $\partial z/\partial y$, we note that the formula for z for $y^2-x^2>0$ is the negative of that for $x^2-y^2>0$. Hence, if either derivative approaches a non-zero limit as $x\to\pm y$, it will be discontinuous across $x=\pm y$. For $x^2-y^2>0$,

$$\frac{\partial z}{\partial x}=3x^2+2xy-y^2=(x+y)(3x-y)$$

and

$$\frac{\partial z}{\partial y}=-3y^2-2xy+x^2=(x+y)(x-3y).$$

Now $\partial z/\partial x$ and $\partial z/\partial y\to 0$ as $x\to -y$ and therefore they are continuous across $x=-y$. $\partial z/\partial x\to 4y^2$ and $\partial z/\partial y\to -4y^2$ as $x\to y$ and therefore they are discontinuous across $x=y$. To sum up, z is continuous everywhere in the x,y-plane and $\partial z/\partial x$ and $\partial z/\partial y$ are continuous everywhere except across the line $x=y$, where they are discontinuous. It is the possibility of discontinuities occurring across characteristics that is a crucial feature of solutions of p.d.e.'s. We shall shortly look into it in more detail.

The last example illustrates another important feature of these p.d.e.'s. It is not always possible to say *a priori* what boundary conditions are needed until after the characteristics are found. When Q/P is independent of z, the characteristics are independent of the boundary conditions. When Q/P depends on z, different boundary conditions produce different sets of characteristics. For example, in (i) of Fig. 5.5, the conditions on AB give the illustrated set of characteristics and hence the solution everywhere over the rectangle; in (ii) the solution is only known to the left of BF and further conditions are needed to solve in BFC.

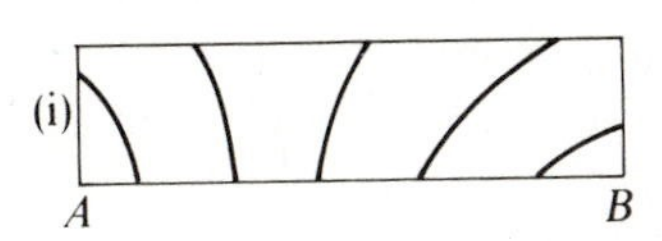

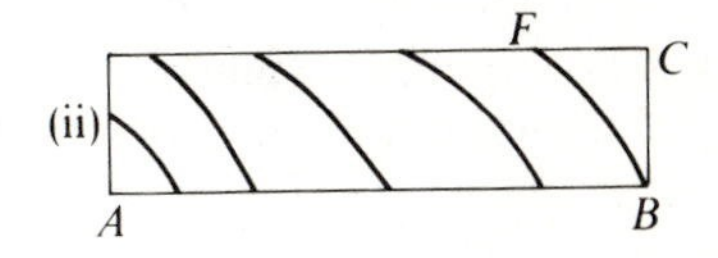

Fig. 5.5

What happens if z is discontinuous at a point A on the boundary and Q/P depends on z? The slope of the characteristics changes discontinuously at A to produce one of the two situations shown in Fig. 5.6. In case (i) there appear to be two characteristics through one point; in case (ii) there is a region in which there appear to be no characteristics. Case (i) leads to shock waves, case (ii) to centred fans of characteristics. Both cases are beyond the scope of this chapter but will be treated in Chapter 9.

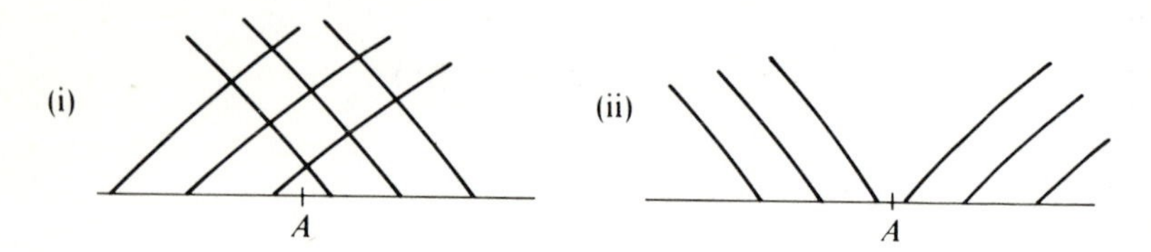

Fig. 5.6

The situation is now investigated where a p.d.e.

$$P\frac{\partial z}{\partial x}+Q\frac{\partial z}{\partial y}=R$$

has a solution in which z is continuous but in which $\partial z/\partial x$ and $\partial z/\partial y$ may be discontinuous. It is supposed that the p.d.e. is solved in a particular region up to but not including a boundary curve C. Can the solution be continued across C? To do this, a Taylor's series expansion about each point $A(x_0, y_0)$ on C (Fig. 5.7) is required of the form

$$z=z_0+(x-x_0)\left(\frac{\partial z}{\partial x}\right)_0+(y-y_0)\left(\frac{\partial z}{\partial y}\right)_0+\ldots.$$

To carry out the calculation just across C, z_0, $(\partial z/\partial x)_0$, and $(\partial z/\partial y)_0$ must be found. Since z is continuous, z_0 is equal to the limit of its value a distance r inside the region of known solution as $r \to 0$. But $(\partial z/\partial x)_0$ and $(\partial z/\partial y)_0$ may be discontinuous, and so the same limiting process cannot be applied. However, if we consider a small element of C about A, then $\mathrm{d}x$, $\mathrm{d}y$ and $\mathrm{d}z$ are known for this element and

$$\mathrm{d}z=\left(\frac{\partial z}{\partial x}\right)_0\mathrm{d}x+\left(\frac{\partial z}{\partial y}\right)_0\mathrm{d}y. \tag{5.4}$$

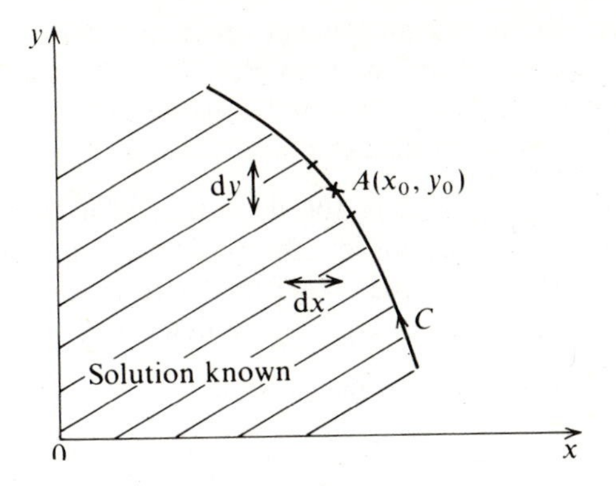

Fig. 5.7

From the p.d.e.,

$$R_0 = \left(\frac{\partial z}{\partial x}\right)_0 P_0 + \left(\frac{\partial z}{\partial y}\right)_0 Q_0 \tag{5.5}$$

where P_0, Q_0, R_0 are functions of x_0, y_0 and z_0 and are therefore known. $(\partial z/\partial x)_0$ and $(\partial z/\partial y)_0$ are the unknowns in the above two equations and can be found unless

$$\Delta = \begin{vmatrix} \mathrm{d}x & \mathrm{d}y \\ P_0 & Q_0 \end{vmatrix} = 0.$$

If $\Delta \neq 0$, $(\partial z/\partial x)_0$ and $(\partial z/\partial y)_0$ can be found and the solution continued across C. If $\Delta = 0$, i.e. $\mathrm{d}x/P_0 = \mathrm{d}y/Q_0$, $(\partial z/\partial x)_0$ and $(\partial z/\partial y)_0$ cannot be found and the solution cannot be continued across C. In other words, the solution can be continued across C unless C is a characteristic. If $\Delta = 0$, the eqns (5.4) and (5.5) are only consistent if

$$\frac{\mathrm{d}z}{R_0} = \frac{\mathrm{d}x}{P_0} \left(= \frac{\mathrm{d}y}{Q_0} \right).$$

The possibility of a discontinuity in $\partial z/\partial x$ and $\partial z/\partial y$ with the consistency condition leads to the same associated equations as previously and hence to the same equation for the characteristics. It may therefore be adopted as the definition of a characteristic. It turns out that this definition is more convenient for higher-order equations.

Definition. A characteristic of a set of nth-order partial differential equations is a curve across which a discontinuity in the nth or higher-order derivatives may occur, all derivatives of lower order being continuous. For the equation $P(\partial z/\partial x)+Q(\partial z/\partial y)=R$, there is one equation of the first order and discontinuities in the first and higher derivatives occur across characteristics.

Discontinuities as defined in the definition of a characteristic are known as 'weak' discontinuities. Discontinuities in derivatives of an order less than that of the p.d.e. are called 'strong' discontinuities, e.g. discontinuities in z itself in eqn (5.1), a first order p.d.e.

Exercise 5.1. For $x \geqslant 0$, solve the equation

$$\frac{\partial z}{\partial x}+x\frac{\partial z}{\partial y}=0,$$

given that $z=\sin y$ on $x=0$. Check your answer.

Exercise 5.2. For $x \geqslant 0$ and $y \geqslant 0$, solve the equation

$$y\frac{\partial z}{\partial x}+x\frac{\partial z}{\partial y}=xy,$$

given that, on $x=0$ for $y>0$, $z=\mathrm{e}^{-y^2}$ and that, on $y=0$ for $x>0$, $z=\mathrm{e}^{-x^2}$. Check your answer.

Exercise 5.3. Solve

$$z\frac{\partial z}{\partial x}-z\frac{\partial z}{\partial y}=z^2+(x+y)^2$$

with $z=1$ when $y=0$. Check your answer.

5.2. Two simultaneous first-order quasi-linear partial differential equations

The equations are

$$P\frac{\partial u}{\partial x}+Q\frac{\partial u}{\partial y}+R\frac{\partial v}{\partial x}+S\frac{\partial v}{\partial y}=T$$

and

$$P'\frac{\partial u}{\partial x}+Q'\frac{\partial u}{\partial y}+R'\frac{\partial v}{\partial x}+S'\frac{\partial v}{\partial y}=T',$$

where $P, Q, \ldots, S', T'$ are functions of the dependent variables u and v and of x and y but not of the partial derivatives of u and v.

Again we ask whether the partial derivatives of u and v can be found on some curve C on which u and v are known. On C

$$\mathrm{d}x\frac{\partial u}{\partial x}+\mathrm{d}y\frac{\partial u}{\partial y}=\mathrm{d}u \quad \text{and} \quad \mathrm{d}x\frac{\partial v}{\partial x}+\mathrm{d}y\frac{\partial v}{\partial y}=\mathrm{d}v.$$

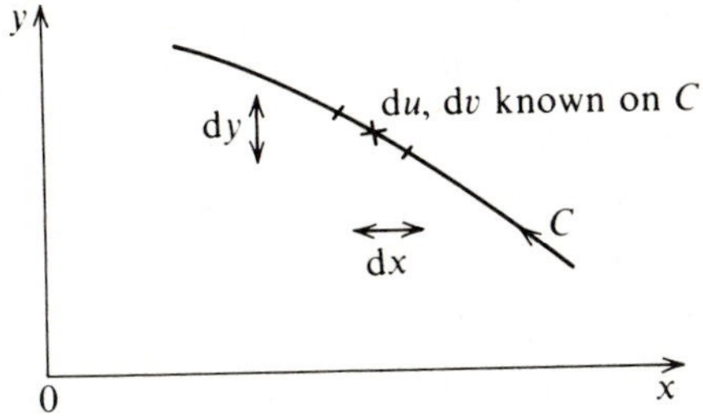

Fig. 5.8

The last four equations can be solved for the four partial derivatives unless

$$\Delta=\begin{vmatrix} P & Q & R & S \\ P' & Q' & R' & S' \\ \mathrm{d}x & \mathrm{d}y & 0 & 0 \\ 0 & 0 & \mathrm{d}x & \mathrm{d}y \end{vmatrix}=0$$

or $\Delta=\lambda(\mathrm{d}x)^2+2\mu\,\mathrm{d}x\,\mathrm{d}y+\nu(\mathrm{d}y)^2=0$, where λ, μ and ν are known functions of $P, Q, \ldots, R', S'$. $\Delta=0$ is a quadratic equation for $\mathrm{d}y/\mathrm{d}x$, i.e. the slope of C. There are three possibilities:

(i) $\mu^2>\lambda\nu$: Two distinct real roots, i.e. two distinct real characteristics. The pair of equations is then said to be hyperbolic.
(ii) $\mu^2=\lambda\nu$: Two coincident real roots, one real (double) characteristic. The pair of equations is then said to be parabolic.
(iii) $\mu^2<\lambda\nu$: Two complex conjugate roots, no real characteristics in the x, y-plane. The equations are then said to be elliptic.

In the elliptic case, no lines of weak discontinuity can appear in the solution of the p.d.e. Discontinuities may appear on the boundary but they do not propagate into the interior. In the hyperbolic case, discontinuities can exist in the interior; when the independent variables are the time t and the cartesian spatial coordinate x, the ratio $\mathrm{d}x/\mathrm{d}t$ represents a characteristic velocity. The parabolic case is intermediate and its treatment is more difficult.

If $\Delta = 0$, then the four equations for the partial derivatives are only consistent if the matrix

$$\begin{pmatrix} P & Q & R & S & T \\ P' & Q' & R' & S' & T' \\ \mathrm{d}x & \mathrm{d}y & 0 & 0 & \mathrm{d}u \\ 0 & 0 & \mathrm{d}x & \mathrm{d}y & \mathrm{d}v \end{pmatrix}$$

is of rank 3. Taken with $\Delta = 0$, this condition is satisfied if the determinant, formed by replacing any column of Δ by the last column of the matrix, is zero, e.g.

$$\begin{vmatrix} P & Q & R & T \\ P' & Q' & R' & T' \\ \mathrm{d}x & \mathrm{d}y & 0 & \mathrm{d}u \\ 0 & 0 & \mathrm{d}x & \mathrm{d}v \end{vmatrix} = 0.$$

In practice, one generally removes that column in Δ which contains the least number of zeros (to make the subsequent algebra as simple as possible). This last equation is between $\mathrm{d}u$ and $\mathrm{d}v$. It holds only when $\Delta = 0$, i.e. on a characteristic. By substituting for $\mathrm{d}y/\mathrm{d}x$, a relation is found between $\mathrm{d}u$ and $\mathrm{d}v$ on each characteristic. This is the two-variable analogue of $\mathrm{d}z/R = \mathrm{d}x/P$.

As an example, we treat the equations governing the two-dimensional irrotational steady adiabatic flow of a compressible fluid, namely,

$$(c^2 - u^2)\frac{\partial u}{\partial x} - uv\left(\frac{\partial u}{\partial y} + \frac{\partial v}{\partial x}\right) + (c^2 - v^2)\frac{\partial v}{\partial y} = 0$$

and

$$\frac{\partial u}{\partial y} - \frac{\partial v}{\partial x} = 0,$$

where u and v are the components of velocity in the x and y directions and $c(>0)$ is the velocity of sound in the fluid. The equations for the characteristics are

$$\Delta = \begin{vmatrix} c^2 - u^2 & -uv & -uv & c^2 - v^2 \\ 0 & 1 & -1 & 0 \\ \mathrm{d}x & \mathrm{d}y & 0 & 0 \\ 0 & 0 & \mathrm{d}x & \mathrm{d}y \end{vmatrix} = 0$$

or

$$(c^2-v^2)(\mathrm{d}x)^2+2uv\,\mathrm{d}x\,\mathrm{d}y+(c^2-u^2)(\mathrm{d}y)^2=0.$$

The characteristics are real and distinct if

$$u^2v^2-(c^2-v^2)(c^2-u^2)>0 \quad \text{or} \quad V^2=u^2+v^2>c^2,$$

where $V\geqslant 0$ is the magnitude of the velocity of the fluid. Hence,

(i) $V>c$: equations hyperbolic; supersonic flow; discontinuities (known as Mach lines) can occur;
(ii) $V<c$: equations elliptic; subsonic flow; no discontinuities;
(iii) $V\simeq c$: transonic flow.

The relations holding along the characteristics will be obtained by replacing the third column of Δ by $(0, 0, \mathrm{d}u, \mathrm{d}v)^{\mathrm{T}}$, so that

$$\begin{vmatrix} c^2-u^2 & -uv & 0 & c^2-v^2 \\ 0 & 1 & 0 & 0 \\ \mathrm{d}x & \mathrm{d}y & \mathrm{d}u & 0 \\ 0 & 0 & \mathrm{d}v & \mathrm{d}y \end{vmatrix}=0$$

or

$$\begin{vmatrix} c^2-u^2 & 0 & c^2-v^2 \\ \mathrm{d}x & \mathrm{d}u & 0 \\ 0 & \mathrm{d}v & \mathrm{d}y \end{vmatrix}=0;$$

i.e. $(c^2-u^2)\,\mathrm{d}u\,\mathrm{d}y+(c^2-v^2)\,\mathrm{d}x\,\mathrm{d}v=0$ on

$$(c^2-v^2)(\mathrm{d}x)^2+2uv\,\mathrm{d}x\,\mathrm{d}y+(c^2-u^2)(\mathrm{d}y)^2=0.$$

Hence,

$$\left(-uv\pm c\sqrt{u^2+v^2-c^2}\right)\mathrm{d}u+(c^2-v^2)\,\mathrm{d}v=0 \quad \text{on} \quad C_{\pm},$$

where $C_{\pm}$ are given respectively by

$$\frac{\mathrm{d}y}{\mathrm{d}x}=\frac{-uv\pm c\sqrt{u^2+v^2-c^2}}{c^2-u^2}.$$

In general, neither the characteristic equations nor the relations holding along them can be integrated exactly. They must be integrated numerically for each different set of boundary conditions. Further study of each particular set of equations is normally carried out within the subject to which it applies.

Exercise 5.4. Under what condition do the equations

$$au_x + bv_y = 0, \quad cu_y + fv_x = 0,$$

where a, b, c and f are real constants, have real distinct characteristics? In this case, determine the relation between u and v which holds on each characteristic.

Exercise 5.5. The equations of motion for one-dimensional isentropic flow of a perfect gas with constant specific heats may be written in the form

$$u_t + uu_x + \frac{2}{\gamma - 1} aa_x = 0, \quad a_t + ua_x + \frac{\gamma - 1}{2} au_x = 0,$$

provided viscosity is negligible, where the variables u and a represent respectively the gas velocity and the local speed of sound. The constant γ is the ratio of specific heats. Find (i) the differential equations for the characteristics, and (ii) the relations between u and a which hold on the characteristics.

5.3. The second-order quasi-linear partial differential equation

A large number of the p.d.e.'s arising in applied mathematics come into this category. It is so important that we treat it in two different but equivalent ways. The equation is

$$a\frac{\partial^2 \phi}{\partial x^2} + 2b\frac{\partial^2 \phi}{\partial x \partial y} + c\frac{\partial^2 \phi}{\partial y^2} = f \tag{5.6}$$

where a, b, c and f are functions of x, y, ϕ, $\partial\phi/\partial x$ and $\partial\phi/\partial y$. If we substitute

$$u = \frac{\partial \phi}{\partial x}, \quad v = \frac{\partial \phi}{\partial y},$$

then

$$a\frac{\partial u}{\partial x} + b\left(\frac{\partial u}{\partial y} + \frac{\partial v}{\partial x}\right) + c\frac{\partial v}{\partial y} = f$$

and

$$\frac{\partial u}{\partial y} - \frac{\partial v}{\partial x} \qquad = 0.$$

a, b, c, f are now functions of ϕ, u, v, x and y. Since u and v are required to be continuous, so are their integrals, in particular ϕ. The characteristics are

$$\begin{vmatrix} a & b & b & c \\ 0 & 1 & -1 & 0 \\ \mathrm{d}x & \mathrm{d}y & 0 & 0 \\ 0 & 0 & \mathrm{d}x & \mathrm{d}y \end{vmatrix} = 0.$$

or

$$c(\mathrm{d}x)^2 - 2b\,\mathrm{d}x\,\mathrm{d}y + a(\mathrm{d}y)^2 = 0. \tag{5.7}$$

The second method makes use of the fact that, since the equation is second order, ϕ and its first derivatives are treated as continuous when we are looking for weak discontinuities. Again consider a curve C across which discontinuities in the second derivatives may occur. Since $\partial\phi/\partial x$ and $\partial\phi/\partial y$ are known on C,

$$\mathrm{d}\left(\frac{\partial\phi}{\partial x}\right) = \frac{\partial^2\phi}{\partial x^2}\,\mathrm{d}x + \frac{\partial^2\phi}{\partial x\partial y}\,\mathrm{d}y$$

and

$$\mathrm{d}\left(\frac{\partial\phi}{\partial y}\right) = \frac{\partial^2\phi}{\partial x\partial y}\,\mathrm{d}x + \frac{\partial^2\phi}{\partial y^2}\,\mathrm{d}y.$$

Also

$$f = \frac{\partial^2\phi}{\partial x^2}\,a + \frac{\partial^2\phi}{\partial x\partial y}\,2b + \frac{\partial^2\phi}{\partial y^2}\,c.$$

These three equations can be solved for the second-order partial derivatives unless

$$\begin{vmatrix} a & 2b & c \\ \mathrm{d}x & \mathrm{d}y & 0 \\ 0 & \mathrm{d}x & \mathrm{d}y \end{vmatrix} = 0$$

or

$$c(\mathrm{d}x)^2 - 2b\,\mathrm{d}x\,\mathrm{d}y + a(\mathrm{d}y)^2 = 0 \text{ as before.} \tag{5.7}$$

The classification is analogous to that for two first-order simultaneous p.d.e.'s:

(i) $b^2 > ac$: two real distinct characteristics; p.d.e. hyperbolic;
(ii) $b^2 = ac$: one real characteristic (double); p.d.e. parabolic;
(iii) if $b^2 < ac$: no real characteristics; p.d.e. elliptic.

The classical equations of mathematical physics in two independent variables are

(i) $\dfrac{\partial^2\phi}{\partial x^2}-\dfrac{1}{c'^2}\dfrac{\partial^2\phi}{\partial t^2}=0$, the wave equation, $a=1$, $b=0$, $c=-\dfrac{1}{c'^2}$,
$b^2-ac=\dfrac{1}{c'^2}>0$, HYPERBOLIC

(ii) $\dfrac{\partial^2\phi}{\partial x^2}=k\dfrac{\partial\phi}{\partial t}$, the heat conduction equation, $a=1$, $b=0$, $c=0$,
$b^2-ac=0$ PARABOLIC

(iii) $\dfrac{\partial^2\phi}{\partial x^2}+\dfrac{\partial^2\phi}{\partial y^2}=0$, Laplace's equation, $a=1$, $b=0$, $c=1$,
$b^2-ac=-1$ ELLIPTIC

Hence, real characteristics and the possibility of discontinuities in the second derivatives occur for the wave equation but not for Laplace's equation.

Example: Show that the characteristics of the p.d.e.

$$(x+y)\frac{\partial^2\phi}{\partial x^2}+(x+2y)\frac{\partial^2\phi}{\partial x\partial y}+y\frac{\partial^2\phi}{\partial y^2}=0$$

are real and distinct except on the y-axis and that the characteristics are mutually orthogonal on a certain straight line passing through the origin. In this problem

$$a=x+y,\quad b=\tfrac{1}{2}x+y,\quad c=y.$$

The characteristics are given by

$$y(\mathrm{d}x)^2-(x+2y)\,\mathrm{d}x\,\mathrm{d}y+(x+y)(\mathrm{d}y)^2=0,$$

and

$$b^2-ac=(\tfrac{1}{2}x+y)^2-(x+y)y=\tfrac{1}{4}x^2.$$

Therefore $b^2-ac>0$ except on $x=0$ where $b^2-ac=0$; the characteristics are real and distinct except on $x=0$, where they are real and coincident. If two straight lines are orthogonal, the product of their slopes is -1. The product of the roots of the quadratic equation for $\mathrm{d}y/\mathrm{d}x$ is $y/x+y$. Hence the characteristics are mutually orthogonal if $y/x+y=-1$ or $x+2y=0$, a straight line passing through the origin.

Exercise 5.6. Classify the equations

(i) $$\frac{\partial^2\phi}{\partial x^2}+\frac{\partial^2\phi}{\partial z^2}+102\frac{\partial\phi}{\partial x}\frac{\partial\phi}{\partial z}+3081\frac{\partial\phi}{\partial x}+5392\phi=0.$$

(ii) $x\dfrac{\partial^2\phi}{\partial x^2}=\dfrac{\partial^2\phi}{\partial t^2}$ for $x>, =$ and <0. This is Tricomi's equation.

Exercise 5.7. Find the characteristics, where real, of the equations

(i) $k\dfrac{\partial^2\phi}{\partial x^2}=\dfrac{\partial\phi}{\partial t}$, the heat conduction equation;

(ii) $$\frac{\partial^2\phi}{\partial x^2}+\frac{\partial^2\phi}{\partial y^2}+k^2\phi=0;$$

(iii) $\left(1+\dfrac{\partial\theta}{\partial x}\right)^2\dfrac{\partial^2\theta}{\partial x^2}=\dfrac{1}{c^2}\dfrac{\partial^2\theta}{\partial t^2}$, a non-linear wave equation.

Exercise 5.8. Show that the characteristics of the p.d.e.

$$x\frac{\partial^2\phi}{\partial x^2}+2(x+y)\frac{\partial^2\phi}{\partial x\partial y}+(2y-1)\frac{\partial^2\phi}{\partial y^2}=\phi^2$$

are real outside a certain circle in the x, y-plane and orthogonal on a certain straight line. Find circle and line.

5.4. Characteristic treatment of the wave equation

The wave equation is the simplest example of a second-order p.d.e. with real distinct characteristics. Solutions have already been found in Chapters 1 and 2. We now find the solution using characteristics.

As both the characteristics and the characteristic relationships are needed, start by writing down the wave equation and the two equations for $\mathrm{d}(\partial\phi/\partial x)$ and $\mathrm{d}(\partial\phi/\partial t)$. In practice, this is as simple as using the formulae of the last section:

$$\frac{\partial^2\phi}{\partial x^2}\qquad\qquad -\frac{1}{c^2}\frac{\partial^2\phi}{\partial t^2}=0, \tag{5.8}$$

$$\mathrm{d}x\frac{\partial^2\phi}{\partial x^2}+\mathrm{d}t\frac{\partial^2\phi}{\partial x\partial t}\qquad\qquad=\mathrm{d}\left(\frac{\partial\phi}{\partial x}\right), \tag{5.9}$$

$$\mathrm{d}x\frac{\partial^2\phi}{\partial x\partial t}+\mathrm{d}t\frac{\partial^2\phi}{\partial t^2}\qquad=\mathrm{d}\left(\frac{\partial\phi}{\partial t}\right). \tag{5.10}$$

Note how the terms in the equations are spaced to facilitate subsequent writing down of the determinants. The condition that one cannot solve for the second derivatives gives the equation for the characteristics:

$$\begin{vmatrix} 1 & 0 & -\dfrac{1}{c^2} \\ \mathrm{d}x & \mathrm{d}t & 0 \\ 0 & \mathrm{d}x & \mathrm{d}t \end{vmatrix} = 0,$$

$$(\mathrm{d}t)^2 - \frac{1}{c^2}(\mathrm{d}x)^2 = 0 \quad \text{or} \quad \frac{\mathrm{d}x}{\mathrm{d}t} = \pm c.$$

On integration,

$$x = ct + \text{constant} \quad \text{and} \quad x = -ct + \text{constant}.$$

The characteristics are called C_+ and C_- characteristics respectively.

The relation holding on a characteristic is found by substituting the r.h.s. of eqns (5.8) to (5.10) for the last column of the determinant:

$$\begin{vmatrix} 1 & 0 & 0 \\ \mathrm{d}x & \mathrm{d}t & \mathrm{d}\left(\dfrac{\partial \phi}{\partial x}\right) \\ 0 & \mathrm{d}x & \mathrm{d}\left(\dfrac{\partial \phi}{\partial t}\right) \end{vmatrix} = 0.$$

Therefore,

$$\mathrm{d}\left(\frac{\partial \phi}{\partial t}\right)\mathrm{d}t - \mathrm{d}\left(\frac{\partial \phi}{\partial x}\right)\mathrm{d}x = 0 \quad \text{on} \quad \frac{\mathrm{d}x}{\mathrm{d}t} = \pm c. \tag{5.11}$$

On a C_+ characteristic, i.e. $\mathrm{d}x/\mathrm{d}t = c$, $\mathrm{d}\left(\dfrac{\partial \phi}{\partial t}\right) - c\,\mathrm{d}\left(\dfrac{\partial \phi}{\partial x}\right) = 0$ or

$$\mathrm{d}\left(\frac{\partial \phi}{\partial t} - c\frac{\partial \phi}{\partial x}\right) = 0 \quad \text{on} \quad C_+.$$

Therefore, $\partial\phi/\partial t - c(\partial\phi/\partial x)$ is constant on $x - ct$ constant, and similarly $\partial\phi/\partial t + c(\partial\phi/\partial x)$ is constant on $x + ct$ constant.
Hence,

$$\frac{\partial \phi}{\partial t} - c\frac{\partial \phi}{\partial x} = -2cf'(x - ct) \quad \text{and} \quad \frac{\partial \phi}{\partial t} + c\frac{\partial \phi}{\partial x} = 2cg'(x + ct),$$

where f' and g' are arbitrary functions of their arguments; the constants $\mp 2c$ and the primes denoting differentiation are included for algebraic convenience. Addition and subtraction in turn of the last two equations gives

$$\frac{\partial\phi}{\partial t}=-cf'(x-ct)+cg'(x+ct) \quad \text{and} \quad \frac{\partial\phi}{\partial x}=f'(x-ct)+g'(x+ct).$$

On integration,

$$\phi=f(x-ct)+g(x+ct)+h(x) \quad \text{and} \quad \phi=f(x-ct)+g(x+ct)+j(t).$$

These equations are only consistent if $h(x)=j(t)=C$, constant. The constant C can be included in either f or g. Hence

$$\phi(x,t)=f(x-ct)+g(x+ct). \tag{5.11}$$

The characteristic treatment of the wave equation has reproduced D'Alembert's solution. The value of $f(x-ct)$, constant on a C_+ characteristic, is determined at the point on this characteristic whose time coordinate is the least which has physical significance in the problem under consideration. If the initial conditions are given at time $t=0$, then this point is the intercept of the C_+ characteristic with the x-axis. Similarly the value of $g(x+ct)$ on a C_- characteristic is determined at the intercept of the C_- characteristic with the x-axis (see Fig. 5.9). When ϕ and $\partial\phi/\partial t$ are given by eqns (1.16) and (1.17) at $t=0$, then $f(x)$ and $g(x)$ are given by eqns (1.18) and (1.19).

The value of ϕ at any point $P(x,t)$ in Fig. 5.9 is determined by the values of $f(x-ct)$ and $g(x+ct)$ on the characteristics $(C_+)_P$ and $(C_-)_P$ respectively which pass through P. But the value of $f(x-ct)$ on

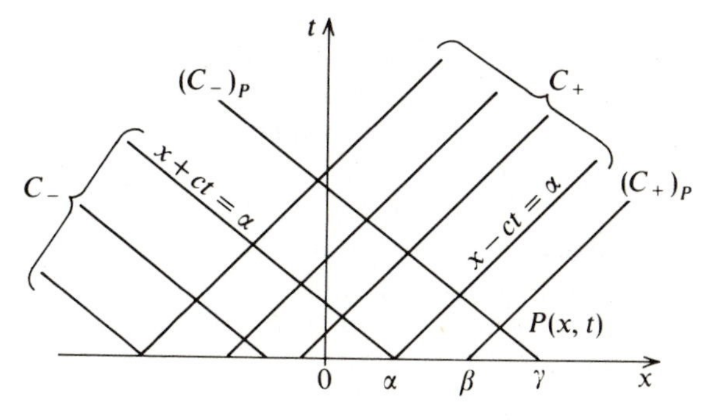

Fig. 5.9

$(C_+)_P$ is determined at $(\beta, 0)$ and that of $g(x+ct)$ on $(C_-)_P$ is determined at $(\gamma, 0)$, $(\beta, 0)$ and $(\gamma, 0)$ being the intercepts of $(C_+)_P$ and $(C_-)_P$ respectively with the x-axis.

If $f(x)$ and $g(x)$ are given for $a \leqslant x \leqslant b$ but not elsewhere on the x-axis, the solution for ϕ will be determined at all points within the triangle ABC (Fig. 5.10) by the construction of the last paragraph. This triangle is called the 'domain of dependence' of ϕ on the initial data given on AB. If D and E are respectively points on C_+ characteristic through B and C_- characteristic through A, then $f(x-ct)$ but not $g(x+ct)$ is known everywhere on the strip bounded by AC produced, CB, and BD produced, and $g(x+ct)$ but not $f(x-ct)$ is known everywhere on the strip bounded by BC produced, CA, and AE produced. The regions formed by these two strips, excluding the common part of the triangle ABC, are called 'domains of influence' of the initial data on AB. The domains of influence are hatched in Fig. 5.10.

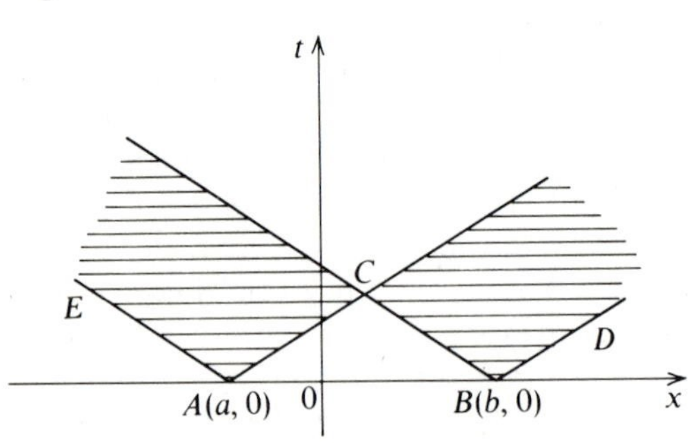

Fig. 5.10

So far in this section, it has been implicitly assumed that the region under consideration has been $-\infty < x < \infty$ and $0 \leqslant t < \infty$. We now restrict x to $0 \leqslant x \leqslant l$, $l > 0$, and assume $f(x)$ and $g(x)$ given for $0 \leqslant x \leqslant l$ at $t = 0$. From the initial data, the triangle $0LC$ (Fig. 5.11) is the domain of dependence and the triangles LCN and $0CM$ are domains of influence. Neither f nor g is known above MCN.

For any point P in LCN, the value of f_P is equal to the known value f_A and the value of g_P is equal to the value of g on the C_- characteristic through P, determined at the point G which corresponds to the first physically significant value of time on this characteristic. g_G is unknown but f_G is equal to f_S and known. Hence

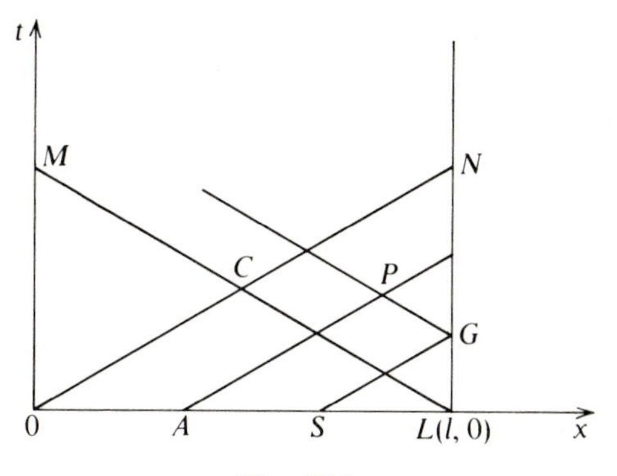

Fig. 5.11

if ϕ is given at G, g_G can be found from

$$g_G = \phi_G - f_G.$$

Since P was any point in LCN, ϕ can be found throughout LCN if ϕ is given on LN. Similarly ϕ can be found throughout $0CM$ if ϕ is given on $0M$. By considering the characteristics through Q, any point in the parallelogram $CMJN$ (Fig. 5.12), it is seen that the information on LN and $0M$ enables ϕ to be found in $CMJN$. To continue the solution further, ϕ must be given on MI and NH.

The boundary condition on $x=0$ and $x=l$ is not always ϕ given. The procedure is treated for the condition $\partial\phi/\partial x$ given on the segment LN of $x=l$ in Fig. 5.13. Take any point G on this segment.

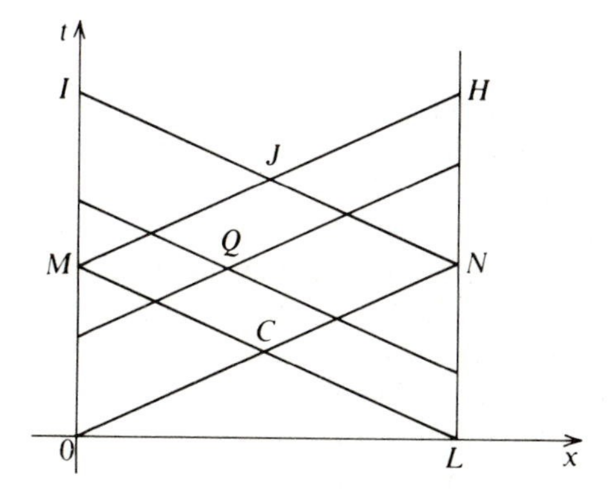

Fig. 5.12

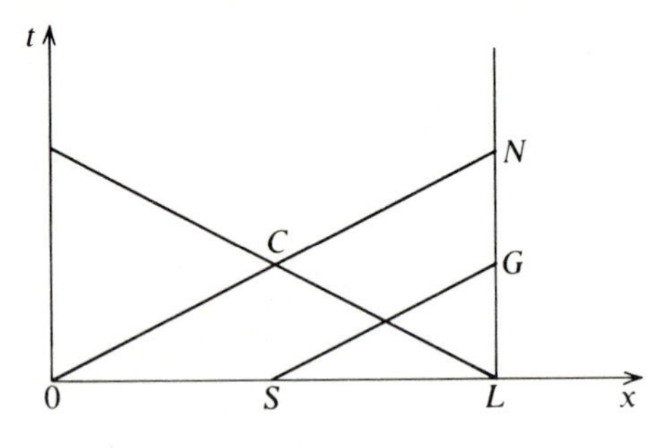

Fig. 5.13

Then as before f_G is known and equal to f_S. Since $f'(x-ct)$ is a function of $x-ct$, $f'_G=f'_S$, known. Since

$$\frac{\partial\phi}{\partial x}=f'(x-ct)+g'(x+ct),\quad \left(\frac{\partial\phi}{\partial x}\right)_G=f'_G+g'_G,$$

an equation for g'_G since $(\partial\phi/\partial x)_G$ is given and f'_G is known. $\partial g/\partial x=g'$ can be found at all points on LN.

Now

$$\frac{\partial g}{\partial t}=\frac{\partial}{\partial t}g(x+ct)=cg'(x+ct)=c\frac{\partial g}{\partial x}.$$

Hence, on integration with respect to t from 0 to t_G,

$$g_G=g_L+c\int_0^{t_G}\frac{\partial g}{\partial x}\,\mathrm{d}t,$$

which determines g_G since g_L is given by the initial conditions. Thus g is found at all points on LN and the solution can be found for ϕ everywhere in the triangle LCN.

It has been shown that the initial and boundary conditions required for the wave equation are two conditions, normally ϕ and $\partial\phi/\partial t$ given, at every point x at the initial time and one condition, normally ϕ or $\partial\phi/\partial x$ given, for all times on each spatial boundary. The difference in type between the conditions on time and spatial boundaries is not caused by the mathematics but by the physics, the causal nature of time. In the next section, these results and ideas will be extended to other partial differential equations.

5.5. Boundary and initial conditions for partial differential equations

Laplace's equation in three dimensions is

$$\frac{\partial^2\phi}{\partial x^2}+\frac{\partial^2\phi}{\partial y^2}+\frac{\partial^2\phi}{\partial z^2}=0.$$

Since x, y and z in physical applications are cartesian spatial co-ordinates, the previous section would suggest that the boundary conditions required are that either ϕ or $\partial\phi/\partial n$, but not both, be given at every point of the boundary. Using vector analysis, it will be proved that the solution to Laplace's equation is unique for such boundary conditions over any volume V bounded by a closed surface S.

Assume the solution is not unique, then there exists at least two distinct solutions ϕ_1 and ϕ_2 which satisfy

(i) $0=\nabla^2\phi=\text{div grad }\phi$ at all points in V, and

(ii) either $\phi=\alpha$ or $\partial\phi/\partial n=\beta$ at all points on S, where α and β are functions given on S.

The function $\psi=\phi_1-\phi_2$ satisfies

(i) div grad $\psi=0$ at all points in V, and

(ii) either $\psi=0$ or $\partial\psi/\partial n=0$ at all points on S.

Hence

ψ grad $\psi\cdot\boldsymbol{n}=0$ at all points on S. Therefore,

$$\int_s \psi \text{ grad } \psi\cdot\boldsymbol{n}=0$$

and by the divergence theorem

$$\int_V \text{div } (\psi \text{ grad } \psi)\, \mathrm{d}V=0.$$

Now

$$\begin{aligned}\text{div } (\psi \text{ grad } \psi) &= \psi \text{ div grad } \psi+\text{grad } \psi\cdot\text{grad } \psi\\ &=\text{grad } \psi\cdot\text{grad } \psi, \quad \text{since div grad } \psi=0.\end{aligned}$$

Therefore,

$$\int_V \text{grad}\,\psi \cdot \text{grad}\,\psi \,\mathrm{d}V = 0$$

or

$$\int_V \left(\left(\frac{\partial \psi}{\partial x}\right)^2 + \left(\frac{\partial \psi}{\partial y}\right)^2 + \left(\frac{\partial \psi}{\partial z}\right)^2\right) \mathrm{d}V = 0.$$

Each term in the above integrand is either positive or zero. But no term can be positive over any region of finite volume, because the integral is zero and there is no negative contribution to balance the positive one. Hence,

$$\frac{\partial \psi}{\partial x} = \frac{\partial \psi}{\partial y} = \frac{\partial \psi}{\partial z} = 0^{\dagger}$$

and $\psi =$ constant throughout V. If ϕ is given at any point of S, then $\psi = 0$, $\phi_1 = \phi_2$ and the two solutions are not distinct, contrary to hypothesis. The assumption was wrong and the solution is unique. If $\partial\phi/\partial n$ is given at all points of S, then the solution is unique apart from an arbitrary additive constant.

Note that it has been proved that there cannot be more than one solution. It has not been proved that there must be at least one solution. Uniqueness but not existence has been proved.

In the case of the wave equation, it was seen that two conditions were given at the initial time and one on each spatial boundary. Is it possible for a p.d.e. with second-order spatial derivatives to have two conditions on one spatial boundary and none on another? The following example of such a situation is due to Hadamard.

The solution of Laplace's equation in two dimensions, namely

$$\frac{\partial^2 \phi}{\partial x^2} + \frac{\partial^2 \phi}{\partial y^2} = 0,$$

over the region $x \geqslant 0$, $-\infty < y < \infty$, when $\phi = 0$ and $\partial\phi/\partial x = (1/n)\sin ny$ on $x = 0$, is given by

$$\phi = \frac{1}{n^2} \sin ny \sinh nx, \tag{5.12}$$

as may be verified by direct substitution. The value of $\partial\phi/\partial x$ on

† Except possibly at isolated points from which no contribution is made to the volume integral.

the boundary $x=0$ may be made less than any prescribed constant ε for all y by making n sufficiently large, $n>n_0$ say. As $n\to\infty$, $(1/n^2)\sinh nx\to\infty$ for all $x>0$.

In a mathematical problem, the boundary conditions can be precisely prescribed; in a physical experiment, the boundary conditions can only be set up to the accuracy of experimental measurement. If the greatest accuracy of measurement of $\partial\phi/\partial x$ on $x=0$ in the above example is ε, then on $x=0$ the boundary conditions $\phi=0$ and $\partial\phi/\partial n=1/n\sin ny$ for $n>n_0$ are indistinguishable experimentally from the conditions $\phi=0$ and $\partial\phi/\partial n=0$. In the former case, the solution to Laplace's equation is eqn (5.12); in the latter it is $\phi=0$. An experimentally indistinguishable change in boundary conditions has produced an arbitrarily large change at almost all interior points.

Since experiments in classical physics are in the main repeatable, the situation just described must be excluded. We require that changes at the interior points in a solution of a p.d.e. must be of the same order as the changes in the boundary conditions. This is a physical requirement—we are asking for stability. Hadamard's example suggests that for this requirement to be satisfied for p.d.e.'s with a second-order spatial derivative and two spatial boundaries, one or both of which may be at infinity, one boundary condition must be given on each boundary and not two on one and none on the other.

The studies of the wave and Laplace equations in the last two sections suggest the following initial and boundary conditions for partial differential equations. Suppose that a p.d.e. for ϕ contains spatial derivatives of order $2n$ and time derivatives of order m, n and m integral. Then one expects to have n conditions on each spatial boundary on ϕ and/or its spatial derivatives for all times and m conditions on ϕ and/or its time derivatives at the initial time at all points. These are only suggestions. In each particular case the boundary conditions and p.d.e. must be satisfied by a solution which is unique where the physical conditions require a unique solution.[†] Existence and uniqueness theorems have been proved in many branches of mathematical physics which are generally, but not always, in line with these suggestions. We have already seen in the simple example of the rectangles in Fig. 5.5 that, for a non-linear

[†] A simple example of a non-unique situation is a column subject to an axial load. The column may either remain straight or become buckled as the load increases.

first-order p.d.e. in two independent variables, whether conditions are required at all on one boundary depends on the conditions given on another boundary.

It has been assumed in the above paragraph that the highest-order spatial derivatives are even. This is the case whenever there is no preferred sense in any spatial direction. In particular, this is true for any phenomenon governed by either the wave equation, Laplace's equation, or the heat conduction equation. However there are exceptions, see for example unidirectional traffic flow considered in Chapter 9, where the highest-order spatial derivative is odd. In this case, the expected initial conditions are the same but the spatial boundary conditions are not.

Answers to exercises

5.1. $z=\sin(y-\tfrac{1}{2}x^2)$.

5.2. $$z=\begin{cases}\tfrac{1}{2}y^2+e^{-(x^2-y^2)}. & \text{for } x>y,\\ \tfrac{1}{2}x^2+e^{-(y^2-x^2)} & \text{for } x<y.\end{cases}$$

5.3. $z=[(1+(x+y)^2)e^{-2y}-(x+y)^2]^{\frac{1}{2}}$.

5.4. Characteristics real and distinct if $abcf>0$.

$$u\pm\left(\frac{bf}{ac}\right)^{\frac{1}{2}}v=\text{constant} \quad \text{on} \quad y=\pm\left(\frac{bc}{af}\right)^{\frac{1}{2}}x+\text{constant}.$$

5.5. $\tfrac{1}{2}(\gamma-1)u\pm a=\text{constant}$ on $\dfrac{dx}{dt}=u\pm a$.

5.6. (i) Elliptic; (ii) $x>0$ hyperbolic; $x=0$ parabolic; $x<0$ elliptic.

5.7. (i) $t=\text{constant}$ (double);
(ii) no real characteristics;
(iii) $\dfrac{dx}{dt}=\pm c\left(1+\dfrac{\partial\theta}{\partial x}\right)$.

5.8. $x^2+y^2+x=0$, $x+2y=1$.

6 Transmission lines

6.1. Use of Laplace transform for partial differential equations with constant coefficients

First consider an ordinary differential equation (o.d.e.) for $f(t)$. If the Laplace transform of $f(t)$ is $\bar{f}(p)$, where

$$\bar{f}(p) = \int_0^\infty \mathrm{e}^{-pt} f(t)\,\mathrm{d}t, \tag{6.1}$$

then the Laplace transforms of $\mathrm{d}f(t)/\mathrm{d}t$ and $\mathrm{d}^2 f(t)/\mathrm{d}t^2$ are $p\bar{f}(p) - f_0$ and $p^2\bar{f}(p) - pf_0 - f_1$ respectively, where f_0 is $f(0)$ and f_1 is $(\mathrm{d}f/\mathrm{d}t)_{t=0}$. As the Laplace transform of the nth derivative of f includes the values of f and the first $n-1$ derivatives at $t=0$, the transform is best applied when all the initial (or boundary) conditions are given at one value of t, which can always be taken as $t=0$.

Next consider a partial differential equation for $f(x,t)$, second order in both its time and space derivatives. In the last chapter, it was seen that the expected initial conditions are f and $\partial f/\partial t$ given for all x at $t = 0$ and one boundary condition for all t on each spatial boundary, $x=0$ and $x=l$ say. Applying the Laplace transform to the variable t brings in both initial conditions and does not introduce an arbitrary function of time to be determined later from the condition at $x=l$. Clearly, it is the time variable to which the Laplace transform generally should be applied.

The result of the application of the transform

$$\bar{f}(x,p) = \int_0^\infty \mathrm{e}^{-pt} f(x,t)\,\mathrm{d}t \tag{6.2}$$

to the p.d.e. is an ordinary differential equation in x for $\bar{f}(x,p)$. Treating p as constant, this equation is solved for $\bar{f}(x,p)$ using as boundary conditions the Laplace transforms of the two spatial boundary conditions.

Finally $\bar{f}(x, p)$ is inverted to obtain $f(x, t)$. This final operation is by far the most difficult step of the whole procedure. It is not always possible to carry out the inversion exactly and various special procedures have been devised for different ranges of the independent variables, often starting from the Bromwich inversion theorem.

As far as the author is aware, no rigorous proof has been given of the correctness of the Laplace transform method of solving p.d.e.'s, in contradistinction to o.d.e.'s. It is therefore essential, and not just advisable, that one checks each solution by substitution back into the p.d.e. and boundary conditions. No-one has found an example where the Laplace transform gives a result that does not check, but this does not constitute a proof!

6.2. The forms of physical laws

When there is dependence on just one space dimension x and time t, many of the laws of mathematical physics can be expressed in the form

$$F(x_1, t) - F(x_2, t) = \frac{\mathrm{d}}{\mathrm{d}t}\int_{x_1}^{x_2} G(x, t)\,\mathrm{d}x + \int_{x_1}^{x_2} H(x, t)\,\mathrm{d}x, \qquad (6.3)$$

where x_1 and x_2 are constant values of x with $x_1 < x_2$ and F, G and H are functions of x and t. Different mathematical forms of eqn (6.3) can be derived for the two cases: (i) when $\partial F/\partial x$ and $\partial G/\partial t$ exist, and (ii) when G has a discontinuity at $x_s(t)$ where $x_1 < x_s(t) < x_2$ but $\partial G/\partial t$ exists elsewhere.

Case (i). Since x_1 and x_2 are constant and $\partial G/\partial t$ exists,

$$\frac{\mathrm{d}}{\mathrm{d}t}\int_{x_1}^{x_2} G(x, t)\,\mathrm{d}x = \int_{x_1}^{x_2} \frac{\partial G}{\partial t}\,\mathrm{d}x.$$

Put $x_2 = x_1 + h$ and divide eqn (6.3) through by h:

$$\frac{1}{h}\{F(x_1, t) - F(x_1 + h, t)\} = \frac{1}{h}\int_{x_1}^{x_1+h} \frac{\partial G}{\partial t}\,\mathrm{d}x + \frac{1}{h}\int_{x_1}^{x_1+h} H\,\mathrm{d}x.$$

Let $h \to 0$. Since $\partial F/\partial x$ exists,

$$\frac{\partial F}{\partial x} + \frac{\partial G}{\partial t} + H = 0. \qquad (6.4)$$

Case (ii). Separate the integral $\int_{x_1}^{x_2} G(x,t)\,\mathrm{d}x$ into $\int_{x_1}^{x_s(t)}$ and $\int_{x_s(t)}^{x_2}$.

$$\frac{\mathrm{d}}{\mathrm{d}t}\int_{x_1}^{x_2} G(x,t)\,\mathrm{d}x = \frac{\mathrm{d}}{\mathrm{d}t}\int_{x_1}^{x_s(t)} G(x,t)\,\mathrm{d}x + \frac{\mathrm{d}}{\mathrm{d}t}\int_{x_s(t)}^{x_2} G(x,t)\,\mathrm{d}x$$

$$= \int_{x_1}^{x_s(t)} \frac{\partial G}{\partial t}\,\mathrm{d}x + G_-(x_s(t),t)\frac{\mathrm{d}x_s(t)}{\mathrm{d}t}$$

$$+ \int_{x_s(t)}^{x_2} \frac{\partial G}{\partial t}\,\mathrm{d}x - G_+(x_s(t),t)$$

where $G_-(x_s(t), t)$ and $G_+(x_s(t), t)$ denote the values of $G(x,t)$ as $x_s(t)$ is approached from below and above respectively. Hence, on putting $x_2 = x_s(t) + h/2$ and $x_1 = x_s(t) - h/2$,

$$F\left(x_s(t)-\frac{h}{2}, t\right) - F\left(x_s(t)+\frac{h}{2}, t\right) = \left(\int_{x_s(t)-h/2}^{x_s(t)} + \int_{x_s(t)}^{x_s(t)+h/2}\right)\frac{\partial G}{\partial t}\,\mathrm{d}x$$

$$+ U[G] + \int_{x_s(t)-h/2}^{x_s(t)+h/2} H\,\mathrm{d}x, \qquad (6.5)$$

where $[G]$ is defined by

$$[G] = G_- - G_+, \qquad (6.6)$$

and U is the velocity of the discontinuity, i.e.

$$U = \frac{\mathrm{d}x_s(t)}{\mathrm{d}t}. \qquad (6.7)$$

Let $h \to 0$ in eqn (6.5):

$$[F] - U[G] = 0. \qquad (6.8)$$

Note first that F has a discontinuity along the same line $x = x_s(t)$ as G, and secondly that in case (i) the relevant equation is divided through by h before proceeding to the limit $h \to 0$ but not in case (ii). Equations (6.3), (6.4) and (6.8) are mathematical expressions of the same physical law, referred to as the integral, differential and jump or shock forms respectively of that law. Laws expressible in the form of eqn (6.3) are sometimes called 'conservation' laws. The jumps in eqn (6.8) are strong discontinuities of the variables in eqn (6.4).

6.3. The transmission line equation

The transmission line is a model of a co-axial cable; the upper line in Fig. 6.1(*a*) represents the centre conductor, the lower the earthed

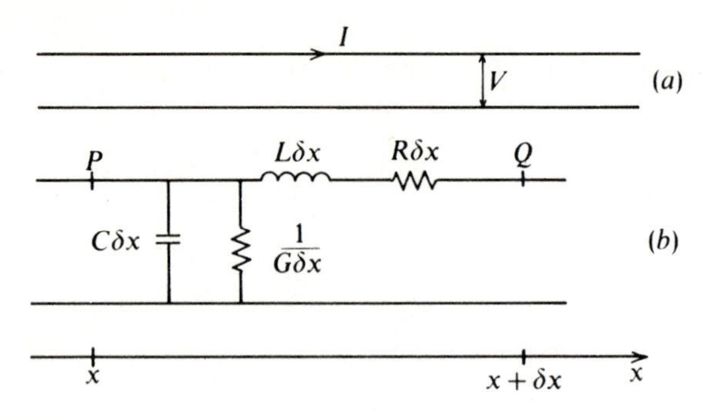

Fig. 6.1

outer sheath. The cable is assumed to have inductance L and resistance R with leakage conductance G and capacitance C between the two wires, all quantities per unit length and constant. Over a small length δx, the effect of these electrical parameters is shown in Fig. 6.1(*b*).

The drop in voltage V between points on the cable with coordinates x_1 and x_2, $x_2 > x_1$, is equal to the sum of the drops in voltage across each element between the two points, i.e.

$$V(x_1, t) - V(x_2, t) = \frac{\mathrm{d}}{\mathrm{d}t} \int_{x_1}^{x_2} LI \,\mathrm{d}x + \int_{x_1}^{x_2} RI \,\mathrm{d}x. \tag{6.9}$$

The fall in current I between x_1 and x_2 is equal to the current loss through the conductances and capacitances between the two points, i.e.

$$I(x_1, t) - I(x_2, t) = \frac{\mathrm{d}}{\mathrm{d}t} \int_{x_1}^{x_2} CV \,\mathrm{d}x + \int_{x_1}^{x_2} GV \,\mathrm{d}x. \tag{6.10}$$

Both eqns (6.9) and (6.10) are of the same form as eqn (6.3). When the partial derivatives of V and I exist, the analogues of eqn (6.4) are

$$\frac{\partial V}{\partial x} + L\frac{\partial I}{\partial t} + RI = 0 \tag{6.11}$$

and

$$\frac{\partial I}{\partial x} + C\frac{\partial V}{\partial t} + GV = 0. \tag{6.12}$$

When V and I are discontinuous, the analogues of eqn (6.8) are

$$[V] - UL[I] = 0 \tag{6.13}$$

and

$$[I]-UC[V]=0, \tag{6.14}$$

where U is the velocity of the discontinuity. From eqns (6.13) and (6.14),

$$U=\pm 1/\sqrt{LC}. \tag{6.15}$$

Strong discontinuities of voltage and current travel with velocity $1/\sqrt{LC}$.

Equations (6.11) and (6.12) are the governing equations for the transmission line in the absence of strong discontinuities of V and I. If I is eliminated between eqns (6.11) and (6.12),

$$LC\frac{\partial^2 V}{\partial t^2}+(LG+RC)\frac{\partial V}{\partial t}-\frac{\partial^2 V}{\partial x^2}+RGV=0. \tag{6.16}$$

Equation (6.16) is called the telegraph equation. With elimination of V between eqns (6.11) and (6.12), I also satisfies eqn (6.16).

The characteristics of eqns (6.11) and (6.12) and the characteristic relations can be found by the methods of Section 5.2:

$$\mathrm{d}\left(V\pm\sqrt{\frac{L}{C}}I\right)\pm G\sqrt{\frac{L}{C}}\left(V\pm\frac{R}{G}\sqrt{\frac{C}{L}}I\right)\mathrm{d}x=0 \tag{6.17}$$

on

$$\frac{\mathrm{d}x}{\mathrm{d}t}=\pm\frac{1}{\sqrt{LC}}. \tag{6.18}$$

Note that the characteristic velocity is the same as that of strong discontinuities. Equations (6.11) and (6.12) are hyperbolic when $LC\neq 0$. Equations (6.17) cannot be integrated for general values of L, C, R and G, only for special values. For very high frequencies, the second-order derivatives dominate and eqn (6.16) approximates

$$LC\frac{\partial^2 V}{\partial t^2}-\frac{\partial^2 V}{\partial x^2}=0,$$

the classical wave equation with velocity $(LC)^{-\frac{1}{2}}$. Note that this velocity is the same as that given by eqn (6.18). This is because it is only the highest-order derivatives that contribute both to the equations for the characteristics and to the high-frequency limit.

The initial conditions for eqns (6.11) and (6.12) are V and I given at $t=0$,

$$V(x,0)=V_0(x) \quad \text{and} \quad I(x,0)=I_0(x). \tag{6.19}$$

When $V_0(x)=0$ and $I_0(x)=0$, the line is spoken of as 'initially dead'. From eqns (6.11) and (6.12),

$$L\left(\frac{\partial I}{\partial t}\right)_{t=0}=-\frac{\partial V_0(x)}{\partial x}-RI_0(x),\quad C\left(\frac{\partial V}{\partial t}\right)_{t=0}=-\frac{\partial I_0(x)}{\partial x}-GV_0(x).$$

Hence, if $V_0(x)$ and $I_0(x)$ are given, $(\partial I/\partial t)_{t=0}$ and $(\partial V/\partial t)_{t=0}$ can be found. If it is convenient to work with eqn (6.16), rather than eqns (6.11) and (6.12), then it is the values of $V_0(x)$ and $(\partial V/\partial t)_{t=0}$ that will be needed.

Apply the Laplace transform to the time variable in eqns (6.11) and (6.12) with the initial conditions (6.19), to give

$$-\frac{\mathrm{d}\bar{V}}{\mathrm{d}x}=(Lp+R)\,\bar{I}-LI_0=Z\bar{I}-LI_0 \tag{6.20}$$

and

$$-\frac{\mathrm{d}\bar{I}}{\mathrm{d}x}=(Cp+G)\,\bar{V}-CV_0=Y\bar{V}-CV_0, \tag{6.21}$$

where

$$Z=Lp+R \quad\text{and}\quad Y=Cp+G. \tag{6.22}$$

Z is the operational form of series impedance per unit length and Y is the operational form of the shunt admittance. Eliminating $\bar{I}$ between eqns (6.20) and (6.21):

$$\frac{\mathrm{d}^2\bar{V}}{\mathrm{d}x^2}=-Z\frac{\mathrm{d}\bar{I}}{\mathrm{d}x}+LI_0'=ZY\bar{V}-CZV_0+LI_0', \tag{6.23}$$

an ordinary differential equation for $\bar{V}$ with complementary functions $\mathrm{e}^{\pm\sqrt{ZY}\,x}$. This leads to a complicated inversion. To get some idea of how to proceed, we look at some simpler cases. The simplest case brings in the error function, so that the next step is to look at its properties.

6.4. The error function; its asymptotic expansion

The error function is so called because it is the integral of the normal or Gaussian distribution in statistics. The error function, denoted by erf x, is defined by

$$\operatorname{erf} x=\frac{2}{\sqrt{\pi}}\int_0^x \mathrm{e}^{-\xi^2}\,\mathrm{d}\xi, \tag{6.24}$$

so that

$$\operatorname{erf} 0=0 \text{ and } \operatorname{erf}(+\infty)=1, \text{ since } \int_0^\infty e^{-x^2}\,dx=\frac{\sqrt{\pi}}{2}.$$

The complementary error function erfc x is defined by

$$\operatorname{erfc} x=\frac{2}{\sqrt{\pi}}\int_x^\infty e^{-\xi^2}\,d\xi, \tag{6.25}$$

so that erfc $x=1-\operatorname{erf} x$, erfc $0=1$ and erfc $(+\infty)=0$.

Now

$$\operatorname{erf}(-x)=\frac{2}{\sqrt{\pi}}\int_0^{-x} e^{-\xi^2}\,d\xi \underset{\xi=-\eta}{=} -\frac{2}{\sqrt{\pi}}\int_0^x e^{-\eta^2}\,d\eta=-\operatorname{erf} x$$

and

$$\operatorname{erfc}(-x)=1-\operatorname{erf}(-x)=1+\operatorname{erf} x=2-\operatorname{erfc} x,$$

so that $\operatorname{erf}(-\infty)=-1$ and $\operatorname{erfc}(-\infty)=2$.

LEMMA.

For $\alpha>0$ *and* $\phi>0$,

$$\frac{2}{\sqrt{\pi}}\int_\phi^\infty e^{-(y^2+\alpha^2/y^2)}\,dy=\frac{1}{2}\left[e^{2\alpha}\operatorname{erfc}\left(\phi+\frac{\alpha}{\phi}\right)+e^{-2\alpha}\operatorname{erfc}\left(\phi-\frac{\alpha}{\phi}\right)\right] \tag{6.26}$$

Proof.

If $u=y\pm\alpha/y$, $du=(1\mp\alpha/y^2)\,dy$, and

$$\begin{aligned}\frac{2}{\sqrt{\pi}}\int_\phi^\infty e^{-(y^2+\alpha^2/y^2)}\,dy&=\frac{1}{\sqrt{\pi}}\int_\phi^\infty\left[\left(1-\frac{\alpha}{y^2}\right)e^{-(y+\alpha/y)^2+2\alpha}\right.\\&\qquad\left.+\left(1+\frac{\alpha}{y^2}\right)e^{-(y-\alpha/y)^2-2\alpha}\right]dy\\&=\frac{1}{\sqrt{\pi}}\left[e^{2\alpha}\int_{\phi+\alpha/\phi}^\infty e^{-u^2}\,du\right.\\&\qquad\left.+e^{-2\alpha}\int_{\phi-\alpha/\phi}^\infty e^{-u^2}\,du\right]\\&=\frac{1}{2}\left[e^{2\alpha}\operatorname{erfc}\left(\phi+\frac{\alpha}{\phi}\right)+e^{-2\alpha}\operatorname{erfc}\left(\phi-\frac{\alpha}{\phi}\right)\right].\end{aligned}$$

CORROLLARY. *Let $\phi \to +0$, then*

$$\frac{2}{\sqrt{\pi}}\int_0^{\infty} e^{-(y^2+\alpha^2/y^2)}\,dy = \tfrac{1}{2}[e^{2\alpha}\,\text{erfc}(+\infty) + e^{-2\alpha}\,\text{erfc}(-\infty)] = e^{-2\alpha}. \tag{6.27}$$

THEOREM. *The Laplace transform of* erf $(a/2\sqrt{t})$ *is* $(1/p)\,(1-e^{-a\sqrt{p}})$. (6.28)

Proof.

$$\int_0^{\infty} e^{-pt}\,\text{erf}\,\frac{a}{2\sqrt{t}}\,dt = \int_0^{\infty} e^{-pt}\left(\frac{2}{\sqrt{\pi}}\int_0^{a/2\sqrt{t}} e^{-y^2}\,dy\right)dt$$

The double integral is over the area shown hatched in Fig. 6.2. Changing the order of integration,

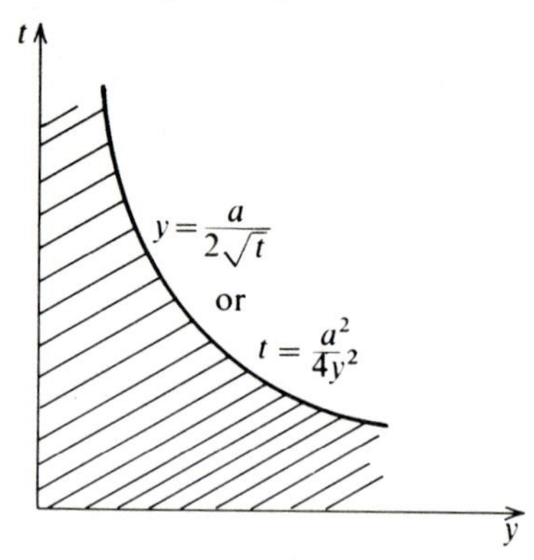

Fig. 6.2

$$\begin{aligned}
\int_0^{\infty} e^{-pt}\,\text{erf}\,\frac{a}{2\sqrt{t}}\,dt &= \frac{2}{\sqrt{\pi}}\int_0^{\infty} e^{-y^2}\left(\int_0^{a^2/4y^2} e^{-pt}\,dt\right)dy \\
&= \frac{2}{\sqrt{\pi}}\int_0^{\infty} e^{-y^2}\,\frac{1}{p}\,(1-e^{-a^2p/4y^2})\,dy \\
&= \frac{1}{p}\,\frac{2}{\sqrt{\pi}}\int_0^{\infty} e^{-y^2}\,dy - \frac{2}{p\sqrt{\pi}}\int_0^{\infty} e^{-(y^2+a^2p/4y^2)}\,dy \\
&= \frac{1}{p} - \frac{1}{p}\,e^{-a\sqrt{p}}
\end{aligned}$$

on putting $\alpha = \tfrac{1}{2}a\sqrt{p}$ in eqn (6.27).

COROLLARY. *The Laplace transform of*

$$\operatorname{erfc}(a/2\sqrt{t}) \text{ is } (1/p)\,\mathrm{e}^{-a\sqrt{p}} \tag{6.29}$$

Proof. The proof follows immediately from $\operatorname{erfc} x = 1 - \operatorname{erf} x$ and from $\int_0^\infty \mathrm{e}^{-pt}\,1\,\mathrm{d}t = 1/p$.

We now consider the form of $\operatorname{erfc} x$ as $x \to +\infty$. Since

$$\frac{\sqrt{\pi}}{2}\operatorname{erfc} x = \int_x^\infty \mathrm{e}^{-y^2}\,\mathrm{d}y = \int_x^\infty \frac{1}{-2y}(-2y\,\mathrm{e}^{-y^2})\,\mathrm{d}y,$$

successive integration by parts gives

$$\frac{\sqrt{\pi}}{2}\operatorname{erfc} x = \frac{\mathrm{e}^{-y^2}}{-2y}\bigg|_x^\infty - \int_x^\infty \frac{\mathrm{e}^{-y^2}}{2y^2}\,\mathrm{d}y = \frac{\mathrm{e}^{-x^2}}{2x} - \int_x^\infty \frac{1}{-4y^3}(-2y\,\mathrm{e}^{-y^2})\,\mathrm{d}y$$

$$= \frac{\mathrm{e}^{-x^2}}{2x} - \frac{\mathrm{e}^{-y^2}}{-4y^3}\bigg|_x^\infty + \frac{1.3}{2^2}\int_x^\infty \frac{\mathrm{e}^{-y^2}}{y^4}\,\mathrm{d}y$$

$$= \frac{\mathrm{e}^{-x^2}}{2x} - \frac{\mathrm{e}^{-x^2}}{4x^3} + \frac{1.3}{2^2}\int_x^\infty \frac{1}{-2y^5}(-2y\,\mathrm{e}^{-y^2})\,\mathrm{d}y$$

$$= \frac{\mathrm{e}^{-x^2}}{2x} - \frac{\mathrm{e}^{-x^2}}{2^2x^3} + \frac{1.3\mathrm{e}^{-x^2}}{2^3x^5} - \frac{1.3.5}{2^4}\int_x^\infty \frac{\mathrm{e}^{-y^2}}{y^6}\,\mathrm{d}y.$$

Further integration by parts yields

$$\operatorname{erfc} x = \frac{\mathrm{e}^{-x^2}}{\sqrt{\pi}}\Bigg[\bigg(\frac{1}{x} - \frac{1}{2x^3} + \frac{1.3}{2^2x^5} - \frac{1.3.5}{2^2x^7} + \ldots$$

$$-(-)^n\frac{1.3\ldots(2n-3)}{2^{n-1}x^{2n-1}}\bigg)$$

$$+(-)^n\frac{1.3\ldots(2n-1)}{2^n}\int_x^\infty \frac{\mathrm{e}^{-y^2}}{y^{2n}}\,\mathrm{d}y\Bigg].$$

The modulus of the ratio of successive terms is $(2n-1)/2x^2$, which tends to $+\infty$ as $n \to \infty$ for all x. The series diverges.

The modulus of the remainder is

$$\frac{1.3\ldots(2n-1)}{2^n}\int_x^\infty \frac{\mathrm{e}^{-y^2}}{y^{2n}}\,\mathrm{d}y < \frac{1.3\ldots(2n-1)}{2^n}\,\mathrm{e}^{-x^2}\int_x^\infty \frac{\mathrm{d}y}{y^{2n}}$$

$$= \frac{1.3\ldots(2n-3)}{2^nx^{2n-1}},$$

i.e. the modulus of the remainder is less than the modulus of the last term included in the series. The sum of the series to n terms has error less than the magnitude of the nth term. Also, each term of the series decreases in magnitude as x increases, hence the error in the sum of the series to any number of terms n can be made arbitrarily small by increasing x. We write

$$\operatorname{erfc} x \sim \frac{1}{\sqrt{\pi}} \mathrm{e}^{-x^2} \left(\frac{1}{x} - \frac{1}{2x^3} + \frac{1.3}{2^2 x^5} - \cdots \right) \quad \text{as } x \to +\infty, \tag{6.30}$$

where the symbol $\sim$ is pronounced 'squiggles'. The series is called the asymptotic expansion of erfc x as $x \to +\infty$. Unlike an ordinary convergent series, whose sum to n terms can be made accurate to within any prescribed positive error ε by increasing n, the error in an asymptotic series to n terms can be arbitrarily small by increasing x.

The magnitude of the terms of the asymptotic series for fixed x decreases initially and then increases. Since the ratio of successive terms is $(2n-1)/2x^2$, the minimum term occurs when n equals one of the integers on either side of $x^2 + \frac{1}{2}$. The modulus of the remainder is of the order of the last term included and therefore this latter modulus can be minimized by cutting off the series at the minimum. There is no way in which the accuracy of the series can be improved for this value of x. A necessary condition for an asymptotic series to be useful in practice is whether this accuracy is acceptable.

Frequently it is permissible to take only one term in an asymptotic series. For example in the series for erfc x, if x is sufficiently large to allow one to neglect $1/x^3$ compared to $1/x$, then

$$\operatorname{erfc} x \sim \frac{1}{\sqrt{\pi x}} \mathrm{e}^{-x^2}; \tag{6.31}$$

because, in the expansion

$$\operatorname{erfc} x \sim \frac{1}{\sqrt{\pi}} \left(\frac{1}{x} - \frac{1}{2x^3} \right) \mathrm{e}^{-x^2},$$

the error is less than the term in $1/x^3$ and this can be neglected compared to the term in $1/x$ for x sufficiently large.

Exercise 6.1. Find the asymptotic expansion for large positive x of the integral

$$f(x) = \int_x^\infty t^{-1} \mathrm{e}^{x-t} \, \mathrm{d}t.$$

Hint: Use successive integrations by parts to show that

$$f(x)=\frac{1}{x}-\frac{1}{x^2}+\frac{2!}{x^3}-\frac{3!}{x^4}+\ \dots\ +\frac{(-)^{n-1}(n-1)!}{x^n}+(-)^n n!\int_x^\infty \frac{e^{x-t}}{t^{n-1}}\,dt.$$

Show that the properties found for the asymptotic series for erfc x also hold for the series for $f(x)$.

6.5. Kelvin's cable and the non-inductive leaky cable

In Kelvin's ideal cable, it is assumed that there is no inductance and that the insulation between the wires is perfect, i.e. $L=0$ and $G=0$. Then $Z=R$ and $Y=Cp$ from eqns (6.22). If the circuit is initially dead, $I_0=V_0=0$, and eqn (6.23) gives

$$\frac{d^2\bar{V}}{dx^2}=CRp\bar{V}$$

with solution

$$\bar{V}(x,p)=A(p)\,e^{-\sqrt{CRp}\,x}+B(p)\,e^{+\sqrt{CRp}\,x}.$$

If the cable is semi-infinite, $x\geqslant 0$, and if $V(0,t)=f(t)$, which corresponds to a signal being fed in at the end $x=0$ for $t>0$, and if $V(x,t)$ is finite as $x\to+\infty$, then $\bar{V}(x,p)$ is finite as $x\to+\infty$ and $B(p)=0$. Hence

$\bar{V}(x,p)=A(p)\,e^{-\sqrt{CRp}\,x}$ and $\bar{f}(p)=\bar{V}(0,p)=A(p)$. Therefore

$$\bar{V}(x,p)=\bar{f}(p)\,e^{-\sqrt{CRp}\,x}. \tag{6.32}$$

When the signal $f(t)=H(t)$, $\bar{f}(p)=1/p$ and

$$\bar{V}(x,p)=\frac{1}{p}\,e^{-\sqrt{CRp}\,x}. \tag{6.33}$$

By eqn (6.29) with $a=\sqrt{CR}\,x$,

$$V(x,t)=\operatorname{erfc}\frac{\sqrt{CR}\,x}{2\sqrt{t}} \tag{6.34}$$

and

$$I(x,t)=-\frac{1}{R}\frac{\partial V}{\partial x}=\sqrt{\frac{C}{\pi Rt}}\,e^{-CRx^2/4t} \tag{6.35}$$

It is seen from eqn (6.34) that, for any $t>0$, however small, $V(x,t)>0$ for all $x \geqslant 0$ although $V(x,t) \to 0$ as $x \to +\infty$. Hence, the signal applied at $t=0$ travels at infinite velocity though its amplitude is very small for large x. For large x, eqn (6.31) gives

$$V(x,t) \sim \frac{2\sqrt{t}}{\sqrt{\pi C R}\, x}\, \mathrm{e}^{-CRx^2/4t} \tag{6.36}$$

The infinite velocity is caused by the neglect of the term $LC(\partial^2 V/\partial t^2)$ in eqn (6.16). In an actual cable, the presence of some inductance will set a limit to the velocity of propagation. Kelvin's ideal cable will form a reasonable approximation at low frequencies, except near the wavefront, provided the insulation is good.

What happens if the assumption of perfect insulation is dropped so that Kelvin's ideal cable is replaced by a non-inductive leaky cable, $L=0$ but $G \neq 0$. The eqn (6.16) for V becomes

$$\frac{\partial^2 V}{\partial x^2} = RC\frac{\partial V}{\partial t} + RGV. \tag{6.37}$$

For an initially dead semi-infinite cable $x \geqslant 0$,

$$\frac{\mathrm{d}^2 \bar{V}}{\mathrm{d}x^2} = (pRC + RG)\,\bar{V}.$$

If $V(0,t)=H(t)$ and $V(x,t)$ finite as $x \to +\infty$,

$$\bar{V}(x,p) = \frac{1}{p}\,\mathrm{e}^{-\sqrt{RC(p+G/C)}\,x}. \tag{6.38}$$

If the straight line $\mathrm{R}[p]=c$ is to the right of all singularities in the p-plane, the inverse transform of $(1/p)\,\bar{f}(p+\alpha)$, $\alpha>0$, by the Bromwich inversion theorem is

$$\frac{1}{2\pi \mathrm{i}}\int_{c-\mathrm{i}\infty}^{c+\mathrm{i}\infty} \frac{\bar{f}(p+\alpha)}{p}\,\mathrm{e}^{pt}\,\mathrm{d}p = \frac{1}{2\pi \mathrm{i}}\int_{\alpha+c-\mathrm{i}\infty}^{\alpha+c+\mathrm{i}\infty} \frac{\bar{f}(p)}{p-\alpha}\,\mathrm{e}^{(p-\alpha)t}\,\mathrm{d}p.$$

Therefore,

$$\frac{1}{p}\bar{f}(p+\alpha) = \mathrm{e}^{-\alpha t}\,\frac{1}{p-\alpha}\bar{f}(p), \quad \alpha>0, \tag{6.39}$$

since the straight line $\mathrm{R}[p]=c+\alpha$ is to the right of all singularities in the p-plane because it is to the right of $\mathrm{R}[p]=c$. Apply eqn (6.39) to the r.h.s. of eqn (6.38):

$$\bar{V}(x,p)=\mathrm{e}^{-Gt/C}\frac{1}{p-G/C}\mathrm{e}^{-\sqrt{RCp}\,x}$$

$$=\mathrm{e}^{-Gt/C}\frac{p}{p-G/C}\frac{\mathrm{e}^{-\sqrt{RCp}\,x}}{p}.$$

By the convolution theorem,

$$V(x,t)=\mathrm{e}^{-Gt/C}\int_0^t\left(\delta(t-u)+\frac{G}{c}\mathrm{e}^{G(t-u)/C}\right)\operatorname{erfc}\frac{x\sqrt{RC}}{2\sqrt{u}}\,\mathrm{d}u$$

where we have used

$$\frac{p}{p-G/C}=1+\frac{G/C}{p-G/C},\quad \int_0^\infty \delta(t)\,\mathrm{e}^{-pt}\,\mathrm{d}t=1$$

and eqn (6.29). Therefore, on integration by parts

$$V(x,t)=\mathrm{e}^{-Gt/C}\left\{\operatorname{erfc}\frac{x\sqrt{RC}}{2\sqrt{t}}-\left[\mathrm{e}^{-G(t-u)/C}\operatorname{erfc}\frac{x\sqrt{RC}}{2\sqrt{u}}\right]_{u=0}^{u=t}\right.$$

$$\left.+\int_0^t \mathrm{e}^{G(t-u)/C}\frac{2}{\sqrt{\pi}}\mathrm{e}^{-RCx^2/4u}\left(-\frac{x\sqrt{RC}}{4u^{\frac{3}{2}}}\right)\mathrm{d}u\right\}$$

$$=\frac{x\sqrt{RC}}{2\sqrt{\pi}}\int_0^t\exp\left(-\frac{Gu}{C}-\frac{RCx^2}{4u}\right)u^{-\frac{3}{2}}\,\mathrm{d}u.$$

Put $u=RCx^2/4\xi^2$, so that $u^{-\frac{3}{2}}\,\mathrm{d}u=-4\,\mathrm{d}\xi/\sqrt{RC}\,x$ and

$$V(x,t)=\frac{2}{\sqrt{\pi}}\int_{\sqrt{RC}\,x/2\sqrt{t}}^{\infty}\mathrm{e}^{-(\xi^2+RGx^2/4\xi^2)}\,\mathrm{d}\xi.$$

Therefore, by eqn (6.26),

$$V(x,t)=\frac{1}{2}\mathrm{e}^{\sqrt{RG}\,x}\operatorname{erfc}\left(\frac{\sqrt{RC}\,x}{2\sqrt{t}}+\sqrt{\frac{G}{C}}\sqrt{t}\right)$$

$$+\frac{1}{2}\mathrm{e}^{-\sqrt{RG}\,x}\operatorname{erfc}\left(\frac{\sqrt{RC}\,x}{2\sqrt{t}}-\sqrt{\frac{G}{C}}\sqrt{t}\right). \quad (6.40)$$

A partial check on eqn (6.40) is to put $G=0$. This reproduces eqn (6.34). The non-inductive leaky cable becomes Kelvin's ideal cable when $G=0$.

Exercise 6.2. Use eqn (6.31) to show from eqn (6.40) that

$$V(x,t) \underset{x\to\infty}{\sim} \frac{\sqrt{RC}\,x}{2\sqrt{\pi t}}\left(\frac{RCx^2}{4t}-\frac{Gt}{C}\right)^{-1}\exp\left(-\frac{RCx^2}{4t}-\frac{Gt}{C}\right), \tag{6.41}$$

and check that eqn (6.41) reduces to eqn (6.36) when $G=0$.

Equation (6.34) for Kelvin's cable and its refinement for leaky cables, eqn (6.40), are easily evaluated for particular x and t by digital computer. However, such cables are subject to practical objection. If Morse code is transmitted by cable, input voltage is of the type (shown in Fig. 6.3),

$$V(0,t)=H(t)-H(t-t_1).$$

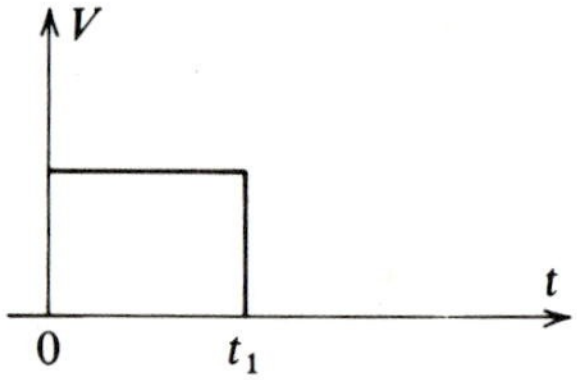

Fig. 6.3

For a Kelvin cable, using eqn (6.34),

$$V(x,t)=\operatorname{erfc}\frac{x\sqrt{RC}}{2\sqrt{t}}\,H(t)-\operatorname{erfc}\frac{x\sqrt{RC}}{2\sqrt{(t-t_1)}}\,H(t-t_1).$$

For fixed x, the first term $\to 1$ as $t\to\infty$ and so does the second term. If, therefore, the first term does not become appreciable before the second term starts growing, the signal will be lost. This means that t_1 must not be too small. Suppose erfc z is appreciable only when $z<z_0$, then the first term is significant as soon as $t=t_0=RCx^2/4z_0^2$ and t_1 must not be small compared to t_0. As x increases, so must t_1 to obtain the same definition. To send a recognizable signal over a submarine cable may require an excessively slow rate of signalling. Similar but more complicated arguments hold for eqn (6.40).

Exercise 6.3. Show that it is possible[†] to maintain a semi-infinite Kelvin ideal cable ($L=G=0$) at constant potential $V_0\neq 0$ (for $t<0$). For $t>0$, the end $x=0$ is short-circuited ($V=0$). Find the potential and current on the line for $t>0$. Check that your result for $V(x,t)$ satisfies the initial and boundary conditions.

[†] Is it possible on a cable when none of L, R, C or G is zero?

6.6. Heaviside cables

It was Oliver Heaviside who suggested that self-inductance L be added to a cable to overcome the lack of definition described in the last paragraph. To see how this might work, we consider harmonic wave solutions of the governing equation. In eqn (6.16), substitute

$$V(x,t)=h(x)\,\mathrm{e}^{\mathrm{i}\omega t} \tag{6.42}$$

to give

$$\frac{\mathrm{d}^2 h}{\mathrm{d}x^2}=-\omega^2 LCh+\mathrm{i}\omega(LG+RC)h+RGh$$

or

$$\frac{\mathrm{d}^2 h}{\mathrm{d}x^2}+\kappa^2 h=0, \tag{6.43}$$

where

$$\kappa^2=\omega^2 LC-RG-\mathrm{i}\omega(LG+RC). \tag{6.64}$$

The solution of eqn (6.43) is

$$h(x)=A\mathrm{e}^{\mathrm{i}\kappa x}+B\mathrm{e}^{-\mathrm{i}\kappa x}. \tag{6.45}$$

Put $\kappa=\alpha-\mathrm{i}\beta$ where $\alpha\geqslant 0$. From eqn (6.44), one root κ must have non-negative real part. Then,

$$\alpha^2-\beta^2=\omega^2 LC-RG \tag{6.46}$$

and

$$2\alpha\beta=\omega(LG+RC). \tag{6.47}$$

From the second equation, $\beta>0$ since $\alpha>0$. For a wave travelling in the x-direction, the phase velocity v is ω/α and the attenuation is β.

In general, both ω/α and β will depend on ω, so that waves of different frequency travel at different speeds and are attenuated differently. For signals to be unaltered in shape as they travel, ω/α and β must be independent of ω. Put $\alpha=\omega/v$ in eqns (6.46) and (6.47):

$$\frac{\omega^2}{v^2}-\beta^2=\omega^2 LC-RG \tag{6.48}$$

and

$$\frac{2\beta}{v}=LG+RC. \tag{6.49}$$

For v and β to be independent of ω, eqn (6.48) requires that

$$\frac{1}{v^2}=LC \quad \text{and} \quad \beta^2=RG. \tag{6.50}$$

For eqn (6.49) to be consistent with eqns (6.50),

$$2\sqrt{RGLC} = LG + RC \quad \text{or} \quad LG = RC. \tag{6.51}$$

Equation (6.51) is the necessary condition for no distortion of signals.

To prove sufficiency, put $G = RC/L$ in eqn (6.44):

$$\kappa^2 = LC\left(\omega^2 - \frac{R^2}{L^2} - 2\mathrm{i}\omega\frac{R}{L}\right) = LC\left(\omega - \mathrm{i}\frac{R}{L}\right)^2,$$

$$\kappa = \sqrt{LC}\left(\omega - \mathrm{i}\frac{R}{L}\right) = \omega\sqrt{LC} - \mathrm{i}\sqrt{RG}. \tag{6.52}$$

Substitute back for κ into eqn (6.42),

$$V(x, t) = A\mathrm{e}^{\mathrm{i}\omega(t - \sqrt{LC}x)}\mathrm{e}^{-\sqrt{RG}x} + B\mathrm{e}^{\mathrm{i}\omega(t + \sqrt{LC}x)}\mathrm{e}^{\sqrt{RG}x}, \tag{6.53}$$

which represents harmonic waves travelling with velocity $1/\sqrt{LC}$ and attenuation $\sqrt{RG}$ for all frequencies ω. Hence, all signals travel with velocity $1/\sqrt{LC}$, with no change of shape but with attenuation $\sqrt{RG}$. A cable for which $LG = RC$ is known as a Heaviside distortionless cable.

To find the signal propagated along an initially dead semi-infinite Heaviside cable for a given input $V(0, t) = f(t)$, we have from eqns (6.16) and (6.51),

$$\frac{\mathrm{d}^2\bar{V}}{\mathrm{d}x^2} = \left(p^2LC + 2pRC + R^2\frac{C}{L}\right)\bar{V} = LC\bar{V}\left(p^2 + 2p\frac{R}{L} + \frac{R^2}{L^2}\right); \tag{6.54}$$

whence

$$\bar{V} = \bar{f}(p)\,\mathrm{e}^{-\sqrt{LC}(p + R/L)x} \tag{6.55}$$

if $V(x, t)$ is finite as $x \to +\infty$. Inversion of eqn (6.55) gives

$$V(x, t) = f(t - \sqrt{LC}\,x)\,H(t - \sqrt{LC}\,x)\,\mathrm{e}^{-\sqrt{RG}x}, \tag{6.56}$$

where in the exponential index $\sqrt{(C/L)}\,R$ has been replaced by $\sqrt{RG}$. Equation (6.56) shows that the signal propagates with velocity $1/\sqrt{LC}$, with attenuation $\sqrt{RG}$ but with no distortion. It follows that signals can be propagated along Heaviside cables over long distances provided booster stations are placed at regular intervals to increase the strength of the signal so as to counteract the effect of the attenuation.

It was pointed out earlier that the characteristic relations (6.17) can only be integrated for special values of constants L, C, R and G. When $LG=RC$, eqns (6.17) become

$$\mathrm{d}\left(V\pm\sqrt{\frac{L}{C}}\,I\right)\pm\sqrt{RG}\left(V\pm\sqrt{\frac{L}{C}}\,I\right)\mathrm{d}x=0$$

or

$$V\pm\sqrt{\frac{L}{C}}\,I=\beta \mathrm{e}^{\mp\sqrt{RG}\,x} \quad \text{on } x=\pm\frac{1}{\sqrt{LC}}\,t+\alpha, \tag{6.57}$$

where β and α are constant on a characteristic.

Let us look at the half-space problem using the characteristic equations (6.57). All C_- characteristics intersect the positive x-axis (Fig. 6.4) on which $V=I=0$, since the line is initially dead. Hence $\beta=0$ on all C_- and therefore $V-\sqrt{(L/C)}\,I=0$ everywhere, i.e.

$$I=\sqrt{\frac{C}{L}}\,V. \tag{6.58}$$

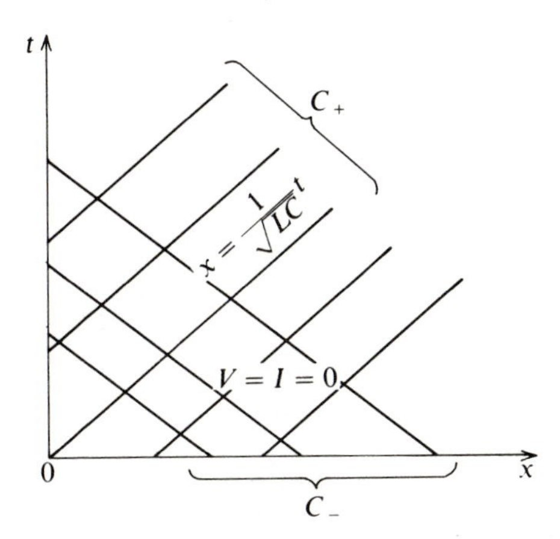

Fig. 6.4

Equation (6.57) for the C_+ characteristics becomes

$$2V=\beta \mathrm{e}^{-\sqrt{RG}\,x} \quad \text{on } x=\frac{1}{\sqrt{LC}}\,t+\alpha. \tag{6.59}$$

If $\alpha \geqslant 0$, then the C_+ characteristic intersects the positive x-axis on which $V=0$. Therefore $\beta=0$ and $V=0$ at all points below the line $x=(1/\sqrt{LC})\,t$, i.e.

$$V=0 \quad \text{for } t-\sqrt{LC}\,x<0. \tag{6.60}$$

If $\alpha<0$, then the C_+ characteristic intersects the positive t-axis, at $t=T$ say. At $t=T$, $x=0$ and $V=f(T)$, therefore $\alpha=-(1/\sqrt{LC})\,T$ and $\beta=2f(T)$. Hence,

$$V=f(T)\mathrm{e}^{-\sqrt{RG}\,x} \quad \text{on } x=\frac{1}{\sqrt{LC}}(t-T).$$

Elimination of T between these last two equations gives

$$V(x,t)=f(t-\sqrt{LC}\,x)\mathrm{e}^{-\sqrt{RG}\,x} \quad \text{for } \alpha<0 \quad \text{or} \quad t-\sqrt{LC}\,x>0. \tag{6.61}$$

Combination of eqns (6.60) and (6.61) gives

$$V(x,t)=f(t-\sqrt{LC}\,x)\mathrm{e}^{-\sqrt{RG}\,x}H(t-\sqrt{LC}\,x),$$

in agreement with eqn (6.56).

6.7. Finite Heaviside line; impedance matching

The signal $f(t)$ is again fed in at $x=0$ (Fig. 6.5) but the line terminates at $x=l$, where the transforms of V and I satisfy

$$\bar{V}=Z\bar{I} \quad \text{at} \quad x=l. \tag{6.62}$$

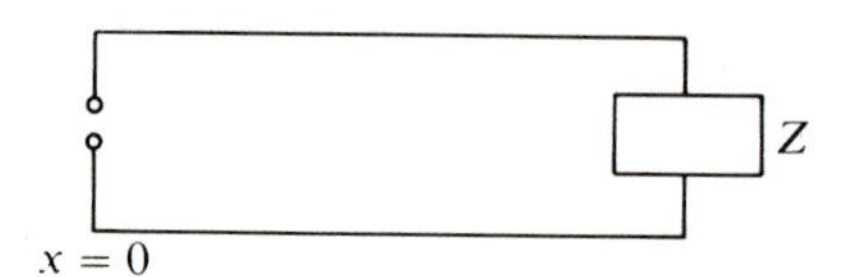

Fig. 6.5

Z is known as the impedance at the end of the line. With the line initially dead the analysis is the same as in the last section until eqn (6.54). Both terms must now be included in the solution of this equation so that

$$\bar{V}(x,p)=A(p)\mathrm{e}^{-\sqrt{LC(p+R/L)}\,x}+B(p)\mathrm{e}^{+\sqrt{LC(p+R/L)}\,x}. \tag{6.63}$$

For
$$\bar{V}(0,p)=\bar{f}(p),$$
$$\bar{f}(p)=A(p)+B(p). \tag{6.64}$$

By eqn (6.20) with $I_0=0$,
$$\bar{I}(x,p)=-(Lp+R)^{-1}\frac{\partial \bar{V}}{\partial x}=\sqrt{\frac{C}{L}}\,(A(p)$$
$$\times\,\mathrm{e}^{-\sqrt{LC(p+R/L)}\,x}-B(p)\,\mathrm{e}^{\sqrt{LC(p+R/L)}\,x}).$$

To satisfy eqn (6.62),
$$A(p)\left(1-\sqrt{\frac{C}{L}}\,Z\right)\mathrm{e}^{-\sqrt{LC(p+R/L)}\,l}+B(p)\left(1+\sqrt{\frac{C}{L}}\,Z\right)$$
$$\times\,\mathrm{e}^{\sqrt{LC(p+R/L)}\,l}=0. \tag{6.65}$$

Equations (6.64) and (6.65) are two equations to solve for $A(p)$ and $B(p)$. $A(p)$ and $B(p)$ are substituted back into eqn (6.63) for $\bar{V}(x,p)$ and then $\bar{V}(x,p)$ is inverted. However, it is easier to treat different end conditions at $x=l$ separately.

Case (i). *Open end, $I=0$ at $x=l$.* Since $\bar{I}=0$ at $x=l$ and $\bar{V}$ remains finite, $1/Z=0$. Divide eqn (6.65) through by Z and then put $1/Z=0$:
$$-A(p)\,\mathrm{e}^{-\sqrt{LC(p+R/L)}\,l}+B(p)\,\mathrm{e}^{\sqrt{LC(p+R/L)}\,l}=0. \tag{6.66}$$

Solving eqns (6.64) and (6.66),
$$A(p)(1+\mathrm{e}^{-2\sqrt{LC(p+R/L)}\,l})=\bar{f}(p)$$
and
$$B(p)(1+\mathrm{e}^{-2\sqrt{LC(p+R/L)}\,l})=\bar{f}(p)\,\mathrm{e}^{-2\sqrt{LC(p+R/L)}\,l},$$
whence
$$\bar{V}(x,p)=\frac{f(p)}{1+\mathrm{e}^{-2\sqrt{LC}(p+R/L)\,l}}\,(\mathrm{e}^{-\sqrt{LC(p+R/L)}\,x}+\mathrm{e}^{\sqrt{LC(p+R/L)}\,(x-2l)}).$$

Since $\mathrm{R}[p]>0$, $|\mathrm{e}^{-2\sqrt{LC(p+R/L)}\,l}|<1$ and the denominator can be expanded by the binomial theorem to give
$$\bar{V}(x,p)=\bar{f}(p)(\mathrm{e}^{-\sqrt{LC(p+R/L)}\,x}+\mathrm{e}^{\sqrt{LC(p+R/L)}\,(x-2l)})$$
$$\times\sum_{n=0}^{\infty}(-)^n\,\mathrm{e}^{-2n\sqrt{LC(p+R/L)}\,l}.$$

Therefore,

$$V(x,t)=\sum_{n=0}^{\infty}(-)^n[e^{-\sqrt{LC}(R/L)(x+2nl)}$$

$$\times f(t-\sqrt{LC}(x+2nl))\,H(t-\sqrt{LC}(x+2nl))$$

$$+e^{\sqrt{LC}(R/L)(x-(2n+2)l)}f(t+\sqrt{LC}(x-(2n+2)l))$$

$$\times H(t+\sqrt{LC}(x-(2n+2)l))]. \tag{6.67}$$

The physical significance of eqn (6.67) is best seen by considering increasing values of t. Remember $0 \leqslant x \leqslant l$. For $t<0$ all Heaviside functions in eqn (6.67) are zero. For $0<t<\sqrt{LC}\,l$, all Heaviside functions except one are zero and, using $LG=RC$,

$$V(x,t)=e^{-\sqrt{RG}\,x}f(t-\sqrt{LC}\,x)\,H(t-\sqrt{LC}\,x). \tag{6.68}$$

Equation (6.68) represents a wave travelling with velocity $1/\sqrt{LC}$ in the positive x-direction and decaying exponentially in amplitude with distance travelled, attenuation constant $\sqrt{RG}$. At time t, the front of the wave is at $x=t/\sqrt{LC}$, where its amplitude is $e^{-(R/L)x}f(0)$.

For $\sqrt{LC}\;l<t<2\sqrt{LC}\,l$,

$$V(x,t)=e^{-\sqrt{RG}\,x}f(t-\sqrt{LC}\,x)+e^{-\sqrt{RG}\,(2l-x)}$$

$$\times f(t-\sqrt{LC}(2l-x))\,H(t-\sqrt{LC}\,(2l-x)). \tag{6.69}$$

The Heaviside function is omitted from the first term in eqn (6.69) because it is always equal to $+1$ for $0 \leqslant x \leqslant l$ and $\sqrt{LC}\,l<t<2\sqrt{LC}\;l$. The first term on the r.h.s. is the continuation of the wave of eqn (6.68), the second term is the reflection of this wave from the end $x=l$. The reflected wave travels in the negative x-direction with velocity $1/\sqrt{LC}$ and attenuation constant $\sqrt{RG}$. This wave reaches the end $x=0$ at time $t=2\sqrt{LC}\;l$ and is reflected back in the positive sense. Hence, for $2\sqrt{LC}\;l<t<3\sqrt{LC}\;l$, three terms will be found to be non-zero in the expression for $V(x,t)$ eqn (6.67), and so on as time increases further.

Case (ii). *Closed end,* $V=0$ *at* $x=l$, *set as Exercise* 6.4.

Exercise 6.4. A finite Heaviside line is initially dead and has input $V(0,t)=f(t)\,H(t)$ at $x=0$. The end $x=l$ is short-circuited ($V=0$). Find the potential

V at all points x on the line at time $t=\frac{5}{2}\sqrt{LC}\,l$ and explain the physical significance of each term in the result.

Case (iii). *Impedance matching.* It has just been seen that after time $t=2\sqrt{LC}\,l$, two waves are travelling in the positive sense on the cable. The twice-reflected wave is not wanted, because its reception at the end $x=l$ starting at time $t=3\sqrt{LC}\,l$ will interfere with the reception of the original incident wave. Since the condition at the end $x=0$ is given by the signal input, it is only the condition at the end $x=l$ which can be used to stop this interference. The interference will not occur if there is no reflected wave from the end $x=l$, i.e. if $B(p)=0$ in eqn (6.63). From eqn (6.65), $B(p)=0$ if

$$Z=\sqrt{\frac{L}{C}}. \tag{6.70}$$

When $B(p)=0$, from eqn (6.64), $A(p)=\bar{f}(p)$ and substitution in eqn (6.63) gives

$$\bar{V}(x,p)=\bar{f}(p)\,\mathrm{e}^{-\sqrt{LC}(p+R/L)x},$$

whence

$$V(x,t)=f(t-\sqrt{LC}\,x)\,H(t-\sqrt{LC}\,x)\,\mathrm{e}^{-\sqrt{RG}\,x}. \tag{6.71}$$

Equation (6.71) is identical with eqn (6.56). Impedance matching means that the signal travels from the end $x=0$ to the end $x=l$ where it is completely absorbed. Substituting back from eqn (6.70) into eqn (6.62),

$$\bar{V}=\sqrt{\frac{L}{C}}\,\bar{I} \quad \text{at} \quad x=l$$

or

$$V=\sqrt{\frac{L}{C}}\,I \quad \text{at} \quad x=l;$$

the impedance matching is equivalent to a resistance of magnitude $\sqrt{(L/C)}$ across the end $x=l$.

6.8. The general semi-infinite transmission line, initially dead

We are now ready to return to eqn (6.23) with V_0 and I_0' zero. Substituting for Z and Y from eqns (6.22),

$$\frac{\mathrm{d}^2\bar{V}}{\mathrm{d}x^2}=LC\left(p^2+p\left(\frac{G}{C}+\frac{R}{L}\right)+\frac{GR}{CL}\right)\bar{V}.$$

Put

$$\frac{1}{LC}=v^2,\quad a=\frac{1}{2}\left(\frac{G}{C}+\frac{R}{L}\right)\quad \text{and}\quad b=\frac{1}{2}\left|\frac{R}{L}-\frac{G}{C}\right|$$

so that

$$\frac{\mathrm{d}^2\bar{V}}{\mathrm{d}x^2}=\frac{1}{v^2}\left[(p+a)^2-b^2\right]\bar{V}; \tag{6.72}$$

b is a measure of how much the line considered differs from the Heaviside line. If $V(0,t)=f(t)$ and $V(x,t)$ remains finite as $x\to+\infty$,

$$\bar{V}(x,p)=\bar{f}(p)\,\mathrm{e}^{-(x/v)\sqrt{[(p+a)^2-b^2]}}. \tag{6.73}$$

We shall make use of tables of Laplace transforms to find $V(t)$. The inverse transform of $\bar{u}(p+a)$ is $\mathrm{e}^{-at}\,u(t)$ and of $\mathrm{e}^{-(x/v)\sqrt{p^2-b^2}}$ is

$$\frac{bx}{v}\left(t^2-\frac{x^2}{v^2}\right)^{-\frac{1}{2}} I_1\left(b\left(t^2-\frac{x^2}{v^2}\right)^{\frac{1}{2}}\right)H\left(t-\frac{x}{v}\right)+\delta\left(t-\frac{x}{v}\right).$$

Hence, with use of the convolution theorem,

$$\begin{aligned}
V(x,t)=&\int_0^t f(t-u)\,\mathrm{e}^{-au}\left[\frac{bx}{v}\left(u^2-\frac{x^2}{v^2}\right)^{-\frac{1}{2}}\right.\\
&\left.\times I_1\left(b\left(u^2-\frac{x^2}{v^2}\right)^{\frac{1}{2}}\right)H\left(u-\frac{x}{v}\right)+\delta\left(u-\frac{x}{v}\right)\right]\mathrm{d}u\\
=&\,\mathrm{e}^{-ax/v}f\left(t-\frac{x}{v}\right)H\left(t-\frac{x}{v}\right)+H\left(t-\frac{x}{v}\right)\frac{bx}{v}\\
&\times\int_{x/v}^t f(t-u)\,\mathrm{e}^{-au}\left(u^2-\frac{x^2}{v^2}\right)^{-\frac{1}{2}} I_1\left(b\left(u^2-\frac{x^2}{v^2}\right)^{\frac{1}{2}}\right)\mathrm{d}u.
\end{aligned} \tag{6.74}$$

The front of the signal travels with velocity $v=1/\sqrt{LC}$ and, if $f(0)\neq 0$, the front has amplitude decreasing as $\mathrm{e}^{-ax/v}$. If $f(t)$ is non-zero only when $0<t<\tau$, then for x fixed $f(t-x/v)$ is only non-zero when $x/v<t<x/v+\tau$, but the integral is non-zero in general for all $t>x/v$. Hence, if a continuing signal is being transmitted, the tail, i.e. the integral, coming from the first part of the signal will interfere with the fronts of signals coming later. However, if $b=0$ the integral vanishes. But $b=0$ when $CR=LG$, the Heaviside line. It is now clear that the Heaviside line is preferable for transmission of signals not only to the Kelvin ideal cable and the non-inductive leaky cable but also to the general cable with C, R, L and G all non-zero with $CR \neq LG$.

6.9. Propagation of discontinuities

In the last section we saw that the solution at the wavefront was relatively simple, but that behind the wavefront the solution was complicated. Can the solution at the wavefront be obtained directly without considering the solution behind the front? The motivation is not only that this might be simpler when the solution behind is complicated but also that it may be possible for equations for which it is not possible to find the general solution.

The following lemma first proved by Jacques Hadamard is fundamental to the analysis of discontinuities.

HADAMARD'S LEMMA. *Let $V(x,t)$ and its partial derivatives $\partial V/\partial x$ and $\partial V/\partial t$ be continuous everywhere in a region of x,t-space except across a curve C. Let a suffix $+$ denote the value of a variable just to the right of and below C, a suffix $-$ to the left and above. Let $[f]=f_- - f_+$. Let $\mathrm{d}/\mathrm{d}t_s$ denote time differentiation following the curve C. Then*

$$\frac{\mathrm{d}}{\mathrm{d}t_s}[V]=\left[\frac{\mathrm{d}V}{\mathrm{d}t_s}\right]. \tag{6.75}$$

Proof. Let $P(x,t)$ be any point on the curve C (Fig. 6.6). Let $Q(x+\mathrm{d}x,\, t+\mathrm{d}t)$ be a neighbouring point on the curve. Then, since V and its derivatives are continuous on either side of C,

$$(V_-)_Q=(V_-)_P+\left(\frac{\partial V}{\partial x}\right)_{-,P}\mathrm{d}z+\left(\frac{\partial V}{\partial t}\right)_{-,P}\mathrm{d}t+\mathrm{O}((\mathrm{d}x)^2+(\mathrm{d}t)^2),$$

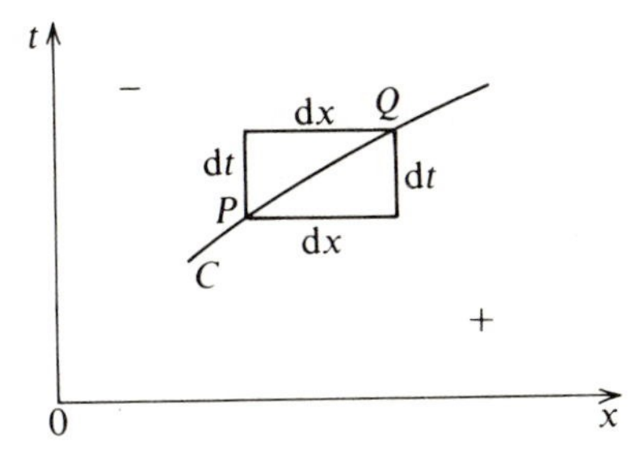

Fig. 6.6

and

$$(V_+)_Q=(V_+)_P+\left(\frac{\partial V}{\partial x}\right)_{+,P}\mathrm{d}x+\left(\frac{\partial V}{\partial t}\right)_{+,P}\mathrm{d}t+\mathrm{O}((\mathrm{d}x)^2+(\mathrm{d}t)^2);$$

on subtraction,

$$[V]_Q=[V]_P+\left[\frac{\partial V}{\partial x}\right]_P\mathrm{d}x+\left[\frac{\partial V}{\partial t}\right]_P\mathrm{d}t+\mathrm{O}((\mathrm{d}x)^2+(\mathrm{d}t)^2). \qquad (6.76)$$

Since $\mathrm{d}t/\mathrm{d}t_s=1$ and $\mathrm{d}x/\mathrm{d}t_s=U$, where U is the velocity of the discontinuity,

$$\frac{\mathrm{d}}{\mathrm{d}t_s}[V]=\lim_{\mathrm{d}t_s\to 0}\frac{[V]_Q-[V]_P}{\mathrm{d}t_s}=\left[\frac{\partial V}{\partial x}\right]U+\left[\frac{\partial V}{\partial t}\right] \qquad (6.77)$$

on substitution from eqn (6.76). Since P is any point, the suffix P has been dropped. Take values of $\mathrm{d}V/\mathrm{d}t_s$ on either side of the discontinuity:

$$\left(\frac{\mathrm{d}V}{\mathrm{d}t_s}\right)_-=\left(\frac{\partial V}{\partial t}\right)_-+U\left(\frac{\partial V}{\partial x}\right)_- \quad\text{and}\quad \left(\frac{\mathrm{d}V}{\mathrm{d}t_s}\right)_+=\left(\frac{\partial V}{\partial t}\right)_++U\left(\frac{\partial V}{\partial x}\right)_+;$$

on subtraction,

$$\left[\frac{\mathrm{d}V}{\mathrm{d}t_s}\right]=\left[\frac{\partial V}{\partial t}\right]+U\left[\frac{\partial V}{\partial x}\right]. \qquad (6.78)$$

Comparison of eqns (6.77) and (6.78) proves the lemma, eqn (6.75). Note that the only derivative of a jump that has meaning is a derivative following the discontinuity, because the jump only exists on the discontinuity.

We now treat the velocity and decay of weak discontinuities. Since the highest-order derivatives occurring in the governing equations

$$L\frac{\partial I}{\partial t}+RI=-\frac{\partial V}{\partial x} \quad (6.11) \quad\text{and}\quad C\frac{\partial V}{\partial t}+GV=-\frac{\partial I}{\partial x} \quad (6.12)$$

are first order, weak discontinuities occur when all derivatives of order less than n, $n\geqslant 1$, are continuous and at least one derivative of order n is discontinuous. The case $n=1$ will be treated. Since V and I are continuous, by Hadamard's lemma

$$0=\frac{\mathrm{d}}{\mathrm{d}t_s}[V]=\left[\frac{\partial V}{\partial t}\right]+U\left[\frac{\partial V}{\partial x}\right] \quad\text{and}\quad 0=\frac{\mathrm{d}}{\mathrm{d}t_s}[I]=\left[\frac{\partial I}{\partial t}\right]+U\left[\frac{\partial I}{\partial x}\right],$$

where U is the velocity of the weak discontinuity. Since the p.d.e.'s hold on both sides of the discontinuity, and since V and I are continuous,

$$L\left[\frac{\partial I}{\partial t}\right]+\left[\frac{\partial V}{\partial x}\right]=0 \quad \text{and} \quad C\left[\frac{\partial V}{\partial t}\right]+\left[\frac{\partial I}{\partial x}\right]=0.$$

For the equations for the four jumps in partial derivatives to have non-zero solutions,

$$\begin{vmatrix} 1 & U & 0 & 0 \\ 0 & 0 & 1 & U \\ 0 & 1 & L & 0 \\ C & 0 & 0 & 1 \end{vmatrix}=0,$$

$$-1+U^2LC=0 \quad \text{or} \quad U=\pm\frac{1}{\sqrt{LC}}. \tag{6.79}$$

The weak discontinuity travels at characteristic velocity. This is consistent with the definition of characteristics, which were defined as curves across which a weak discontinuity might occur. The jumps across a characteristic are related by

$$-\frac{1}{U}\left[\frac{\partial V}{\partial t}\right]=\left[\frac{\partial V}{\partial x}\right]=-L\left[\frac{\partial I}{\partial t}\right]=UL\left[\frac{\partial I}{\partial x}\right]. \tag{6.80}$$

The rate of decay of $[\partial V/\partial x]$ is given by

$$\begin{aligned}\frac{\mathrm{d}}{\mathrm{d}t_s}\left[\frac{\partial V}{\partial x}\right]&=\left[\frac{\mathrm{d}}{\mathrm{d}t_s}\frac{\partial V}{\partial x}\right] \quad \text{by Hadamard's lemma}\\ &=\left[\frac{\partial^2 V}{\partial t\partial x}\right]+U\left[\frac{\partial^2 V}{\partial x^2}\right].\end{aligned} \tag{6.81}$$

Differentiate equations (6.11) and (6.12) with respect to x, take jumps and substitute in eqn (6.81):

$$\frac{\mathrm{d}}{\mathrm{d}t_s}\left[\frac{\partial V}{\partial x}\right]=-UL\left[\frac{\partial^2 I}{\partial t\partial x}\right]-UR\left[\frac{\partial I}{\partial x}\right]-\frac{G}{C}\left[\frac{\partial V}{\partial x}\right]-\frac{1}{C}\left[\frac{\partial^2 I}{\partial x^2}\right].$$

Also

$$\frac{\mathrm{d}}{\mathrm{d}t_s}\left[\frac{\partial V}{\partial x}\right]=UL\frac{\mathrm{d}}{\mathrm{d}t_s}\left[\frac{\partial I}{\partial x}\right]=UL\left[\frac{\partial^2 I}{\partial t\partial x}\right]+U^2L\left[\frac{\partial^2 I}{\partial x^2}\right],$$

using eqns (6.79) and (6.80). Add the last two equations and use $U\left[\frac{\partial I}{\partial x}\right]=\frac{1}{L}\left[\frac{\partial V}{\partial x}\right]$:

$$2\frac{\mathrm{d}}{\mathrm{d}t_s}\left[\frac{\partial V}{\partial x}\right]=-\left(\frac{R}{L}+\frac{G}{C}\right)\left[\frac{\partial V}{\partial x}\right]$$

or

$$\left[\frac{\partial V}{\partial x}\right]=\left[\frac{\partial V}{\partial x}\right]_0 \exp\left(-\frac{1}{2}\left(\frac{R}{L}+\frac{G}{C}\right)t\right), \tag{6.82}$$

where $[\partial V/\partial x]_0$ is the value of $[\partial V/\partial x]$ at $t=0$.

The rate of decay of weak discontinuities is exponential, the 'half-time' being

$$2\left(\frac{R}{L}+\frac{G}{C}\right)^{-1}.$$

When a discontinuity in voltage V travels along the cable, the discontinuity is strong because derivatives of V occur in the governing partial differential equations (6.11) and (6.12). The jump conditions satisfied by V and I were derived in Section 6.3 as

$$LU[I]=[V] \tag{6.13}$$

and

$$CU[V]=[I], \tag{6.14}$$

with U, the shock velocity, given by

$$U=\pm\frac{1}{\sqrt{CL}}. \tag{6.15}$$

The velocity of a strong discontinuity is the same as that of a weak discontinuity, the characteristic velocity. This result holds only for linear theory. The jumps in V and I are not independent; they are related by either of eqns (6.13) and (6.14). Just as in the continuous solution of linear homogeneous p.d.e.'s, the dependent variables in both weakly and strongly discontinuous solutions of these same equations are undetermined to within an arbitrary multiplicative constant. Strong discontinuities are frequently referred to as 'shocks'. This nomenclature comes from fluid mechanics, the branch of applied mathematics in which strong discontinuities were first studied in detail.

In addition to the two jump conditions $LU[I]=[V]$ and $CU[V]=[I]$, eqns (6.11) and (6.12) hold on both sides of the discontinuity. If the equation on one side is subtracted from the equation on the other,

$$L\left[\frac{\partial I}{\partial t}\right]+R[I]=-\left[\frac{\partial v}{\partial x}\right] \tag{6.83}$$

and

$$C\left[\frac{\partial v}{\partial t}\right]+G[V]=-\left[\frac{\partial I}{\partial x}\right]. \tag{6.84}$$

All these jump conditions and Hadamard's lemma are needed to calculate the rates of decay of the jumps in potential and current along the discontinuity.

The rate of decay of the jump in potential is $\mathrm{d}[V]/\mathrm{d}t_s$. By Hadamard's lemma,

$$\frac{\mathrm{d}}{\mathrm{d}t_s}[V]=\left[\frac{\mathrm{d}V}{\mathrm{d}t_s}\right]=\left[\frac{\partial V}{\partial t}+U\frac{\partial V}{\partial x}\right]=\left[\frac{\partial V}{\partial t}\right]+U\left[\frac{\partial V}{\partial x}\right]$$
$$=-\frac{1}{C}\left[\frac{\partial I}{\partial x}\right]-\frac{G}{C}[V]-UL\left[\frac{\partial I}{\partial t}\right]-UR[I]$$

using eqns (6.83) and (6.84). Using eqn (6.15),

$$\frac{\mathrm{d}}{\mathrm{d}t_s}[V]=-UL\left[\frac{\partial I}{\partial t}+U\frac{\partial I}{\partial x}\right]-UR[I]-\frac{G}{C}[V]$$
$$=-UL\left[\frac{\mathrm{d}I}{\mathrm{d}t_s}\right]-UR[I]-\frac{G}{C}[V];$$

hence $$\frac{\mathrm{d}}{\mathrm{d}t_s}[V]+UL\frac{\mathrm{d}}{\mathrm{d}t_s}[I]=-\frac{G}{C}[V]-UR[I]$$

by Hadamard's lemma

and

$$2\frac{\mathrm{d}}{\mathrm{d}t_s}[V]=-\left(\frac{G}{C}+\frac{R}{L}\right)[V] \quad \text{on using } LU\,[I]=[V];$$

Hence

$$[V]=[V]_0\exp\left(-\frac{1}{2}\left(\frac{G}{C}+\frac{R}{L}\right)t_s\right), \tag{6.85}$$

where $[V]_0$ is the value of $[V]$ at $t_s=0$. The 'half-time' for the decay of the potential jumps is $2(G/C+R/L)^{-1}$. The rate of decay for strong discontinuities is the same as for weak discontinuities.

The values of $[I]$ and the jumps in the first-order partial derivatives can now be found.

$$[\mathrm{I}]=\sqrt{\frac{L}{C}}[V]=\sqrt{\frac{L}{C}}[V]_0 \exp\left(-\frac{1}{2}\left(\frac{G}{C}+\frac{R}{L}\right)t_s\right), \tag{6.86}$$

$$\frac{\mathrm{d}}{\mathrm{d}t_s}[V]=-\frac{1}{2}\left(\frac{G}{C}+\frac{R}{L}\right)[V]_0 \exp\left(-\frac{1}{2}\left(\frac{G}{C}+\frac{R}{L}\right)t_s\right) \tag{6.87}$$

and

$$\frac{\mathrm{d}}{\mathrm{d}t_s}[I]=-\frac{1}{2}\left(\frac{G}{C}+\frac{R}{L}\right)\sqrt{\frac{L}{C}}[V]_0 \exp\left(-\frac{1}{2}\left(\frac{G}{C}+\frac{R}{L}\right)t_s\right). \tag{6.88}$$

Equations (6.83) and (6.84),

$$\frac{\mathrm{d}}{\mathrm{d}t_s}[V]=\left[\frac{\partial V}{\partial t}\right]+U\left[\frac{\partial V}{\partial x}\right] \quad \text{and} \quad \frac{\mathrm{d}}{\mathrm{d}t_s}[I]=\left[\frac{\partial I}{\partial t}\right]+U\left[\frac{\partial I}{\partial x}\right]$$

provide four equations for $[\partial V/\partial t]$, $[\partial V/\partial x]$, $[\partial I/\partial t]$ and $[\partial I/\partial x]$ on substitution from eqns (6.87) and (6.88). When these equations have been solved, the jumps in the second derivatives of V and I can be found from the equations obtained by differentiating the p.d.e.'s with respect to x and t, taking jumps and applying Hadamard's lemma to the first derivatives of V and I.

When the discontinuity propagates into a region of zero V and I, then since V, I, and all their derivatives are zero ahead, the value of the jump of any quantity is equal to the value of the quantity just behind the discontinuity. This was the case in Section 6.8.

There is a very important consequence of the fact that both weak and strong discontinuities travel with finite velocity on a transmission line, in fact with the same velocity $(CL)^{-\frac{1}{2}}$. If the line is initially dead and if the voltage V becomes non-zero for $t>0$ at the point taken as $x=0$, in such a way that either V or one of its time derivatives is discontinuous at $x=0$ for $t=0$, then the discontinuities at the front of the voltage waves travel along the transmission line in either direction with velocity $(CL)^{-\frac{1}{2}}$ and the voltage remains zero at any point on the line until one or other of the discontinuities reaches this point. If the point has coordinate a, then

$$V(a,t)=0 \quad \text{for} \quad t<(CL)^{\frac{1}{2}}|a|. \tag{6.89}$$

If $x=0$ is one end of the line $x>0$, then the discontinuity travels in the positive sense only, $a>0$, and eqn (6.89) becomes

$$V(a,t)=0 \quad \text{for} \quad t<(CL)^{\frac{1}{2}}a. \tag{6.90}$$

The same arguments apply to discontinuities suddenly generated at time $t=0$ at position $x=0$ in initially dead media governed by hyperbolic partial differential equations with maximum characteristic or shock velocity U. The media at section $x=a$ remains undisturbed for times t when $t<U^{-1}|a|$. Equation (6.74) is an illustration of these arguments; they will be used again, in particular in Section 7.3.

Exercise 6.5. On a transmission line, governed by

$$L\frac{\partial I}{\partial t}+RI=-\frac{\partial V}{\partial x} \quad \text{and} \quad C\frac{\partial V}{\partial t}+bV=-\frac{\partial I}{\partial x},$$

V, I and their first derivatives are continuous, but some or all of their second derivatives are discontinuous. Find the velocity of propagation of the discontinuities and any equations relating the jumps in the various second derivatives. Determine the rate of decay of $[\partial^2 V/\partial x^2]$.

Answers to exercises

6.3. $V=V_0 \operatorname{erf}\dfrac{x\sqrt{RC}}{2\sqrt{t}}, \quad I=-V_0\sqrt{\dfrac{C}{\pi Rt}}\,e^{-x^2RC/4t}$

6.4. $$V\left(x,\frac{5l}{2v}\right)=e^{-ax/v}f\left(\frac{5l/2-x}{v}\right)+e^{-a(x+2l)/v}f\left(\frac{\frac{1}{2}l-x}{v}\right)H\left(\frac{l}{2}-x\right)$$
$$-e^{-a(2l-x)/v}f\left(\frac{l/2+x}{v}\right).$$

6.5. Velocity of propagation $=\pm 1/\sqrt{LC}$,

$$\left[\frac{\partial^2 V}{\partial x^2}\right]=CL\left[\frac{\partial^2 V}{\partial t^2}\right]=-\sqrt{CL}\left[\frac{\partial^2 V}{\partial x\partial t}\right]=\sqrt{\frac{L}{C}}\left[\frac{\partial^2 I}{\partial x^2}\right]$$
$$=L\sqrt{CL}\left[\frac{\partial^2 I}{\partial t^2}\right]=-L\left[\frac{\partial^2 I}{\partial x\partial t}\right],$$

$$\left[\frac{\partial^2 V}{\partial x^2}\right]=\left[\frac{\partial^2 V}{\partial x^2}\right]_{t=0}\exp\left(-\frac{1}{2}\left(\frac{R}{L}+\frac{G}{C}\right)t\right).$$

7 Linear isentropic isotropic elastodynamics

7.1. Governing equations stated; dilatational and rotational waves

The governing equations of elasticity, which includes thermo-elasticity, come from the analysis of strain, the analysis of stress, and from thermodynamics applied to reversible changes in solids. The analyses of strain and stress are common to all continuous media, solids, liquids and gases. The formulation of the equations is in cartesian tensors.

From the analysis of strain the relation between strain ε_{ij} and displacement u_i is given by the six independent equations

$$\varepsilon_{ij}=\tfrac{1}{2}(u_{j,i}+u_{i,j}), \tag{7.1}$$

where i denotes $\partial/\partial x_i$ and the displacement gradients are sufficiently small to neglect their squares and products compared to their first powers. From the analysis of stress, which uses Newton's laws of motion, the stress tensor σ_{ij} is symmetric and the three equations of momentum become

$$\sigma_{ij,j}+\rho X_i=\rho\frac{\partial^2 u_i}{\partial t^2}, \tag{7.2}$$

where X_i is the body force per unit mass and ρ is the density.

An elastic solid is defined by the two hypotheses that (i) the six independent components of strain and either the temperature T or the specific entropy S form a complete and independent set of variables of state, and that (ii) a thermodynamically reversible path exists between any two states. Taken with the first two laws of thermodynamics, these two hypotheses lead to the seven constitutive equations which express σ_{ij} and T in terms of ε_{ij} and S. The heat conduction equation provides the seventeenth equation for the seventeen unknowns, six ε_{ij}, three u_i, six σ_{ij}, T and S.

In isentropic linear elasticity, changes of entropy are taken to be zero, i.e. $S=0$, and the heat conduction equation is not satisfied.

Elimination of T leads to the six independent stress–strain equations, which for isotropic[†] elastic solids are

$$\sigma_{ij} = \lambda\varepsilon_{kk}\delta_{ij} + 2\mu\varepsilon_{ij}, \tag{7.3}$$

where λ and μ are constants, named after Lamé who first introduced them. The accuracy of the isentropic approximation will be investigated in the next chapter. The derivation of eqns (7.1), (7.2) and (7.3) will be found in text books on elasticity and in introductions to continuum mechanics.

In most applications, the body force X_i does not influence wave motion and it will not be included henceforth. σ_{ij} and ε_{ij} can be eliminated from eqns (7.1) to (7.3) to give

$$\rho\frac{\partial^2 u_i}{\partial t^2} = (\lambda\varepsilon_{kk}\delta_{ij} + 2\mu\varepsilon_{ij})_{,j} = \lambda\varepsilon_{kk,i} + 2\mu\varepsilon_{ij,j}$$

$$= \lambda u_{k,ki} + \mu(u_{j,ij} + u_{i,jj})$$

or

$$\rho\frac{\partial^2 u_i}{\partial t^2} = (\lambda + \mu)u_{j,ji} + \mu u_{i,jj}. \tag{7.4}$$

Equations (7.4) are three equations for the three components of u_i. They can be written in vector notation,

$$\rho\frac{\partial^2 \boldsymbol{u}}{\partial t^2} = (\lambda + \mu)\text{grad div } \boldsymbol{u} + \mu\nabla^2\boldsymbol{u}.$$

Since $\nabla^2\boldsymbol{u} = \text{grad div } \boldsymbol{u} - \text{curl curl } \boldsymbol{u}$,

$$\rho\frac{\partial^2 \boldsymbol{u}}{\partial t^2} = (\lambda + 2\mu)\text{grad div } \boldsymbol{u} - \mu \text{ curl curl } \boldsymbol{u}. \tag{7.5}$$

Equations (7.4) and (7.5) are equivalent. The reader must be prepared to use whichever is the more appropriate in particular circumstances.

Consider the equation

$$\phi = A\mathrm{e}^{\mathrm{i}\omega(t - \alpha n_j x_j)}, \tag{7.6}$$

where A, ω and α are real constants with ω and α positive, and n_j is a real unit vector; $n_j x_j$ is the projection of the vector x_i in the direction

[†] An isotropic material has intrinsic properties which are independent of the direction in which they are measured.

of n_j. If the $0x'_1$ axis is oriented in the n_i direction, then $n_j x_j = x'_1$ and eqn (7.6) becomes

$$\phi = A\mathrm{e}^{\mathrm{i}\omega(t - \alpha x'_1)}. \tag{7.7}$$

Equation (7.7) represents a sinusoidal wave of radian frequency ω and amplitude A travelling in the positive sense in the x'_1 direction with velocity $1/\alpha$. Since all points in any plane whose normal is parallel to $0x'_1$ have the same coordinate x'_1, eqn (7.7) represents a plane wave in three-dimensional space because ϕ is the same for all points in a plane x'_1 constant. It follows that eqn (7.6) represents a plane wave of radian frequency ω and amplitude A travelling in the direction and sense of n_j with velocity $1/\alpha$, the value of ϕ being the same at all points in any plane $n_j x_j = $constant. Similarly

$$u_i = A_i \mathrm{e}^{\mathrm{i}\omega(t - \alpha n_j x_j)} \tag{7.8}$$

represents a displacement vector whose ith component has amplitude A_i travelling as a plane wave of radian frequency ω in the direction and sense of n_j with velocity $1/\alpha$, each component u_i being the same at all points in any plane $n_j x_j = $constant. The letter i is frequently used both as a tensor suffix and to denote $\sqrt{-1}$, distinguished typographically by italic and roman fount respectively.

Look for plane wave solutions of eqn (7.4). On substitution from eqn (7.8) into eqn (7.4),

$$-\omega^2 \rho A_i = -\alpha^2 \omega^2 (\lambda + \mu) A_j n_j n_i - \alpha^2 \omega^2 \mu A_i n_j n_j$$

or

$$(\rho - \alpha^2 \mu) A_i = \alpha^2 (\lambda + \mu) A_j n_j n_i. \tag{7.9}$$

Equation (7.9) is a vector equation. There are two possible solutions: either (i) A_i is parallel to n_i, i.e. $A_i = A n_i$, where A is a scalar; or (ii) A_i is perpendicular to n_i, in which case $A_j n_j = 0$.

Case (i). $A_i = A n_i$, $A_j n_j n_i = A n_i$, $\rho - \alpha^2 \mu = \alpha^2 (\lambda + \mu)$, $\alpha = \sqrt{\rho/(\lambda + 2\mu)}$, and the wave velocity is

$$\frac{1}{\alpha} = \sqrt{\frac{\lambda + 2\mu}{\rho}}.$$

If $0x'_1$ is oriented parallel to n_i, $A'_1 = A$, $A'_2 = A'_3 = 0$, so that

$$u'_1 = A\mathrm{e}^{\mathrm{i}\omega(t - \alpha x'_1)}, \quad u'_2 = u'_3 = 0. \tag{7.10}$$

The dilatation

$$\varepsilon_{ii} = u_{i,\,i} = u'_{j,\,j} = \frac{\partial u'_1}{\partial x'_1} \neq 0,$$

but the components of rotation $\frac{1}{2}(\partial u'_j/\partial x'_i - \partial u'_i/\partial x'_j)$ are all zero. Substituting $A_i = An_i$ back into eqn (7.8), it is seen that

$$u_i = An_i \mathrm{e}^{\mathrm{i}\omega(t - x_j n_j/c_1)}, \tag{7.11}$$

where

$$c_1 = \sqrt{\frac{\lambda + 2\mu}{\rho}} \tag{7.12}$$

represents a sinusoidal wave in which (i) the displacement is parallel to n_j, the direction of propagation of the wave; (ii) the velocity is c_1; and (iii) the dilatation is non-zero but the rotation is zero. The wave is non-dispersive because c_1 is independent of ω. It is usually called a 'dilatational' wave, sometimes 'longitudinal'.

Case (*ii*). $A_j n_j = 0$. Since not all A_i are zero for a wave to exist, from eqn (7.9) $\alpha = \sqrt{\rho/\mu}$. The wave velocity is $\sqrt{\mu/\rho}$. If $0x'_1$ is oriented parallel to n_i, then $A'_1 = 0$, A'_2 and A'_3 are arbitrary and

$$u'_1 = 0, \quad u'_2 = A'_2\, \mathrm{e}^{\mathrm{i}\omega(t - \alpha x'_1)}, \quad u_3 = A'_3 \mathrm{e}^{\mathrm{i}\omega(t - \alpha x'_1)}. \tag{7.13}$$

The dilatation $u'_{j,\,j} = 0$, but the components of rotation

$$\frac{1}{2}\left(\frac{\partial u'_1}{\partial x'_3} - \frac{\partial u'_3}{\partial x'_1}\right) \quad \text{and} \quad \frac{1}{2}\left(\frac{\partial u'_2}{\partial x'_1} - \frac{\partial u'_1}{\partial x'_2}\right)$$

are non-zero. Equation (7.8) with $\alpha = 1/c_2$, where

$$c_2 = \sqrt{\frac{\mu}{\rho}}, \tag{7.14}$$

and $A_i n_i = 0$ represents a sinusoidal wave in which (i) the displacement is in any direction perpendicular to n_i, the direction of propagation, (ii) the velocity is c_2, and (iii) the rotation is non-zero but the dilatation is zero. The wave is again non-dispersive. It is called a 'rotational' or 'transverse' wave or, less frequently, an 'equivoluminal' wave. Note that there are two constants left undetermined in the equation for the rotational wave, not only the amplitude but also the direction of the displacement in the plane of the wave.

Elastic waves, either dilatational or rotational, need not be plane. Using a vector formulation, which is independent of any coordinate system, we start from eqns (7.5). Since the dilatation is given by div $\boldsymbol{u}$, take the divergence of both sides of eqn (7.5):

$$\rho\frac{\partial^2 \operatorname{div} \boldsymbol{u}}{\partial t^2}=(\lambda+2\mu)\operatorname{div}\operatorname{grad}\operatorname{div}\boldsymbol{u}$$

$$=(\lambda+2\mu)\nabla^2\operatorname{div}\boldsymbol{u}, \tag{7.15}$$

using $\operatorname{div}\operatorname{curl}\equiv 0$. Equation (7.15) shows that $\operatorname{div}\boldsymbol{u}$ satisfies the classical wave equation (in three dimensions) with velocity $c_1=\sqrt{(\lambda+2\mu)/\rho}$.

The components of rotation are given by $\operatorname{curl}\boldsymbol{u}$ in cartesian coordinates. This suggests that the operator curl be applied to both sides of eqn (7.5):

$$\rho\frac{\partial^2 \operatorname{curl}\boldsymbol{u}}{\partial t^2}=-\mu\operatorname{curl}\operatorname{curl}\boldsymbol{u}, \tag{7.16}$$

since $\operatorname{curl}\operatorname{grad}\equiv 0$. Since $\operatorname{curl}\operatorname{curl}\boldsymbol{a}=\operatorname{grad}\operatorname{div}\boldsymbol{a}-\nabla^2\boldsymbol{a}$, eqn (7.16) can be written

$$\rho\frac{\partial^2 \operatorname{curl}\boldsymbol{u}}{\partial t^2}=\mu\nabla^2\operatorname{curl}\boldsymbol{u}. \tag{7.17}$$

Each rectangular cartesian component of $\operatorname{curl}\boldsymbol{u}$ satisfies the classical wave equation with velocity $c_2=\sqrt{\mu/\rho}$. Since the classical wave equation can be referred to curvilinear coordinate systems, solutions representing non-plane dilatational and rotational waves can be found, see for example Section 7.3.

Equation (5) can also be treated by means of the Helmholtz resolution of a vector, which expresses any vector $\boldsymbol{u}$ in the form

$$\boldsymbol{u}=\operatorname{grad}\phi+\operatorname{curl}\boldsymbol{H} \tag{7.18}$$

with

$$\operatorname{div}\boldsymbol{H}=0. \tag{7.19}$$

ϕ is known as the Lamé scalar potential, $\boldsymbol{H}$ as the Lamé vector potential. Substitute from eqn (7.18) into eqn (7.5):

$$\rho\operatorname{grad}\frac{\partial^2\phi}{\partial t^2}+\rho\operatorname{curl}\frac{\partial^2\boldsymbol{H}}{\partial t^2}=(\lambda+2\mu)\operatorname{grad}\nabla^2\phi-\mu\operatorname{curl}\operatorname{curl}\operatorname{curl}\boldsymbol{H}, \tag{7.20}$$

since div curl$\equiv 0$ and curl grad$\equiv 0$. Equation (7.20) is satisfied if

$$\rho\frac{\partial^2\phi}{\partial t^2}=(\lambda+2\mu)\nabla^2\phi \tag{7.21}$$

and

$$\frac{\partial^2 \boldsymbol{H}}{\partial t^2}+\frac{\mu}{\rho}\operatorname{curl}\operatorname{curl}\boldsymbol{H}=0.$$

The last equation, since div $\boldsymbol{H}=0$, is equivalent to

$$\frac{\partial^2 \boldsymbol{H}}{\partial t^2}=\frac{\mu}{\rho}\nabla^2\boldsymbol{H}. \tag{7.22}$$

The potentials ϕ and $\boldsymbol{H}$ are much used in elastodynamics. Once ϕ and $\boldsymbol{H}$ are found, $\boldsymbol{u}$ follows from eqn (7.18), ε_{ij} from eqns (7.1), and σ_{ij} from eqns (7.2) to complete the solution.

Elastic waves are generally caused by stresses applied to or displacements imposed on the boundaries.† One of the simplest examples is the semi-infinite elastic solid $x>0$ at rest, unstrained for $t<0$, and subject to a given normal pressure $p(t)$ on its boundary $x=0$ for $t>0$. The boundary conditions can be written

$$\sigma_{xx}=-p(t)\,H(t)\quad\text{and}\quad\sigma_{xy}=\sigma_{xz}=0\quad\text{for } x=0. \tag{7.23}$$

Since the boundary conditions are independent of y and z and the stress vector on the boundary is in the x-direction, look for a solution for the displacement vector (u, v, w) of the form

$$u=u(x,t),\quad v=w=0. \tag{7.24}$$

The dilatation is $\partial u/\partial x$, the rotation is zero. The easiest approach to a solution is to substitute from eqns (7.24) into eqns (7.4). The equations for $i=2$ and 3 are satisfied identically and the equation for $i=1$ gives

$$\rho\frac{\partial^2 u}{\partial t^2}=(\lambda+2\mu)\frac{\partial^2 u}{\partial x^2}. \tag{7.25}$$

The solution of eqn (7.25) which represents a wave moving in the positive sense is

$$u=f\left(t-\frac{x}{c_1}\right),\quad c_1=\sqrt{\frac{\lambda+2\mu}{\rho}}. \tag{7.26}$$

---

† The stress (vector) $\sigma_{ij}n_j$ acting on a boundary, outward normal n_i, is frequently referred to as the 'traction'.

The form of f is chosen to fit the boundary condition at $x=0$. The non-zero components of strain and stress by eqns (7.1) and (7.2) are

$$\varepsilon_{xx}=\frac{\partial u}{\partial x}=-\frac{1}{c_1}f'\left(t-\frac{x}{c_1}\right), \tag{7.27}$$

$$\sigma_{xx}=-\frac{\lambda+2\mu}{c_1}f'\left(t-\frac{x}{c_1}\right), \quad \sigma_{yy}=\sigma_{zz}=-\frac{\lambda}{c_1}f'\left(t-\frac{x}{c_1}\right). \tag{7.28}$$

The boundary conditions, eqns (7.23), are satisfied if

$$-\frac{\lambda+2\mu}{c_1}f'(t)=-p(t)\,H(t). \tag{7.29}$$

Hence[†]

$$f(t)=\frac{c_1}{\lambda+2\mu}\int_0^t p(\tau)\mathrm{d}\tau\, H(t) \tag{7.30}$$

and

$$u(x,t)=[\rho(\lambda+2\mu)]^{-\frac{1}{2}}\int_0^{t-x/c_1} p(\tau)\mathrm{d}\tau\, H\left(t-\frac{x}{c_1}\right). \tag{7.31}$$

From eqns (7.28) and (7.29),

$$\begin{aligned}\sigma_{xx}(x,t)&=-p\left(t-\frac{x}{c_1}\right)H\left(t-\frac{x}{c_1}\right),\\ \sigma_{yy}=\sigma_{zz}&=-\frac{\lambda}{\lambda+2\mu}p\left(t-\frac{x}{c_1}\right)H\left(t-\frac{x}{c_1}\right)\end{aligned} \tag{7.32}$$

The normal stress acting on the boundary propagates through the solid at velocity c_1, the dilatational velocity; so do the 'reaction' stresses σ_{yy} and σ_{zz} which are necessary to maintain $\varepsilon_{yy}=\varepsilon_{zz}=0$.

7.2. Rayleigh waves

The dilatational and rotational waves are frequently referred to as body waves. They travel effectively within an infinite elastic body, as no conditions on any boundary of the body have been imposed. For waves travelling in a semi-infinite elastic solid, $z>0$, of which the

---

[†] The arbitrary constant which can be added to $f(t)$ represents a fixed rigid body displacement, not a wave.

surface $z=0$ is unloaded, i.e. stress-free, conditions are imposed on the surface $z=0$. However there is a compensating factor; terms of the form e^{-az}, $\mathrm{R}[a]>0$, may be included in the equations for the displacement components u_i because e^{-az} only tends to infinity as $z\to-\infty$ and this is impossible within the solid.

Consider waves travelling in the x-direction on the surface of the solid $z\geqslant 0$ (Fig. 7.1). There is no dependence on the y-coordinate. Look for a solution in which the Lamé potentials ϕ and $\boldsymbol{H}$ are of the forms

$$\phi=\phi(x,z,t) \quad \text{and} \quad \boldsymbol{H}\equiv(0,-\psi(x,z,t),0). \tag{7.33}$$

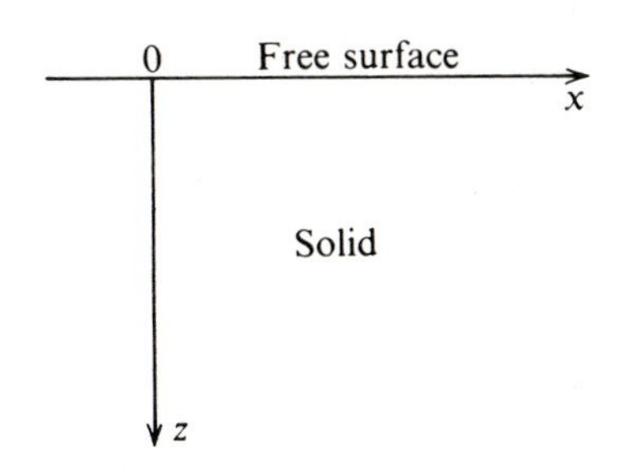

Fig. 7.1

Equation (7.19), $\operatorname{div}\boldsymbol{H}=0$, is satisfied. The displacement components, by eqn (7.18), are

$$u=\frac{\partial\phi}{\partial x}+\frac{\partial\psi}{\partial z}, \quad v=0 \quad \text{and} \quad w=\frac{\partial\phi}{\partial z}-\frac{\partial\psi}{\partial x}, \tag{7.34}$$

and eqns (7.21) and (7.22) for ϕ and for $\psi(=-H_y)$ become

$$\frac{\partial^2\phi}{\partial t^2}-c_1^2\left(\frac{\partial^2\phi}{\partial x^2}+\frac{\partial^2\phi}{\partial z^2}\right)=0 \tag{7.35}$$

and

$$\frac{\partial^2\psi}{\partial t^2}-c_2^2\left(\frac{\partial^2\psi}{\partial x^2}+\frac{\partial^2\psi}{\partial z^2}\right)=0. \tag{7.36}$$

Look for solutions which represent a sinusoidal wave in the x-direction, i.e.

$$\phi=A\,\mathrm{e}^{-qz}\,\mathrm{e}^{\mathrm{i}(\omega t-kx)} \quad \text{and} \quad \psi=B\,\mathrm{e}^{-sz}\,\mathrm{e}^{\mathrm{i}(\omega t-kx)}. \tag{7.37}$$

ω and k are both real and positive and are the same for both ϕ and ψ as a solution representing a single wave is sought. However, there is no reason why ϕ and ψ should have the same decay in the z-direction. It is only required that

$$\mathrm{R}[q]>0 \quad \text{and} \quad \mathrm{R}[s]>0. \tag{7.38}$$

Substitution from eqn (7.37) into eqns (7.35) and (7.36) gives

$$q^2-k^2=-\omega^2/c_1^2 \tag{7.39}$$

and

$$s^2-k^2=-\omega^2/c_2^2. \tag{7.40}$$

The surface $z=0$ is stress-free. Therefore $\sigma_{xz}=\sigma_{yz}=\sigma_{zz}=0$ on $z=0$. $\sigma_{yz}=2\mu\varepsilon_{yz}=\mu(\partial w/\partial y+\partial v/\partial z)=0$ everywhere, in particular on $z=0$.

$$\begin{aligned}\sigma_{zz}&=\lambda(\varepsilon_{xx}+\varepsilon_{yy}+\varepsilon_{zz})+2\mu\varepsilon_{zz}\\&=\lambda\left(\frac{\partial u}{\partial x}+\frac{\partial w}{\partial z}\right)+2\mu\frac{\partial w}{\partial z}\\&=\lambda\left(\frac{\partial^2\phi}{\partial x^2}+\frac{\partial^2\phi}{\partial z^2}\right)+2\mu\left(\frac{\partial^2\phi}{\partial z^2}-\frac{\partial^2\psi}{\partial x\partial z}\right)\end{aligned} \tag{7.41}$$

and

$$\begin{aligned}\sigma_{xz}&=2\mu\varepsilon_{xz}=\mu\left(\frac{\partial w}{\partial x}+\frac{\partial u}{\partial z}\right)\\&=\mu\left(2\frac{\partial^2\phi}{\partial x\partial z}-\frac{\partial^2\psi}{\partial x^2}+\frac{\partial^2\psi}{\partial z^2}\right).\end{aligned} \tag{7.42}$$

Substitute for ϕ and ψ from eqns (7.37) into the right-hand sides of eqns (7.41) and (7.42); equate to zero and put $z=0$ (i.e. $\sigma_{zz}=\sigma_{xz}=0$ on $z=0$):

$$A[(\lambda+2\mu)q^2-\lambda k^2]-B\,2\mu iks=0$$

and

$$A\,2ikq+B(s^2+k^2)=0. \tag{7.43}$$

Note that the factor $e^{i(\omega t-kx)}$ has been divided out from eqns (7.43). This would not have been possible if ϕ and ψ had not had the same x and t dependence. For there to exist solutions for A and B from eqns (7.43) other than $A=B=0$,

$$[(\lambda+2\mu)q^2-\lambda k^2]\,(s^2+k^2)=4\mu k^2sq.$$

Square both sides and substitute for q^2 and s^2 from eqns (7.39) and (7.40):

$$[2\mu k^2-(\lambda+2\mu)\omega^2/c_1^2]^2\,(2k^2-\omega^2/c_2^2)^2$$
$$=16\mu^2k^4(k^2-\omega^2/c_1^2)(k^2-\omega^2/c_2^2).$$

Divide through by μ^2k^8 and use $(\lambda+2\mu)/\mu=c_1^2/c_2^2$:

$$\left(2-\frac{\omega^2}{c_2^2k^2}\right)^2\left(2-\frac{\omega^2}{c_2^2k^2}\right)^2=16\left(1-\frac{\omega^2}{c_1^2k^2}\right)\left(1-\frac{\omega^2}{c_2^2k^2}\right). \quad (7.44)$$

Put

$$\frac{c_2^2}{c_1^2}=\alpha^2,\quad \xi^2=\frac{\omega^2}{c_2^2k^2}: \quad (7.45)$$

$$(2-\xi^2)^4=16(1-\alpha^2\xi^2)(1-\xi^2)$$

or

$$\xi^6-8\xi^4+(24-16\alpha^2)\,\xi^2+16\alpha^2-16=0, \quad (7.46)$$

which is a cubic for ξ^2, independent of ω.

The wave velocity $=c_s=\omega/k=\xi c_2$. Since ξ is independent of ω, c_s is independent of ω and the waves are non-dispersive. The l.h.s. of eqn (7.46) is negative for $\xi^2=0$, since $\alpha^2<1$, and positive for $\xi^2=1$. Hence there is one root ξ_1 of eqn (7.46), for which $0<\xi_1<1$. From eqns (7.39), (7.40) and (7.45),

$$q^2/k^2=1-\alpha^2\xi^2 \quad (7.47) \quad \text{and} \quad s^2/k^2=1-\xi^2 \quad (7.48)$$

Since $\alpha^2<1$, q^2 and s^2 are positive for $0<\xi_1<1$ and q and s can be chosen positive.

The values of ξ_1 and the other two roots ξ_2^2 and ξ_3^2 are given in Table 7.1 for $\nu=0(0.05)0.5$.

$$\alpha^2=\frac{c_2^2}{c_1^2}=\frac{\mu}{\lambda+2\mu}=\frac{1-2\nu}{2(1-\nu)}. \quad (7.49)$$

When ξ_2^2 and ξ_3^2 are real, they are greater than unity, and then by eqn (7.48) s^2 is negative and so s cannot be chosen with negative real part. When ξ_2^2 and ξ_3^2 are complex, by the second of eqns (7.45) k is complex, contrary to assumption. There is therefore one and only one root ξ for each ν, $0\leqslant\nu<0.5$, corresponding to a sinusoidal wave travelling in the x-direction with amplitude decreasing with depth, i.e. one and only one surface (Rayleigh) wave. Since frequency, amplitude, and phase are arbitrary, non-sinusoidal Rayleigh waves are possible; their velocity depends only on the Poisson ratio of the solid.

Table 7.1

ν	0	0.05	0.10	0.15	0.20	0.25	0.30	0.35	0.40	0.45	0.50
ξ_1	0.874	0.884	0.893	0.902	0.911	0.919	0.927	0.935	0.942	0.949	0.955
$R[\xi_2^2]$	2.00	2.11	2.25	2.43	2.69	3.15	3.57	3.56	3.56	3.55	3.54
$I[\xi_2^2]$	0	0	0	0	0	0	0.74	1.18	1.54	1.88	2.23
$R[\xi_3^2]$	5.24	5.11	4.95	4.75	4.48	4.00	3.57	3.56	3.56	3.55	3.54
$I[\xi_3^2]$	0	0	0	0	0	0	−0.74	−1.18	−1.54	−1.88	−2.23

7.3. Uniform pressure on spherical cavity

Consider a spherical cavity of radius a in an infinite elastic solid subject to an internal pressure $p_0 H(t)$ at all points on the cavity surface; p_0 constant. The problem is to find the displacement and stress throughout the solid for all positive times t. The problem is spherically symmetric, so a solution is sought in which the displacement is radial of the form $(U(r, t),\ 0,\ 0)$ and in which the principal stresses are σ_{rr}, $\sigma_{\theta\theta}$ and $\sigma_{\phi\phi}$ with $\sigma_{\theta\theta}=\sigma_{\phi\phi}$, and both displacement and stress are referred to spherical polar coordinates with origin at the centre of the cavity.

Since spherically symmetric motions involve dilatation but not rotation, look for a solution in which the scalar Lamé potential ϕ is of the form $\phi(r, t)$ and the vector potential $\boldsymbol{H}$ is identically zero. From eqn (7.21), under spherical symmetry,

$$\frac{\partial^2\phi}{\partial t^2}=c^2\left(\frac{\partial^2\phi}{\partial r^2}+\frac{2}{r}\frac{\partial\phi}{\partial r}\right), \quad c=\sqrt{\frac{\lambda+2\mu}{\rho}}. \tag{7.50}$$

The substitution $\phi=(1/r)\chi$ reduces the p.d.e. to

$$\frac{\partial^2\chi}{\partial t^2}=c^2\frac{\partial^2\chi}{\partial r^2} \tag{7.51}$$

with solution $\chi=g(t-(r-a)/c)$ for an outgoing wave.† Hence

$$\phi=\frac{1}{r}g\left(t-\frac{r-a}{c}\right), \tag{7.52}$$

where g is an arbitrary function of its argument. From eqn (7.18) in spherical polars, with ϕ given by eqn (7.52) and with $\boldsymbol{H}\equiv 0$,

$$U(r, t)=\frac{\partial\phi}{\partial r}=-\frac{1}{rc}g'\left(t-\frac{r-a}{c}\right)-\frac{1}{r^2}g\left(t-\frac{r-a}{c}\right). \tag{7.53}$$

The characteristics of eqn (7.50) are $r=\pm ct+\text{constant}$. No signal can travel with velocity greater than c. Hence $U(r, t)=0$ for $r-a>ct$. Since U is continuous everywhere provided there is no break in the solid, $U=0$ for $r-a=ct$. Substitution of $t=(r-a)/c$ in eqn (7.53)

---

† Since a boundary condition is applied at $r=a$, it is convenient to choose an argument of g which reduces to t at $r=a$. It also reduces to zero at $r-a=ct$! The solution for χ is D'Alembert's solution again.

gives

$$0 = -\frac{1}{rc}g'(0) - \frac{1}{r^2}g(0),$$

and for this equation to be true for all r,

$$g(0) = g'(0) = 0. \tag{7.54}$$

The components of stress, σ_{rr} and $\sigma_{\theta\theta}$, are found by observing that $\sigma_{rr}(r, t) = \sigma_{xx}(r, 0, 0, t)$ and $\sigma_{\theta\theta}(r, t) = \sigma_{yy}(r, 0, 0, t)$. The cartesian components of displacement are $(x/r)U$, $(y/r)U$ and $(z/r)U$. Hence

$$\varepsilon_{xx} = \frac{\partial}{\partial x}\left(\frac{x}{r}U\right) = \left(\frac{1}{r} - \frac{x^2}{r^3}\right)U + \frac{x^2}{r^2}\frac{\partial U}{\partial r}$$

with similar expressions for ε_{yy} and ε_{zz}. Since $\sigma_{xx} = \lambda\Delta + 2\mu\varepsilon_{xx}$ and $\sigma_{yy} = \lambda\Delta + 2\mu\varepsilon_{yy}$,

$$\sigma_{rr}(r, t) = \lambda\Delta + 2\mu\frac{\partial U}{\partial r} \quad \text{and} \quad \sigma_{\theta\theta}(r, t) = \lambda\Delta + 2\mu\frac{U}{r}. \tag{7.55}$$

In spherical symmetry,

$$\begin{aligned}\Delta = \operatorname{div} \boldsymbol{u} &= \frac{\partial U}{\partial r} + 2\frac{U}{r} \\ &= \frac{1}{rc}g''\left(t - \frac{r-a}{c}\right),\end{aligned} \tag{7.56}$$

on substitution from eqn (7.53). Substitute from eqns (7.53) and (7.56) into eqns (7.55):

$$\sigma_{rr}(r, t) = \frac{\lambda + 2\mu}{rc^2}g''\left(t - \frac{r-a}{c}\right) + \frac{4\mu}{r^2c}g'\left(t - \frac{r-a}{c}\right) + \frac{4\mu}{r^3}g\left(t - \frac{r-a}{c}\right) \tag{7.57}$$

and

$$\sigma_{\theta\theta}(r, t) = \frac{\lambda}{rc^2}g''\left(t - \frac{r-a}{c}\right) - \frac{2\mu}{r^2c}g'\left(t - \frac{r-a}{c}\right) - \frac{2\mu}{r^3}g\left(t - \frac{r-a}{c}\right). \tag{7.58}$$ †

† From eqns (7.57), (7.58) and (7.56), $\sigma_{rr} + 2\sigma_{\theta\theta} = (3\lambda + 2\mu)\Delta$, a partial check on the algebra so far.

Since

$$\sigma_{rr}(a,t) = -p_0 H(t),$$

$$\frac{\lambda+2\mu}{ac^2} g''(t) + \frac{4\mu}{a^2 c} g'(t) + \frac{4\mu}{a^3} g(t) + p_0 H(t) = 0, \tag{7.59}$$

an o.d.e. with constant coefficients for $g(t)$ with boundary conditions given by eqn (7.54).

Introduce a dimensionless time τ, where

$$\tau = \frac{ct}{a}, \quad \text{so that} \quad \frac{\mathrm{d}}{\mathrm{d}t} = \frac{c}{a}\frac{\mathrm{d}}{\mathrm{d}\tau}, \tag{7.60}$$

a dimensionless constant α such that

$$\alpha = \frac{4\mu}{\lambda+2\mu} = 4 - \frac{2}{1-\nu}, \tag{7.61}$$

where ν is Poisson's ratio ($3 > \alpha > 0$ for $-1 < \nu < \frac{1}{2}$), and put

$$\beta = \frac{p_0 a^3}{\lambda + 2\mu}. \tag{7.62}$$

Equation (7.59), on putting $g(a\tau/c) = G(\tau)$, becomes

$$G''(\tau) + \alpha G'(\tau) + \alpha G(\tau) + \beta H(\tau) = 0, \tag{7.63}$$

with the boundary conditions (7.54) becoming

$$G(0) = G'(0) = 0. \tag{7.64}$$

The general solution of eqn (7.63) is

$$G(\tau) = \left(A\,\mathrm{e}^{\gamma\tau} + B\,\mathrm{e}^{\bar{\gamma}\tau} - \frac{\beta}{\alpha} \right) H(\tau),$$

where

$$\gamma = -\frac{\alpha}{2} + \mathrm{i}\sqrt{\alpha - \frac{\alpha^2}{4}} \quad \text{and} \quad \bar{\gamma} = -\frac{\alpha}{2} - \mathrm{i}\sqrt{\alpha - \frac{\alpha^2}{4}}. \tag{7.65}$$

Since $3 \geqslant \alpha > 0$, $\alpha > \alpha^2/4$. To satisfy the boundary conditions (7.54),

$$A = \frac{\beta}{\alpha}\frac{\bar{\gamma}}{\bar{\gamma}-\gamma} \quad \text{and} \quad B = \frac{\beta}{\alpha}\frac{\gamma}{\gamma-\bar{\gamma}}.$$

Hence,

$$G(\tau) = \frac{\beta}{\alpha}\left[\frac{1}{\bar{\gamma}-\gamma}(\bar{\gamma}\mathrm{e}^{\gamma\tau} - \gamma\mathrm{e}^{\bar{\gamma}\tau}) - 1 \right] H(\tau), \tag{7.66}$$

with

$$G'(\tau)=\frac{\beta}{\alpha}\frac{\gamma\bar{\gamma}}{\bar{\gamma}-\gamma}(e^{\gamma\tau}-e^{\bar{\gamma}\tau})\ H(\tau) \tag{7.67}$$

and

$$G''(\tau)=\frac{\beta}{\alpha}\frac{\gamma\bar{\gamma}}{\bar{\gamma}-\gamma}(\gamma e^{\gamma\tau}-\bar{\gamma}e^{\bar{\gamma}\tau})\ H(\tau). \tag{7.68}$$

Back-substitution into eqns (7.53), (7.57) and (7.58) with use of eqns (7.60), (7.61), (7.62) and (7.65) determines U, σ_{rr} and $\sigma_{\theta\theta}$ as functions of r and t.

The stresses just behind the wavefront are found by substituting $r=a+ct$ into eqns (7.57) and (7.58). The arguments of g, g' and g'' all become zero. Then $g(0)=G(0)=0$, $g'(0)=(c/a)G'(0)=0$ and $g''(0)=(c^2/a^2)\ G''(0)$. From eqns (7.68), (7.62) and (7.65),

$$g''(0)=\frac{c^2}{a^2}\frac{\beta}{\alpha}(-\gamma\bar{\gamma})=-\frac{c^2}{a^2}\beta=-\frac{p_0 a}{\rho},$$

$$\sigma_{rr}\left(r,\frac{r-a}{c}\right)=-\frac{p_0 a}{r},$$

and

$$\sigma_{\theta\theta}\left(r,\frac{r-a}{c}\right)=-\frac{\lambda}{\lambda+2\mu}\frac{p_0 a}{r}.$$

There is a strong discontinuity of stress at the wavefront, since the stress just ahead of the wavefront is zero.

As $t\to\infty$ at fixed r, $g(t)=G(\tau)\to-\beta/\alpha=-p_0a^3/4\mu$, $g'(t)\to 0$ and $g''(t)\to 0$, as the real parts of γ and $\bar{\gamma}$ are negative. From eqns (7.43), (7.57) and (7.58),

$$U(r,t)\underset{t\to\infty}{\to}\frac{p_0a^3}{4\mu r^2},$$

$$\sigma_{rr}(r,t)\underset{t\to\infty}{\to}-\frac{p_0a^3}{r^3},$$

and

$$\sigma_{\theta\theta}(r,t)\underset{t\to\infty}{\to}\frac{p_0a^3}{2r^3},$$

which is the static solution for the spherical hole in an infinite elastic solid under internal pressure p_0.

Comparison of the equations for σ_{rr} and $\sigma_{\theta\theta}$ $(=\sigma_{\phi\phi})$ derived in this section with eqns (7.32) shows that the fact that the waves in this section are spherical, instead of plane as in Section 7.1, not only causes the stress amplitudes at the wavefront to decay as $1/r$ but also causes the stresses behind the wavefront to vary. This is sometimes described as a 'tail'. For discussion of the detailed structure behind the wavefront, the reader is referred to texts on elastodynamics.†

7.4. Reflection of plane sinusoidal waves at a free plane boundary

It was seen in Section 7.1 that two types of plane elastic body wave exist, the dilatational and the transverse. When one of these waves is incident on a plane boundary, there can be reflected a wave of the same type, or a wave of the other type, or waves of both types. The last possibility will be assumed to be true, unless symmetry arguments suggest otherwise, because it includes the first two possibilities when zero amplitudes occur. The plane boundary will be taken as $x=0$ with the elastic solid occupying the half-space $x>0$.

Case (*i*). *Incident dilatational wave.* Choose the y-axis to lie in the plane of the normal I to the incident wave and of the x-axis. Let α be the angle between I and the x-axis (Fig. 7.2). The normals to the reflected dilatational and rotational waves, denoted by D and R respectively, are assumed to lie in the $0xy$ plane, because there is no reason why they should have positive rather than negative z-com-

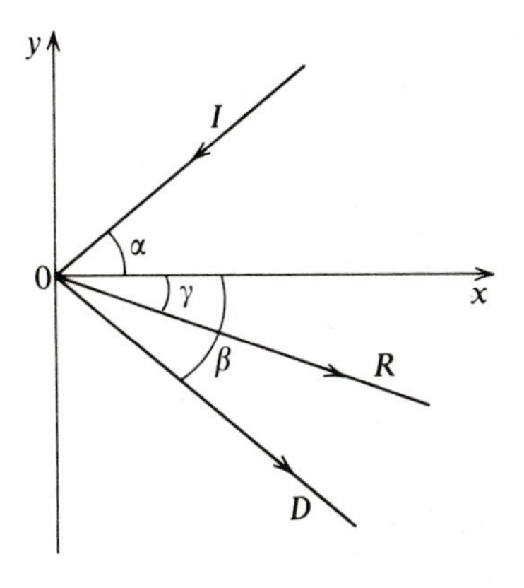

Fig. 7.2

† See, e.g., Eringen, A. C. and Suhubi, E. S. *Elastodynamics*, section 6.8.

ponents or vice versa. Let the angles made by D and R with the x-axis be β and γ respectively. The displacement vector of the incident wave is parallel to its normal I and therefore has no component in the z-direction. Again using the principle of insufficient reason, the displacement vectors of both reflected waves are taken to have no components in the z-direction. We look for a solution of the form described. The reader should note that if such a solution could not be found, the next step would be to look for a more complex solution system such as two reflected dilatational and/or rotational waves with equal and opposite z-components. A free boundary is stress-free so that the boundary conditions are

$$\sigma_{xx}=\sigma_{xy}=\sigma_{xz}=0 \quad \text{on} \quad x=0. \tag{7.69}$$

Since the direction cosines of I, the normal to the incident wave and the direction in which it is travelling, are $(-\cos\alpha, -\sin\alpha, 0)$ and since the wave displacement is parallel to the normal, the displacement components are

$$(u^I, v^I)=A(-\cos\alpha, -\sin\alpha)\exp(i\omega(t-(-x\cos\alpha-y\sin\alpha)/c_1)), \tag{7.70}$$

where ω is the radian frequency and A is the amplitude. Since the wave is dilatational, its velocity is c_1. All the constants of the incident wave are known. The reflected dilatational wave with normal D has components

$$(u^D, v^D)=B(\cos\beta, -\sin\beta)\exp(i\omega_D(t-(x\cos\beta-y\sin\beta)/c_1)), \tag{7.71}$$

where B, β and ω_D are constants to be determined. The normal R to the reflected rotational wave has direction cosines $(\cos\gamma, -\sin\gamma, 0)$; the displacement is perpendicular to R and lies in the x, y-plane by assumption, and the wave velocity is c_2. Hence,

$$(u^R, v^R)=C(\sin\gamma, \cos\gamma)\exp(i\omega_R(t-(x\cos\gamma-y\sin\gamma)/c_2)), \tag{7.72}$$

where C, γ and ω_R are constants to be determined.

The boundary conditions (7.69) apply at all points $(0, y, z)$ on $x=0$ for all times t. A necessary condition for this to be possible is that the exponentials on the r.h.s. of eqns (7.70), (7.71) and (7.72) must have the same form in t, y and z (they then divide out from the boundary conditions), i.e.

$$\omega_D=\omega_R=\omega, \tag{7.73}$$

$$\frac{1}{c_1}\sin\alpha = \frac{1}{c_1}\sin\beta = \frac{1}{c_2}\sin\gamma \tag{7.74}$$

and there is no z-dependence. Equations (7.73) mean that the angular frequencies of both reflected waves are the same as that of the incident wave. From eqns (7.74),

$$\beta = \alpha \tag{7.75}$$

and

$$\gamma = \arcsin\left(\frac{c_2}{c_1}\sin\alpha\right). \tag{7.76}$$

The angle of reflection β of the reflected dilatational wave is equal to the angle of incidence α. Since $0 < c_2/c_1 < 1$, $0 < \gamma < \alpha$ for $\alpha > 0$; $\gamma = 0$ if $\alpha = 0$.

Since the total displacement

$$u_i = u_i^I + u_i^D + u_i^R \tag{7.77}$$

and

$$\sigma_{i1} = \lambda u_{j,j}\,\delta_{i1} + \mu(u_{i,1} + u_{1,i}), \tag{7.78}$$

the condition $\sigma_{xx} = 0$ on $x = 0$ gives

$$\lambda\left(\frac{-\mathrm{i}\omega A}{c_1} - \frac{\mathrm{i}\omega B}{c_1}\right) + 2\mu\left(-\mathrm{i}\omega A\cos^2\frac{\alpha}{c_1} - \mathrm{i}\omega B\cos^2\frac{\beta}{c_1} - \mathrm{i}\omega C\sin\gamma\cos\frac{\gamma}{c_2}\right) = 0$$

or

$$(A+B)\,\frac{\lambda + 2\mu\cos^2\alpha}{c_1} + 2C\,\frac{\mu\sin\gamma\cos\gamma}{c_2} = 0. \tag{7.79}$$

For $\sigma_{xy} = 0$ on $x = 0$, $\partial u/\partial y + \partial v/\partial x = 0$ on $x = 0$, i.e.

$$A\,\frac{-2\sin\alpha\cos\alpha}{c_1} + B\,\frac{2\sin\beta\cos\beta}{c_1} + C\,\frac{\sin^2\gamma - \cos^2\gamma}{c_2} = 0$$

or

$$\frac{B-A}{c_1}\sin 2\alpha - \frac{C}{c_2}\cos 2\gamma = 0. \tag{7.80}$$

σ_{xz} is identically zero. Equations (7.79) and (7.80) determine B/A and C/A, both real since their coefficients in eqns (7.79) and (7.80) are real. Therefore, at the boundary both reflected waves are either in phase or exactly out of phase with the incident wave.

In general, both reflected waves are present. For no reflected rotational wave, $C=0$. Equations (7.79) and (7.80) are only satisfied if $B=-A$ and $\alpha=0$ $(=\beta)$, i.e. normal incidence and the reflected dilatational wave of equal amplitude to and exactly out of phase with the incident wave. The conditions $C=0$, $\alpha=\pi/2$ $(=\beta)$ and $B=-A$ does not represent a reflection—the total displacement is identically zero! There are no angles of incidence α for which $B=0$.

Case (*ii*). *Incident rotational wave with displacement in z-direction.* The displacement in an incident rotational wave is normal to the direction of propagation I in the x, y-plane. It can be resolved into the sum of two displacements, one in the z-direction and one in the x, y-plane normal to I. We first treat the former displacement. It has the form

$$u^I=v^I=0, \quad w^I=A\exp(\mathrm{i}\omega(t-(-x\cos\alpha-y\sin\alpha)/c_2), \qquad (7.81)$$

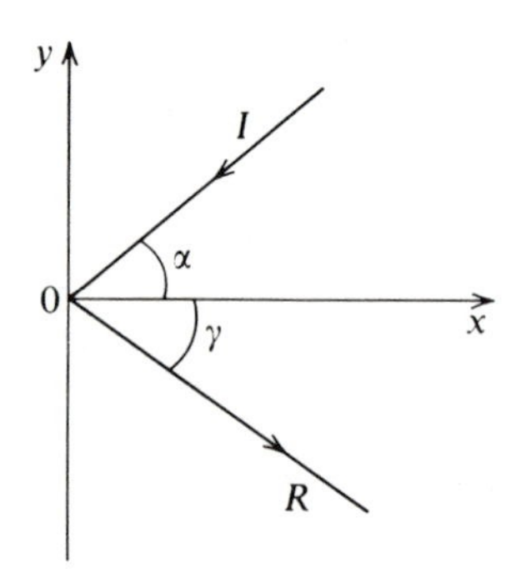

Fig. 7.3

where α is the angle of incidence. Since a dilatational wave travelling in the x, y-plane has zero z-component of displacement, look for a solution in which the only reflected wave is rotational with displacement in the z-direction, i.e.

$$u^R=v^R=0, \quad w^R=C\exp(\mathrm{i}\omega_R(t-(x\cos\gamma-y\sin\gamma)/c_2), \qquad (7.82)$$

where γ is the angle of reflection. For the exponentials to have the same form in t, y and z on $x=0$, $\omega_R=\omega$ and $(\sin\alpha)/c_2=(\sin\gamma)/c_2$ or $\alpha=\gamma$. The angle of incidence equals the angle of reflection. Using eqns (7.77) and (7.78), σ_{xx} and σ_{xy} are identically zero and $\sigma_{xz}=\mu\partial w/\partial x$. Put $\partial w/\partial x=0$ on $x=0$, to give $A\cos\alpha - C\cos\gamma=0$ or $C=A$. The reflected wave is equal in magnitude to an in phase with the incident wave.

Case (iii). Incident rotational wave with displacement in x, y-plane. If α is the angle of incidence, the incident rotational wave has displacement

$$u^I, v^I = A(\sin\alpha, -\cos\alpha)\exp(i\omega(t-(-x\cos\alpha - y\sin\alpha)/c_2)). \qquad (7.83)$$

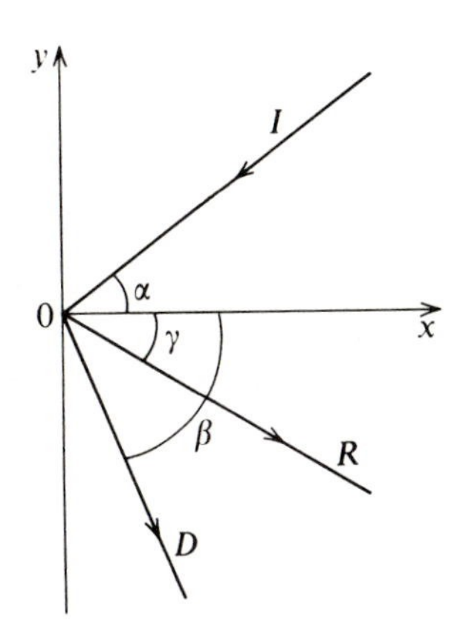

Fig. 7.4

The reflected rotational and dilatational wave displacements are respectively

$$u^R, v^R = C(\sin\gamma, \cos\gamma)\exp(i\omega_R(t-(x\cos\gamma - y\sin\gamma)/c_2)) \qquad (7.84)$$

and

$$u^D, v^D = B(\cos\beta, -\sin\beta)\exp(i\omega_D(t-(x\cos\beta - y\sin\beta)/c_1)). \qquad (7.85)$$

For the exponentials to have the same dependence on t, y and z on $x=0$,

$$\omega_D=\omega_R=\omega \quad \text{and} \quad \frac{\sin\alpha}{c_2}=\frac{\sin\gamma}{c_2}=\frac{\sin\beta}{c_1},$$

or

$$\gamma=\alpha \quad \text{and} \quad \beta=\arcsin\left(\frac{c_1}{c_2}\sin\alpha\right). \qquad (7.86)$$

The frequencies of both reflected waves equal that of the incident wave and the angle of reflection of the reflected rotational wave is equal to the angle of incidence. Since $c_1 > c_2$, β is only real if $(c_1/c_2)\sin\alpha < 1$. Initially we assume this to be the case.

Using eqns (7.77) and (7.78), the conditions $\sigma_{xx}=0$ and $\sigma_{xy}=0$ on $x=0$ give

$$\frac{-\lambda B}{c_1}+2\mu\left(\sin\alpha\cos\alpha\,\frac{A}{c_2}-\cos^2\beta\,\frac{B}{c_1}-\cos\gamma\sin\gamma\,\frac{C}{c_2}\right)=0 \tag{7.87}$$

and

$$A\frac{\sin^2\alpha-\cos^2\alpha}{c_2}+2B\frac{\cos\beta\sin\beta}{c_1}+C\frac{\sin^2\gamma-\cos^2\gamma}{c_2}=0 \tag{7.88}$$

respectively; the condition $\sigma_{xz}=0$ is satisfied everywhere. Equations (7.87) and (7.88) can be written

$$\mu\sin 2\alpha\frac{A-C}{c_2}-(\lambda+2\mu\cos^2\beta)\frac{B}{c_1}=0 \tag{7.89}$$

and

$$-\mu\cos 2\alpha\frac{A+C}{c_2}+\sin 2\beta\,\frac{B}{c_1}=0. \tag{7.90}$$

Equations (7.89) and (7.90) are two equations with real coefficients for the ratios B/A and C/A, which ratios are therefore real. Both reflected waves are either in phase or exactly out of phase with the incident wave. For no reflected dilatational wave, $B=0$; then from eqns (7.89) and (7.90) either $C=A$ and $\alpha=\pi/4$ or $C=-A$ and $\alpha=0$.

Now consider $(c_1/c_2)\sin\alpha>1$. The above analysis remains valid provided a physical interpretation of the results exists; $\sin\beta=(c_1/c_2)\sin\alpha>1$ and $\cos\beta=(1-\sin^2\beta)^{\frac{1}{2}}$ and $\sin 2\beta$ are complex. Put $\cos\beta=\pm i\theta$, $\theta>0$; then $\sin\beta=(1+\theta^2)^{\frac{1}{2}}$. Equation (7.85) becomes

$$u^D, v^D=B(\pm i\theta,\ -\sin\beta)\exp(i\omega(t+y\sin\beta/c_1)\exp(\pm\omega\theta x/c_1). \tag{7.91}$$

For eqn (7.91) to represent a displacement which remains finite as $x\to+\infty$, the lower signs must be taken so that components of displacement decay exponentially in the x-direction. The wave travels in the negative y-direction parallel to the boundary with velocity $c_1\operatorname{cosec}\beta$. Note that $c_1\operatorname{cosec}\beta<c_1$, since $\operatorname{cosec}\beta=1/\sin\beta<1$.

The two components u^D and v^D are $\pi/2$ out of phase. Since $\sin 2\beta = -2i\theta(1+\theta^2)^{\frac{1}{2}}$, a pure imaginary, the coefficient of B in eqn (7.90) is imaginary and hence the ratios B/A and C/A are complex. The reflected waves are no longer in phase, or exactly out of phase, with the incident wave.

The analysis of this section can be extended to the reflection and transmission of plane elastic waves at a plane interface between two elastic solids. To be specific, an incident dilatational wave is treated (see Fig. 7.5). If the interface is $x=0$ and the incident wave I travels in $x>0$ towards the interface, there will be in general two reflected waves D and R and two transmitted waves D' and R'. I, D and R have the forms of eqns (7.70), (7.71) and (7.72). D' and R' have the forms

$$u^{D'}, v^{D'} = B'(-\cos\beta', -\sin\beta')\exp(i\omega_{D'}(t-(-x\cos\beta'-y\sin\beta')/c_1')) \tag{7.92}$$

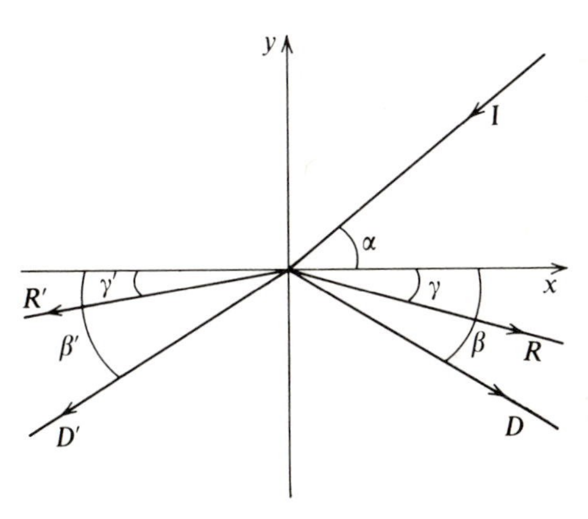

Fig. 7.5

and

$$u^{R'}, v^{R'} = C'(\sin\gamma', -\cos\gamma')\exp(i\omega_{R'}(t-(-x\cos\gamma'-y\sin\gamma')/c_2')), \tag{7.93}$$

where $c_1' = \sqrt{(\lambda'+2\mu')/\rho'}$, $c_2' = \sqrt{\mu'/\rho'}$ and λ', μ' and ρ' are the Lamé constants and density of the elastic medium $x<0$. For eqns (7.70), (7.71), (7.72), (7.92) and (7.93) to have the same dependence on t, y and z for $x=0$,

$$\omega = \omega_D = \omega_R = \omega_{D'} = \omega_{R'} \tag{7.94}$$

and

$$\frac{1}{c_1}\sin\alpha = \frac{1}{c_1}\sin\beta = \frac{1}{c_2}\sin\gamma = \frac{1}{c_1'}\sin\beta' = \frac{1}{c_2'}\sin\gamma'. \tag{7.95}$$

Equations (7.94) show that all five waves have the same frequency ω and eqns (7.95) determine β, γ, β' and γ' in terms of α. If $(c_1'/c_1)\sin\alpha > 1$, then D' travels parallel to the y-axis and decays exponentially as $x \to -\infty$; similarly R' if $(c'_2/c_1)\sin\alpha > 1$. The four equations for the ratios B/A, C/A and B'/A and C'/A are given by the conditions that σ_{xx}, σ_{xy}, u and v are continuous across $x=0$; σ_{xz} and w are identically zero everywhere.

It is possible to give a descriptive account of the propagation of the waves caused by an earthquake. A crude model of the earth is an inner liquid core surrounded by an outer solid mantle. The liquid core transmits only dilatational waves, not rotational ones. The solid mantle transmits both. In seismology, the waves are known as P (dilatational) and S (rotational).†

Suppose an earthquake occurs at some point Q on the earth's surface (Fig. 7.6). Rayleigh waves (R) are transmitted along the surface from Q, and P and S waves are transmitted in all directions

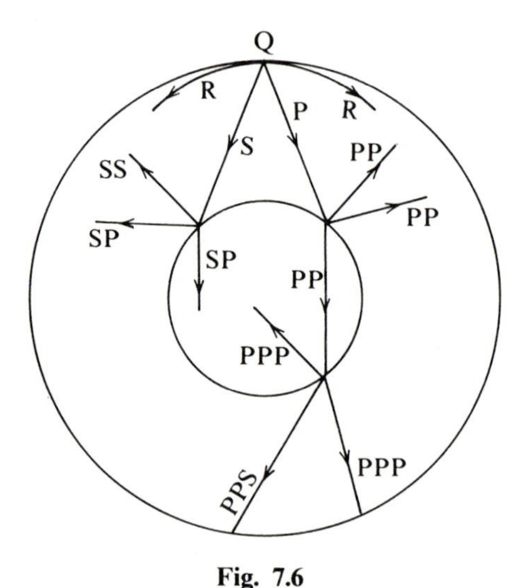

Fig. 7.6

† P for primary, S for secondary; classified by order of arrival.

into the interior. When the P wave strikes the interface between mantle and core, a P wave is transmitted into the core and P and S waves are reflected back into the mantle; these waves are labelled PP, PP and PS respectively in the figure. Similarly, waves labelled SP, SP and SS arise when the original S wave strikes the interface. When the PP wave in the core again strikes the interface, it gives rise to the PPP, PPP, PPS waves shown; and so on. Around the surface of the earth are situated seismological stations in which instruments record the time of arrival, the magnitude and the direction of wave transmitted around or through the earth. One of the main purposes of seismology is to deduce properties of the inside of the earth from these records. The actual structure is, of course, far more complicated in detail than the simple one illustrated in the figure above. In fact it is believed now that there is a solid inner core within the liquid core, as illustrated in Fig. 7.7.

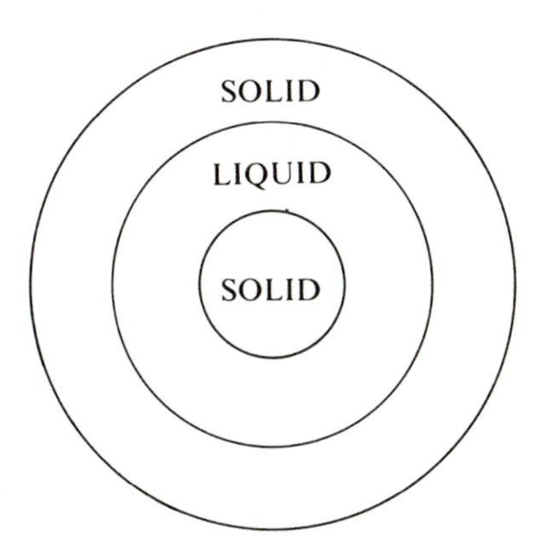

Fig. 7.7

Exercise 7.1. Two semi-infinite elastic media are rigidly joined along their common plane boundary $x=0$ so that the displacement vector is continuous. The density and Lamé constants are ρ, λ and μ for $x>0$, and ρ', λ' and μ' for $x<0$. A wave with displacement components

$$u=v=0, \quad w=A\exp(i\omega(t+(x\cos\alpha+y\sin\alpha)/c)),$$

where $A>0$, $0<\alpha<\frac{1}{2}\pi$ and $c=\sqrt{\mu/\rho}$, travels through the medium in $x>0$ and is incident on the boundary. Given that σ_{xx}, σ_{xy} and σ_{xz} are continuous and that there is one reflected and one transmitted wave, both with $u=v=0$, determine the displacements in both media in the two cases $c \geqslant c'$ and $c < c'$, where $c' = \sqrt{\mu'/\rho'}$.

Exercise 7.2. Love waves. A surface layer occupying the region $h < x_2 < 0$ overlies a semi-infinite solid $x_2 > 0$. The moduli of rigidity and the densities of the materials of the layer and the semi-infinite solid are μ, ρ and μ', ρ' respectively. Assume that $u_1 = 0 = u_2$ in both solid and layer, and show that

$$u_3 = (A\cos sx_2 + B\sin sx_2)\cos(k(x_1 - ct)), \quad h < x_2 < 0,$$

$$u_3 = C\mathrm{e}^{-s'x_2} \quad \cos(k(x_1 - ct)), \qquad x_2 > 0,$$

are possible solutions in the two regions provided

$$\frac{s^2}{k^2} = \frac{c^2}{c_2^2} - 1, \quad \frac{s'^2}{k^2} = 1 - \frac{c^2}{c_2'^2},$$

where

$$c_2^2 = \mu/\rho \quad \text{and} \quad c_2'^2 = \mu'/\rho'.$$

Suppose the surface $x_2 = -h$ of the layer is stress-free and that $\sigma_{12}, \sigma_{22}, \sigma_{32}$ and u_3 are continuous across the interface $x_2 = 0$. Show that the above solution satisfies these conditions if

$$A = -B\cot sh = C$$

and

$$\tan^2 sh = \left(\frac{\mu'}{\mu}\right)^2 \left(1 - \frac{c^2}{c_2'^2}\right) \Bigg/ \left(\frac{c^2}{c_2^2} - 1\right).$$

This last equation gives possible values of the wave velocity c for Love waves. Show that $c_2 < c < c_2'$.

7.5. Longitudinal waves in plates; Rayleigh–Lamb theory

A plate is a solid body bounded by two parallel surfaces, taken as $y = \pm h$. Plane waves propagate in the plate in the x-direction. Plane strain is assumed in the z-direction, i.e. $w = 0$ and u and v are independent of z. Take the Lamé potentials as

$$\phi = \phi(x, y, t) \quad \text{and} \quad \boldsymbol{H} \equiv (0, 0, \psi(x, y, t)),$$

then by eqns (7.18), (7.21) and (7.22),

$$u = \frac{\partial \phi}{\partial x} + \frac{\partial \psi}{\partial y} \quad \text{and} \quad v = \frac{\partial \phi}{\partial y} - \frac{\partial \psi}{\partial x}, \tag{7.96}$$

$$\frac{1}{c_1^2}\frac{\partial^2\phi}{\partial t^2}=\frac{\partial^2\phi}{\partial x^2}+\frac{\partial^2\phi}{\partial y^2} \tag{7.97}$$

and

$$\frac{1}{c_2^2}\frac{\partial^2\psi}{\partial t^2}=\frac{\partial^2\psi}{\partial x^2}+\frac{\partial^2\psi}{\partial y^2}. \tag{7.98}$$

If the waves sought are sinusoidal and travel in the positive x-direction, then ϕ and ψ depend on x and t through $e^{i(\omega t-kx)}$, where ω and k are both real and positive. Look for solutions of the form

$$\phi=Y_1(y)\,e^{i(\omega t-kx)} \quad \text{and} \quad \psi=Y_2(y)\,e^{i(\omega t-kx)}. \tag{7.99}$$

Substitution in eqns (7.97) and (7.98) gives

$$\frac{d^2Y_1}{dy^2}=\alpha^2 Y_1, \tag{7.100}$$

where

$$\alpha^2=k^2-c_1^{-2}\omega^2=k^2-\frac{\rho\omega^2}{\lambda+2\mu}, \tag{7.101}$$

and

$$\frac{d^2Y_2}{dy^2}=\beta^2 Y_2, \tag{7.102}$$

where

$$\beta^2=k^2-c_2^{-2}\omega^2=k^2-\frac{\rho\omega^2}{\mu}. \tag{7.103}$$

The two faces of the plate, $y=\pm h$, are taken to be unloaded, so that

$$\sigma_{xy}=\sigma_{yy}=\sigma_{zy}=0 \quad \text{on} \quad y=\pm h. \tag{7.104}$$

By eqns (7.1) and (7.2), $\sigma_{ij}=\lambda u_{k,k}\delta_{ij}+\mu(u_{j,i}+u_{i,j})$; hence

$$\sigma_{xy}=\mu\left(\frac{\partial u}{\partial y}+\frac{\partial v}{\partial x}\right)=\mu\left(2\frac{\partial^2\phi}{\partial x\partial y}+\frac{\partial^2\psi}{\partial y^2}-\frac{\partial^2\psi}{\partial x^2}\right), \tag{7.105}$$

$$\sigma_{yy}=\lambda\left(\frac{\partial u}{\partial x}+\frac{\partial v}{\partial y}\right)+2\mu\frac{\partial v}{\partial y}=\lambda\left(\frac{\partial^2\phi}{\partial x^2}+\frac{\partial^2\phi}{\partial y^2}\right)$$
$$+2\mu\left(\frac{\partial^2\phi}{\partial y^2}-\frac{\partial^2\psi}{\partial x\partial y}\right), \tag{7.106}$$

and

$$\sigma_{zy}=\mu\left(\frac{\partial u}{\partial z}+\frac{\partial w}{\partial x}\right)=0.$$

A longitudinal wave travelling in the x-direction has u even in y and v odd in y. This requires ϕ even in y and ψ odd in y (see Fig. 7.7). Hence, from eqns (7.100), (7.102) and (7.99),

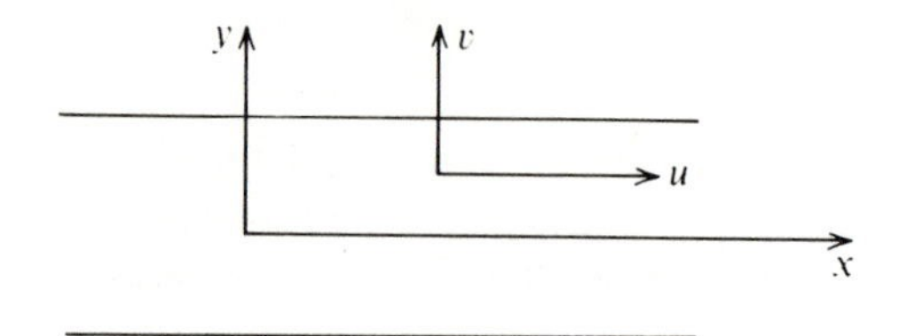

Fig. 7.7

and
$$\left.\begin{aligned}\phi &= A\cosh\alpha y\,\mathrm{e}^{\mathrm{i}(\omega t-kx)}\\ \psi &= B\sinh\beta y\,\mathrm{e}^{\mathrm{i}(\omega t-kx)}\end{aligned}\right\}\quad\text{longitudinal wave.}\tag{7.107}$$

A flexural wave travelling in the x-direction has u odd in y and v even in y. This requires ϕ odd in y and ψ even in y. Hence, from eqns (7.100), (7.102) and (7.99),

and
$$\left.\begin{aligned}\phi &= C\sinh\alpha y\,\mathrm{e}^{\mathrm{i}(\omega t-kx)}\\ \psi &= \mathrm{D}\cosh\beta y\,\mathrm{e}^{\mathrm{i}(\omega t-kx)}\end{aligned}\right\}\quad\text{flexural wave.}\tag{7.108}$$

The longitudinal wave will now be treated. Substitute from eqns (7.107) into the r.h.s. of eqns (7.105) and (7.106). Put $y=\pm h$ so that by eqns (7.104), the l.h.s. are zero. Hence

$$0=\frac{(\sigma_{xy})_{y=h}}{\mu}=(-2\mathrm{i}k\alpha A\sinh\alpha h+(\beta^2+k^2)B\sinh\beta h)\mathrm{e}^{\mathrm{i}(\omega t-kx)}$$

and

$$0=\frac{(\sigma_{yy})_{y=h}}{\mu}=\left(\frac{\lambda}{\mu}(\alpha^2-k^2)A\cosh\alpha h+2(\alpha^2 A\cosh\alpha h\right.$$
$$\left.+\mathrm{i}k\beta B\cosh\beta h)\right)\mathrm{e}^{\mathrm{i}(\omega t-kx)}.$$

Now

$$\frac{\lambda}{\mu}(\alpha^2-k^2)+2\alpha^2=\frac{\lambda+2\mu}{\mu}\alpha^2-\frac{\lambda}{\mu}k^2=\frac{\lambda+2\mu}{\mu}\left(k^2-\frac{\rho\omega^2}{\lambda+2\mu}\right)-\frac{\lambda}{\mu}k^2$$
$$=2k^2-\frac{\rho\omega^2}{\mu}=k^2+\beta^2.$$

Hence,

$$-2ik\alpha \sinh \alpha h\, A + (\beta^2 + k^2) \sinh \beta h\, B = 0 \tag{7.109}$$

and

$$(\beta^2 + k^2) \cosh \alpha h\, A + 2ik\beta \cosh \beta h\, B = 0. \tag{7.110}$$

For there to exist a solution to eqns (7.109) and (7.110) other than $A = B = 0$,

$$4k^2\alpha\beta \sinh \alpha h \cosh \beta h = (\beta^2 + k^2)^2 \sinh \beta h \cosh \alpha h$$

or

$$\frac{\tanh \beta h}{\tanh \alpha h} = \frac{4k^2\alpha\beta}{(\beta^2 + k^2)^2}, \tag{7.111}$$

where α and β are given by eqns (7.101) and (7.103). Equation (7.111) gives the relation between ω and k, the dispersion equation for longitudinal waves in plates. The wave velocity c is given by $c = \omega/k$. Once $\omega = \omega(k)$ has been found, the ratio B/A can be found by back-substitution into either eqn (7.109) or eqn (7.110). If α^2 is negative, put $\alpha = ia$, so that $\tanh \alpha h = i \tan ah$; if β^2 is negative, put $\beta = ib$, so that $\tanh \beta h = i \tan bh$. In all cases eqn (7.111) is real. The behaviour of the dispersion equation for limiting values of ω can be treated analytically.

(i) $|\alpha h|$ *and* $|\beta h| \ll 1$. From eqns (7.101) and (7.103), kh and $\sqrt{(\rho/\mu)}\omega h \ll 1$, since ω and k are both real and positive. This is the low-frequency approximation. For $|\alpha h|$ and $|\beta h|$ small enough to neglect their squares compared to unity, eqn (7.111) becomes $(\beta^2 + k^2)^2 = 4k^2\alpha^2$, on replacing $\tanh \alpha h$ by αh and $\tanh \beta h$ by βh. Substitute for α^2 and β^2 from eqns (7.101) and (7.103):

$$\left(2k^2 - \frac{\rho\omega^2}{\mu}\right)^2 = 4k^2\left(k^2 - \frac{\rho\omega^2}{\lambda + 2\mu}\right)$$

or

$$\frac{\rho\omega^2}{\mu} = 4k^2 \frac{\lambda + \mu}{\lambda + 2\mu}.$$

Therefore,

$$c = \frac{\omega}{k} = 2\sqrt{\frac{\mu(\lambda + \mu)}{\rho(\lambda + 2\mu)}} = \sqrt{\frac{E}{\rho(1 - v^2)}} = c_0. \tag{7.112}$$

The wave velocity c tends to c_0 as $\omega \to 0$. It is easily shown that

$$c_2 < c_0 < c_1. \tag{7.113}$$

(ii) *A high frequency approximation* αh, βh, kh and $\sqrt{(\rho/\mu)}\omega h$ *all* $\gg 1$. Put $\tanh \alpha h$ and $\tanh \beta h$ equal to unity in eqn (7.11), square and substitute for α^2 and β^2 from eqns (7.101) and (7.103) to give

$$16k^4\left(k^2-\frac{\omega^2}{c_1^2}\right)\left(k^2-\frac{\omega^2}{c_2^2}\right)=\left(2k^2-\frac{\omega^2}{c_2^2}\right)^4. \tag{7.114}$$

But eqn (7.114), after division through by k^8, is identical to eqn (7.44). Hence, the high-frequency limit is that

$$c=\frac{\omega}{k}\to c_s, \quad \text{the Rayleigh wave velocity,} \tag{7.115}$$

as $\omega\to\infty$. Since $c_s<c_2<c_1$, $\alpha^2>0$ and $\beta^2>0$ as $\omega\to\infty$, consistent with assumption.

For general values of ω, eqn (7.111) must be solved numerically. $c(\omega)$ decreases monotonically from c_0 to c_s as ω increases from 0 to $+\infty$.

The dispersion curve $c(\omega)$ is sketched in Fig. 7.8 as a full line. Since

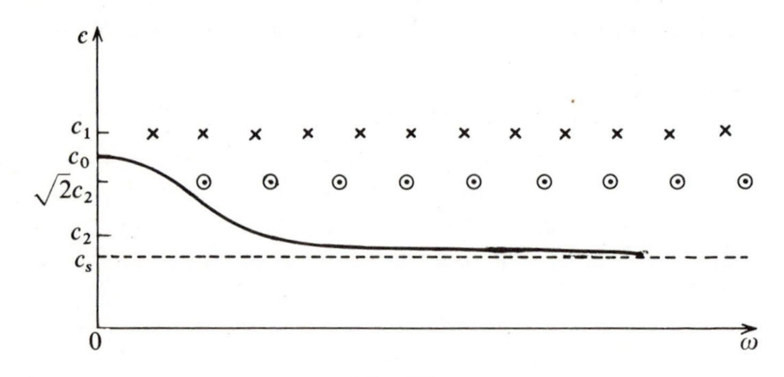

Fig. 7.8

$$\frac{\alpha^2}{k^2}=1-\frac{c^2}{c_1^2} \quad \text{and} \quad \frac{\beta^2}{k^2}=1-\frac{c^2}{c_2^2}, \tag{7.116}$$

$$\alpha^2 \gtreqless 0 \quad \text{for } c \lesseqgtr c_1 \quad \text{and} \quad \beta^2 \gtreqless 0 \quad \text{for} \quad c \lesseqgtr c_2, \tag{7.117}$$

provided k is real.

The displacement components u and v for the longitudinal wave are given by substitution from eqns (7.107) into eqn (7.96) as

$$u=(-\mathrm{i}kA\cosh\alpha y+\beta B\cosh\beta y)\mathrm{e}^{\mathrm{i}(\omega t-kx)} \tag{7.118}$$

and

$$v = (\alpha A \sinh \alpha y + ikB \sinh \beta y)\mathrm{e}^{\mathrm{i}(\omega t - kx)}. \tag{7.119}$$

As $\omega \to 0$, k, α and β are all of order ω and

$$u \underset{\omega \to 0}{\to} (-ikA + \beta B)\mathrm{e}^{\mathrm{i}(\omega t - kx)} \tag{7.120}$$

and

$$v \underset{\omega \to 0}{\to} (\alpha^2 A + ik\beta B) y\, \mathrm{e}^{\mathrm{i}(\omega t - kx)}. \tag{7.121}$$

Therefore, choosing A, and consequently B by eqns (7.109) or (7.110), to be order ω^{-1} as $\omega \to 0$, u is effectively independent of y and $|v/u| \to 0$ as $\omega \to 0$.

As $\omega \to \infty$, α and $\beta \to \infty$, and hence the magnitude of all the hyperbolic functions in eqns (7.118) and 7.119) in the neighbourhood of $y = \pm h$ are exponentially large compared to their respective magnitudes elsewhere in the range $-h < y < +h$. The main part of the wave travels in the surface layers on either side of the plate as $\omega \to \infty$. This phenomenon also occurs in electromagnetism. It is known as the 'skin' effect.

Are there any other branches of the dispersion curve? We first ask whether c is ever equal to c_1, which does not occur on the full curve sketched in Fig. 7.8. If $c = c_1$, $\alpha = 0$ and $\beta^2 < 0$. Put $\beta = ib$, $b > 0$; the dispersion equation (7.111) becomes

$$\tan bh = 0 \quad \text{or} \quad b = \frac{n\pi}{h},$$

where n is a positive integer ($n = 0$ is inconsistent with $\beta^2 < 0$). Since $\beta^2/k^2 = 1 - c_1^2/c_2^2$,

$$k = \frac{c_2 b}{\sqrt{c_1^2 - c_2^2}} = \frac{n\pi c_2}{h\sqrt{c_1^2 - c_2^2}}$$

and

$$\omega = kc_1 = \frac{n\pi c_1 c_2}{h\sqrt{c_1^2 - c_2^2}}. \tag{7.122}$$

These values of ω are represented by the diagonal crosses in Fig. 7.8. The relation between A and B when $\alpha = 0$ is determined from eqn (7.110); eqn (7.109) is satisfied identically, since $\sinh \beta h = \mathrm{i} \sin bh = \mathrm{i} \sin n\pi = 0$. Equation (7.110), after dividing through by k^2, putting

$\alpha=0$ and $\beta=\mathrm{i}b$, becomes

$$\left(1-\frac{b^2}{k^2}\right)A-2\frac{b}{k}\cos n\pi\, B=0$$

or

$$A=-2(-)^n\frac{c_2\sqrt{c_1^2-c_2^2}}{c_1^2-2c_2^2}B. \tag{7.123}$$

It is seen from eqns (7.109) and (7.110) that another special case arises when $\beta^2+k^2=0$. Then from eqn (7.103), $0=2k^2-c_2^{-2}\omega^2$ or $c=\omega/k=\sqrt{2}c_2$, and from eqn (7.101), $\alpha^2/k^2=1-2c_2^2/c_1^2=\lambda/(\lambda+2\mu)>0$. From eqns (7.109) and (7.110), $\sinh\alpha h\, A=0$ and $\cos bh\, B=0$. Hence,

$$A=0 \quad \text{and} \quad bh=(n-\tfrac{1}{2})\pi, \tag{7.124}$$

and

$$\omega=k\sqrt{2}c_2=b\sqrt{2}c_2=\sqrt{2}c_2(n-\tfrac{1}{2})\pi/h. \tag{7.125}$$

These values of ω are represented by dots within circles in Fig. 7.8.

Are there any values of ω for which $c=c_2$ other than that on the full line in Fig. 7.8? If $c=c_2$, $\beta=0$ and eqn (7.111) becomes, on using $c=c_2$ and the second of eqns (7.116),

$$\frac{\tanh\alpha h}{\alpha h}=\frac{k^2}{4\alpha^2}=\frac{c_1^2}{4(c_1^2-c_2^2)}=\frac{\lambda+2\mu}{4(\lambda+\mu)},$$

an equation with only one root for αh positive. However, if the points marked on the figure are points on higher branches of the dispersion curve, then it is possible that on each higher branch $c(\omega)$ tends to c_2 from above as $\omega\to\infty$. This possibility is now investigated.

Put $c=c_2(1+\varepsilon)$, where $0<\varepsilon\ll 1$. Then

$$k=\frac{\omega}{c}=\omega c_2^{-1}(1+\varepsilon)^{-1}, \tag{7.126}$$

and

$$\beta^2=k^2-\omega^2/c_2^2=\frac{\omega^2((1+\varepsilon)^{-2}-1)}{c_2^2}=\frac{-2\varepsilon\omega^2}{c_2^2}+\mathrm{O}(\varepsilon^2).$$

Hence,

$$b=-\mathrm{i}\beta=\sqrt{2}\,\varepsilon^{\frac{1}{2}}\frac{\omega}{c_2}+\mathrm{O}(\varepsilon^{\frac{3}{2}}). \tag{7.127}$$

$$\alpha^2=k^2\left(1-\frac{c^2}{c_1^2}\right)=\frac{\omega^2}{c_2^2}(1+\varepsilon)^{-2}-\frac{\omega^2}{c_1^2}$$

$$=\omega^2\left(\frac{1}{c_2^2}-\frac{1}{c_1^2}-\frac{2\varepsilon}{c_2^2}\right)+\mathrm{O}(\varepsilon^2);$$

hence,

$$\alpha = \frac{\omega\sqrt{c_1^2 - c_2^2}}{c_1 c_2}\left(1 - \frac{c_1^2\varepsilon}{c_1^2 - c_2^2} + \mathrm{O}(\varepsilon^2)\right), \tag{7.128}$$

and

$$\tanh \alpha h \to 1 \quad \text{as } \omega \to \infty. \tag{7.129}$$

Substitution into eqn (7.111) from eqns (7.126) to (7.129) gives

$$\tan bh = \frac{4\omega^2 c_2^{-2}[\omega\sqrt{c_1^2 - c_2^2}/(c_1 c_2)]\sqrt{2}\varepsilon^{\frac{1}{2}}\,\omega/c_2}{\omega^4 c_2^{-4}} + \mathrm{O}(\varepsilon^{\frac{3}{2}})$$

$$\simeq 4\sqrt{2}\sqrt{c_1^2 - c_2^2}\,\varepsilon^{\frac{1}{2}}/c_1 \quad \text{as } \varepsilon \to 0,$$

or

$$bh \sim n\pi + 4\sqrt{2}\sqrt{c_1^2 - c_2^2}\,\varepsilon^{\frac{1}{2}}/c_1 \quad \text{as } \varepsilon \to 0, \quad n \text{ +ve integer.}$$

Therefore, from eqn (7.127),

$$\omega \sim \frac{c_2 b}{\sqrt{2}\,\varepsilon^{\frac{1}{2}}} \sim \frac{4\sqrt{c_1^2 - c_2^2}\,c_2}{c_1 h} + \frac{c_2 n\pi}{\sqrt{2\varepsilon}\,h} \quad \text{as } \varepsilon \to 0. \tag{7.130}$$

There are an (enumerably) infinite number of values of ω for each value of ε and all values of $\omega \to \infty$ as $\varepsilon \to 0$.

On the assumption that all branches of the dispersion curve $c(\omega)$ are monotonic decreasing, the branches can now be sketched for $c(\omega) < c_1$. They are the solid curves in Fig. 7.9.

Since there was only one solution, $c = c_0$, as $\omega \to 0$ for ω and k real and positive, we look for solutions to the dispersion equation in which $c \to +\infty$ for positive ω. If $c \to \infty$, $k = \omega/c = 0$ in the limit. If

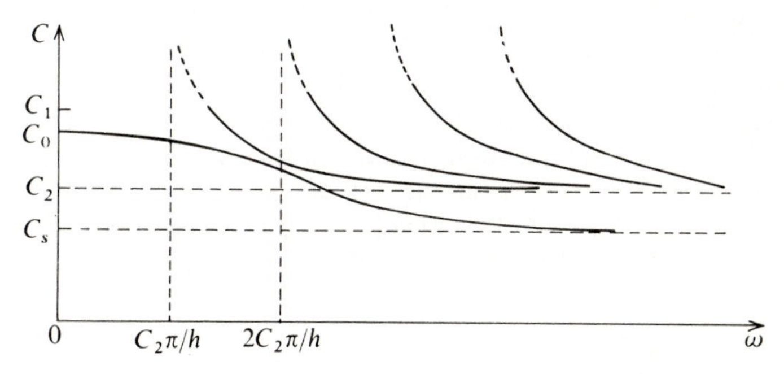

Fig. 7.9

$k=0$, $\alpha^2=-\omega^2/c_1^2$ and $\beta^2=-\omega^2/c_2^2$. Hence $a=-\mathrm{i}\alpha=\omega/c_1$ and $b=-\mathrm{i}\beta=\omega/c_2$. Equations (7.109) and (7.110) reduce respectively to

$$B\sin\frac{\omega h}{c_2}=0 \quad \text{and} \quad A\cos\frac{\omega h}{c_1}=0.$$

Therefore, either

$$A=0 \quad \text{and} \quad \omega=\frac{n\pi c_2}{h}, \qquad n \text{ positive integer} \tag{7.131}$$

or

$$B=0 \quad \text{and} \quad \omega=\frac{(n-\frac{1}{2})\pi c_1}{h}, \qquad n \text{ positive integer.} \tag{7.132}$$

If all branches are to be monotonic decreasing, then the spacing of the values of ω at which $c\to+\infty$ must be less than the spacing of the values of ω for which $c=c_1$. From eqn (7.122), the latter spacing is

$$\frac{\pi c_1 c_2}{h\sqrt{c_1^2-c_2^2}}.$$

Since

$$c_2<\frac{c_1 c_2}{\sqrt{c_1^2-c_2^2}}<c_1,$$

provided $\lambda>0$ or Poisson's ratio lies between 0 and $\frac{1}{2}$, only eqn (7.131) is admissible. Using eqn (7.131), the higher branches of the dispersion curve are completed with dashed lines in Fig. 7.9.

What is the significance of eqn (7.132)? Throughout the analysis of this section it has been assumed that ω and k are real; ω complex corresponds to motions (waves or vibrations) increasing or decreasing in magnitude with time, and k complex corresponds to waves, in motion or stationary, increasing or decreasing in magnitude with distance. We have not been considering such motions. However, $k=0$ is also the limit of complex values of k as $|k|\to 0$ and eqn (7.132) could be the limit of a complex set of values of k. For detailed investigations of eqns (7.111) for complex values of k, the reader is referred to advanced texts on elastodynamics.†

---

† For example, Graff, K. F. *Wave motion in elastic solids*, chapter 8.

7.6. Approximate theories for longitudinal waves in plates

With a dispersion equation as complicated as eqn (7.111), it is difficult to construct solutions to particular problems of longitudinal waves in plates other than those represented by eqns (7.107) or by the sum of such solutions. Because in practice an approximate solution is better than no solution at all, approximate theories have been constructed for longitudinal waves in plates. An engineer often requires estimates of magnitude for a real situation—an exact mathematical solution for a somewhat different problem is not always of value.

The simplest approximate theory assumes that the components of displacement and of stress are functions of x and t only and that only those components of stress are non-zero which are non-zero for static longitudinal tension in the x-direction with plane strain in the z-direction, namely σ_{xx} and σ_{zz}. The equation of motion in the x-direction is

$$\rho \frac{\partial^2 u}{\partial t^2} = \frac{\partial \sigma_{xx}}{\partial x},$$

and the relevant strain–stress equations are

$$\frac{\partial u}{\partial x} = \varepsilon_{xx} = \frac{1}{E}\sigma_{xx} - \frac{\nu}{E}\sigma_{zz}$$

and

$$0 = \varepsilon_{zz} = \frac{1}{E}\sigma_{zz} - \frac{\nu}{E}\sigma_{xx}.$$

Hence,

$$\sigma_{zz} = \nu\sigma_{xx}, \quad \sigma_{xx} = \frac{E}{1-\nu^2}\frac{\partial u}{\partial x} \quad \text{and} \quad \rho\frac{\partial^2 u}{\partial t^2} = \frac{E}{1-\nu^2}\frac{\partial^2 u}{\partial x^2},$$

or, by eqn (7.112),

$$\frac{\partial^2 u}{\partial t^2} = c_0^2 \frac{\partial^2 u}{\partial x^2}; \tag{7.133}$$

the classical wave equation with velocity c_0.

This approximate theory agrees with the Rayleigh–Lamb theory in the limit as $\omega \to 0$ both in the predicted value of the velocity and in the form of u, namely $u = u(x, t)$, as is shown by eqn (7.120). The theory is of value in situations where in the Fourier analysis in time of the imposed boundary conditions the low-frequency components dominate. For larger values of ω, both the wave velocity and the dependence of u upon y cease to be even approximately accurate when predicted by the simplest theory.

Equation (7.133) is a second-order p.d.e. containing one wave operator of the form $\partial^2/\partial t^2 - c_0^2\,\partial^2/\partial x^2$. Now consider a fourth-order p.d.e. containing three such operators of the form

$$\left(\frac{\partial^2}{\partial t^2}-\alpha^2\frac{\partial^2}{\partial x^2}\right)u+\kappa\left(\frac{\partial^2}{\partial t^2}-\beta^2\frac{\partial^2}{\partial x^2}\right)\left(\frac{\partial^2}{\partial t^2}-\gamma^2\frac{\partial^2}{\partial x^2}\right)u=0, \qquad (7.134)$$

where α, β, γ and κ are disposable positive constants with $\gamma>\beta$. To find the dispersion equation corresponding to eqn (7.134), substitute $u=A\,\mathrm{e}^{\mathrm{i}(\omega t-kx)}$:

$$\alpha^2k^2-\omega^2+\kappa(\beta^2k^2-\omega^2)(\gamma^2k^2-\omega^2)=0. \qquad (7.135)$$

The purpose of the analysis that follows is to choose α, β, γ and κ so that the dispersion equation (7.135) reproduces some of the principal features of Fig. 7.9.

Substitute $k=\omega/c$ in eqn (7.135), multiply through by $c^4\omega^{-2}$ to give

$$c^4(\kappa\omega^2-1)-c^2[\kappa(\beta^2+\gamma^2)\omega^2-\alpha^2]+\kappa\beta^2\gamma^2\omega^2=0, \qquad (7.136)$$

with solution

$$c^2(\omega)=\{\tfrac{1}{2}[\kappa(\beta^2+\gamma^2)\omega^2-\alpha^2]\pm\{\tfrac{1}{4}[\kappa(\beta^2+\gamma^2)\omega^2-\alpha^2]^2 - \kappa\beta^2\gamma^2\omega^2(\kappa\omega^2-1)\}^{\frac{1}{2}}\}(\kappa\omega^2-1)^{-1}. \qquad (7.137)$$

As

$$\omega\to 0, \quad c\sim\pm\alpha, \quad \text{or } c\sim\pm\mathrm{i}\beta\gamma\sqrt{\kappa}(\omega/\alpha), \qquad (7.138)$$

as

$$\omega\to+\infty, \quad c\sim\pm\beta, \quad \text{or } c\sim\pm\gamma, \qquad (7.139)$$

and as

$$\omega\to\kappa^{-\frac{1}{2}}, \quad c\sim\pm\infty, \quad \text{or } c\sim\pm\beta\gamma(\beta^2+\gamma^2-\alpha^2)^{-\frac{1}{2}}. \qquad (7.140)$$

The first set of values of c in eqns (7.138) and (7.140) and both sets in eqn (7.139) are easily obtained from eqn (7.137). The second set in eqn (7.138) then follows from the formula for the product of the roots of a quadratic and the second set in eqn (7.140) by putting $\omega^2=1/\kappa$ in eqn (7.136). The roots of the equation for $c^2(\omega)$ are distinct for all values of ω if the discriminant Δ in eqn (7.137) is positive for all positive ω. Since the discriminant is a quadratic in ω^2, this condition

is satisfied if $\Delta(\Delta)$, the discriminant of the discriminant is negative.†
From eqn (7.117),

$$4\Delta=(\beta^2-\gamma^2)^2(\kappa\omega^2)^2-2((\beta^2+\gamma^2)\alpha^2-2\beta^2\gamma^2)\kappa\omega^2+\alpha^4.$$

Hence,

$$\Delta(\Delta)=-\tfrac{1}{4}\beta^2\gamma^2(\gamma^2-\alpha^2)(\alpha^2-\beta^2).$$

The roots are distinct if, using $\gamma>\beta$,

$$\gamma>\alpha>\beta. \tag{7.141}$$

We are now ready to choose the constants α, β and κ so that eqn (7.134) 'approximates' the dispersion curves in Fig. 7.9. Choose $\alpha=c_0$ and $\beta=c_s$ to reproduce the behaviour of the fundamental mode as $\omega\to 0$ and as $\omega\to\infty$. All the higher modes in the figure are replaced by one mode in the approximation. Choose the cut-off frequency of the first high mode, $\pi c_2/h$, to be the cut-off frequency of the approximation, i.e. $\kappa=h^2/(\pi c_2)^2$; γ has yet to be determined. For any value of γ greater than c_2, the curve of the higher mode of the approximation will intersect all the higher modes of the Rayleigh–Lamb theory. The curves of the two modes of the approximation will not intersect if the roots of eqn (7.137) are distinct. By eqn (7.141), this will be the case if

$$\gamma>\alpha=c_0. \tag{7.142}$$

Equation (7.134) has now become

$$\left(\frac{\partial}{\partial t^2}-c_0^2\frac{\partial^2}{\partial x^2}\right)u+\frac{h^2}{\pi^2 t_2^2}\left(\frac{\partial^2}{\partial t^2}-c_s^2\frac{\partial^2}{\partial x^2}\right)\left(\frac{\partial^2}{\partial t^2}-\gamma^2\frac{\partial^2}{\partial x^2}\right)u=0,\ \gamma>c_0. \tag{7.143}$$

There are many examples in applied mathematics where physical phenomena can only be modelled by an approximate theory. Such a theory does not aim to reproduce all the details but it must describe the main physical features. One such feature is stability. If the phenomenon modelled is stable, then the theory must predict stability and vice versa. Longitudinal waves set up in plates do not increase indefinitely in magnitude; does eqn (7.143) predict the same consequence?

---

† $x^2-2bx+c=(x-b)^2+c-b^2$ is >0 for all real x if $b^2<c$. If $b^2>c$ then $x^2-2bx+c=0$ for $x=b\pm\sqrt{b^2-c}$.

Consider $u=\mathrm{R}[\mathrm{e}^{\mathrm{i}(\omega t-kx)}]$. At time $t=0$, $u=\mathrm{R}[\mathrm{e}^{-\mathrm{i}kx}]=\cos kx$ for real k. This is a sinusoidal displacement at time $t=0$ with wave-number k. For u to remain finite at all later times, $\mathrm{I}[\omega]>0$ for all positive k for all possible roots $\omega(k)$. Substitution of $u=\mathrm{R}[\mathrm{e}^{\mathrm{i}(\omega t-kx)}]$ into eqn (7.143) gives the dispersion equation again, but it is now treated as an equation for ω as a function of k:

$$\omega^4-\left(k^2(c_s^2+\gamma^2)+\frac{\pi^2c_2^2}{h^2}\right)\omega^2+c_s^2\gamma^2k^4+\frac{\pi^2c_2^2}{h^2}c_0^2k^2=0. \tag{7.144}$$

The discriminant Δ of this equation for ω^2 is given by

$$4\Delta=k^4(\gamma^2-c_s^2)^2+2k^2\frac{\pi^2c_2^2}{h^2}(\gamma^2+c_s^2-2c_0^2)+\frac{\pi^4c_2^4}{h^4},$$

and the discriminant $\Delta(\Delta)$ of this quadratic form in $k^2h^2/\pi^2c_2^2$ is

$$\Delta(\Delta)=-\tfrac{1}{4}(\gamma^2-c_0^2)(c_0^2-c_s^2).$$

Hence, the discriminant Δ is positive for all k and, hence, from eqn (7.144)†, both roots ω^2 are positive if

$$\gamma>c_0. \tag{7.145}$$

Then $\mathrm{I}[\omega]=0$ for all four roots and the criterion for u to remain finite at all times is satisfied. There are four roots, because eqn (7.143) is fourth-order and for each value of ω, $\mathrm{e}^{\mathrm{i}(\omega t-kx)}$ reduces to the given initial displacement $\mathrm{e}^{-\mathrm{i}kx}$ at $t=0$. Equation (7.145) is the condition for stability. Since eqns (7.145) and (7.143) are identical, the requirement that eqn (7.143) is stable ensures that the two branches of the dispersion curve are distinct.

Faute de mieux, γ is chosen equal to c_1. Equation (7.142) becomes

$$\left(\frac{\partial^2}{\partial t^2}-c_0^2\frac{\partial^2}{\partial x^2}\right)u+\frac{h^2}{\pi^2c_2^2}\left(\frac{\partial^2}{\partial t^2}-c_s^2\frac{\partial^2}{\partial x^2}\right)\left(\frac{\partial^2}{\partial t^2}-c_1^2\frac{\partial^2}{\partial x^2}\right)u=0. \tag{7.146}$$

The branches of the dispersion curve corresponding to eqn (7.146) are plotted in Fig. 7.10).

Comparison of eqns (7.133) and (7.146) shows that the l.h.s. of eqn (7.146) consists of a second-order term identical with the l.h.s. of eqn (7.133) plus a fourth-order term, the latter term being significant at all frequencies except the lowest. Also the constant c_0^2 in eqn (7.133)

---

† The roots of $x^2-bx+c=0$ are both positive when $b^2>4c>0$ and $b>0$.

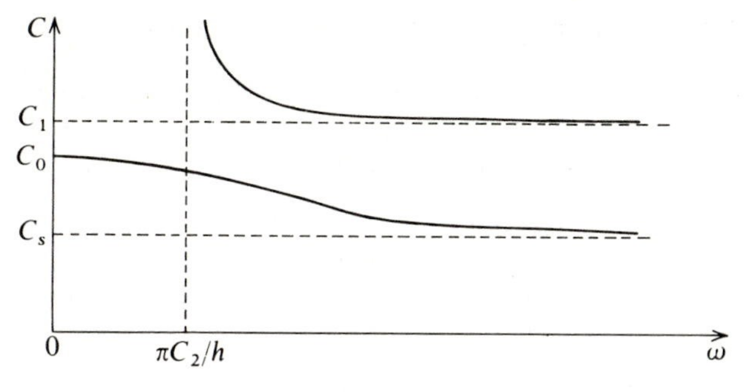

Fig. 7.10

lies in magnitude between the two constants c_1^2 and c_s^2 in the fourth-order term. These properties constitute the first example in this book of two equations forming a wave hierarchy. Further examples arise in Sections 8.4 and 10.3.

Let us return to the dispersion equation (7.137) and look at the second root for $c^2(\omega)$ when $\omega<\kappa^{-\frac{1}{2}}$. For $\omega<\kappa^{-\frac{1}{2}}$, $(\kappa\omega^2-1)^{-1}<0$, $|\frac{1}{4}(\kappa(\beta^2+\gamma^2)\omega^2-\alpha^2)^2-\kappa\beta^2\gamma^2\omega^2(\kappa\omega^2-1)|>|\frac{1}{2}\kappa(\beta^2+\gamma^2)\omega^2-\alpha^2|^2$, and one root of $c^2(\omega)$ is positive and the other negative. Hence, one pair of values of $c(\omega)$ are real, equal and opposite, and the second pair pure imaginary, equal and opposite; their limiting values as $\omega\to 0$ are given by eqns (7.138). Let the second pair be $\pm i\phi(\omega)$ where $\phi(\omega)$ is positive. Then $k=\omega/c=\pm i\omega/\phi(\omega)$ and $u=A\,e^{i(\omega t-kx)}$ becomes

$$u=A\,e^{i\omega t}\,e^{\pm x\omega/\phi(\omega)}, \tag{7.147}$$

which represents a standing wave whose amplitude is exponentially dependent on x. This type of wave also arises in the more detailed analysis of eqn (7.111), referred to at the end of Section 7.5.

The characteristics of eqn (7.146) are determined by the fourth-order terms in that equation and are $dx/dt=\pm c_s$ and $dx/dt=\pm c_1$. Since $c_1>c_s$, the greatest velocity at which any discontinuity can travel is c_1. Is there any contradiction between this fact and the fact that the higher mode in the $c(\omega)$ diagram (Fig. 7.10) has wave velocity greater than c_1? The answer is no, because the dispersion equation is derived from waves of the form $u=A\,e^{i(\omega t-kx)}$, which extend to infinity in both directions and contain no discontinuities.

7.7. Other waves in bounded elastic media

The theory of flexural waves in plates in the main follows that of longitudinal waves. Some of the results are now set as exercises, details being given only where significant differences occur.

Exercise 7.3. Substitute from eqns (7.108) into the r.h.s. of eqns (7.105) and (7.106) to obtain equations for σ_{xy} and σ_{yy}. Apply the boundary conditions (7.104) to obtain two equations between C and D and hence find the dispersion equation for flexural waves, namely

$$\frac{\tanh \alpha h}{\tanh \beta h} = \frac{4k^2\alpha\beta}{(\beta^2+k^2)^2}. \tag{7.148}$$

If, to obtain the low-frequency limit of the dispersion equation, $\tanh \alpha h$ and $\tanh \beta h$ are replaced by αh and βh in eqn (7.148), then all the terms in k vanish from the resulting equation. To find the limit, more terms must be retained. Replace $\tanh \alpha h$ by $\alpha h - \frac{1}{3}(\alpha h)^3$ and $\tanh \beta h$ by $\beta h - \frac{1}{3}(\beta h)^3$ to give, after substitution for α^2 and β^2,

$$\left(2k^2 - \frac{\rho\omega^2}{\mu}\right)^2 \left[1 - \frac{h^2}{3}\left(k^2 - \frac{\rho\omega^2}{\lambda+2\mu}\right)\right] - 4\left(k^2 - \frac{\rho\omega^2}{\mu}\right) k^2 \left[1 - \frac{h^2}{3}\left(k^2 - \frac{\rho\omega^2}{\mu}\right)\right] = 0, \tag{7.149}$$

where, if $\omega h/c_2$ and kh are first order, terms of fourth and sixth order are retained and terms of eighth order are neglected. The only term of fourth order that remains after cancellation is $(\rho\omega^2/\mu)^2$. The remaining terms of sixth order are $k^4(\rho\omega^2/\mu)$, $k^2(\rho\omega^2/\mu)^2$ and $(\rho\omega^2/\mu)^3$.

The term $(\rho\omega^2/\mu)^2$ can only be balanced if k is of different order to ω as $\omega \to 0$. Let $k \sim C\omega^\alpha$ as $\omega \to 0$ where C is a constant. Then the orders in ω of the 'sixth' order terms are $2+4\alpha$, $4+2\alpha$ and 6 respectively; α must be chosen so that the order of at least one of these terms is 4 and no term is of lower order (otherwise that term could not be balanced). There are two possibilities, either (i) $2+4\alpha=4$ or (ii) $4+2\alpha=4$. In case (i), $\alpha=\frac{1}{2}$ and the orders of the 'sixth' order terms are 4, 5 and 6. In case (ii), $\alpha=0$ and the orders are 2, 4 and 6. Case (i) is admissible, but case (ii) is inadmissible because $k^4(\rho\omega^2/\mu)$ could not then be balanced.

It is now shown that treating k as of order $\omega^{\frac{1}{2}}$ does not require more terms to be included in the derivation of eqn (7.149) from

(7.148). Terms of eighth order in k and ω were neglected and eventually terms of fourth order in ω are retained, treating k as of order $\omega^{\frac{1}{2}}$. The only term of eighth order originally neglected which becomes significant in the new ordering is k^8. To determine the coefficient of this term if it had been included in eqn (7.149), yet another term must be retained in the expansions of $\tanh \alpha h$ and $\tanh \mathrm{b}h$, i.e.

$$\tanh \alpha h = \alpha h - \tfrac{1}{3}(\alpha h)^3 + (\alpha h)^5 + \mathrm{O}(\alpha h)^7$$

and

$$\tanh \beta h = \beta h - \tfrac{1}{3}(\beta h)^3 + (\beta h)^5 + \mathrm{O}(\beta h)^7.$$

The reader can verify that the term in k^8 vanishes on substitution of these two equations for $\tanh \alpha h$ and $\tanh \beta h$ into eqn (7.148), and hence the new ordering does not require additional terms to be added to eqn (7.149).

In eqn (7.149), treat k as of order $\omega^{\frac{1}{2}}$ and retain terms of order ω^4, i.e. the terms proportional to ω^4 and $k^4\omega^2$.

Exercise 7.4. Show that as $\omega \to 0$, the dispersion eqn (7.148) gives

$$k \sim \left(\frac{3\rho(\lambda+2\mu)}{4\mu(\lambda+\mu)}\right)^{\frac{1}{4}} \left(\frac{\omega}{h}\right)^{\frac{1}{2}} \tag{7.150}$$

and

$$c = \frac{\omega}{k} \sim \left(\frac{4\mu(\lambda+\mu)}{3\rho(\lambda+2\mu)}\right)^{\frac{1}{4}} (h\omega)^{\frac{1}{2}}. \tag{7.151}$$

The balancing procedure just used is of frequent occurrence in applied mathematics. In particular, it is employed in the scaling to find inner-region solutions in asymptotics (e.g. as in Chapter 8).

The high-frequency approximation is virtually identical to that for longitudinal waves.

Exercise 7.5. When αh, βh, kh and $\sqrt{(\rho/\mu)}\,\omega h$ are all $\gg 1$, show that

$$c \to c_s \quad \text{as} \quad \omega \to \infty. \tag{7.152}$$

The fundamental mode of the dispersion equation from eqns (7.151) and (7.152) and from numerical solution of eqn (7.148) is sketched as a full line in Fig. 7.11.

Exercise 7.6. The approximate equation for the flexural vibration of plates is

$$\frac{\partial^2 v}{\partial t^2} + \frac{D}{2\rho h}\frac{\partial^4 v}{\partial x^4}, \quad D = \frac{8h^3}{3}\mu(\lambda+\mu)(\lambda+2\mu)^{-1};$$

D is known as the flexural rigidity of the plate.

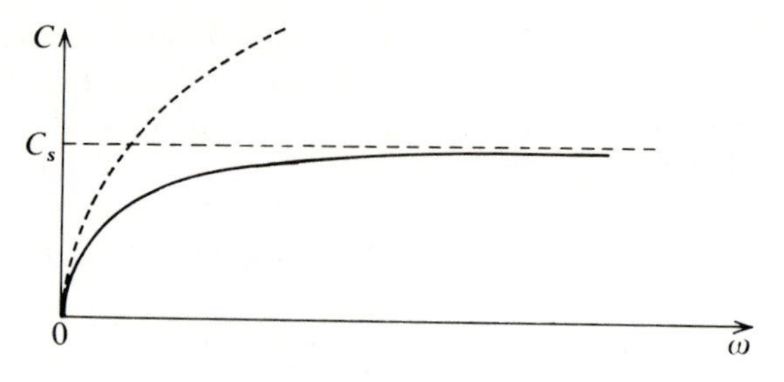

Fig. 7.11

Show that the equation for the dispersion relation $c(\omega)$ is eqn (7.151). Hence the approximate equation is of value for low frequencies only. Equation (7.151) is plotted as a broken curve in Fig. 7.11.

The propagation of waves in circular elastic rods was first studied by Pochhammer and Chree. The theory treats longitudinal, torsional and flexural waves and is similar to the Rayleigh–Lamb theory for plates, except that Bessel functions replace the hyperbolic functions. Approximate 'engineering' theories exist for all three types of wave.†

Answers to Exercises

(i) $c \geqslant c'$: The displacement components w^R (reflected) and w^T (transmitted) are

$$w^R = A\left[2\left(1+\frac{c\mu' \cos\gamma}{c'\mu\cos\alpha}\right)^{-1} - 1\right] e^{i\omega(t-(x\cos\alpha - y\sin\alpha)/c)}$$

and

$$w^T = 2A\left(1+\frac{c\mu'\cos\gamma}{c'\mu\cos\alpha}\right)^{-1} e^{i\omega(t-(-x\cos\gamma - y\sin\gamma)/c')},$$

where

$$\sin\gamma = \frac{c'}{c}\sin\alpha.$$

(ii) $c' > c$: For $\sin\alpha < c/c'$, solution as above. For $\sin\alpha > c/c'$, put $\sin\gamma = (c'/c)\sin\alpha = 1+\eta^2$, $\eta > 0$, $\cos\gamma = \pm i\eta$. For $w^T \to 0$ as $x \to -\infty$, the lower sign must be taken: $\cos\gamma = -i\eta$. The transmitted wave has velocity $c' \operatorname{cosec}\gamma$ and decays exponentially with $-x$. Reflected and transmitted waves are no longer in phase, or exactly out of phase, with the incident wave.

† See Kolsky, H. (1953). *Stress waves in solids*, Clarendon Press.

8 Linear dilatational waves in thermoelastic and Voigt solids

8.1. The derivation of governing equations

This chapter and Chapter 10 treat plane dilatational waves in different solid media. We start by deriving the governing mechanical and thermal equations in an elastic medium. The displacement is normal to the wavefront and is independent of the coordinates in the plane of the front. If the x-axis is normal to the front and if u is the component of displacement along the normal, $u = u(x, t)$ and the other two components of displacement are zero. Plane surface elements of particles, with normals initially in the x-direction, remain with normals in the x-direction during deformation and their areas remain constant. Consequently, the normal stress in the x-direction, defined as the normal force per unit area on a surface whose normal is in the x-direction, is the same whether the initial or current area of the surface of particles is taken.

Consider a cylinder of unit cross-sectional area with generators in the x-direction bounded by the surfaces of particles which in their initial position lie in planes $x = a$ and $x = b$. By symmetry, the principal stresses are σ_{xx}, denoted by σ, σ_{yy} and σ_{zz} with $\sigma_{yy} = \sigma_{zz}$. The equation of motion in the x-direction, in the absence of body force, is

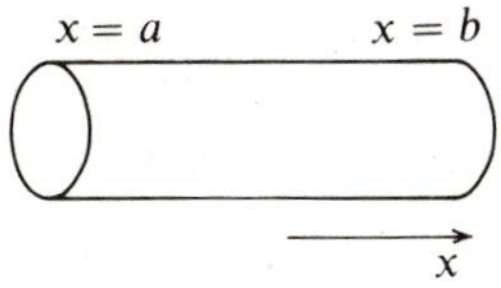

$x = a$ $x = b$

x

Fig. 8.1

$$\sigma(b, t) - \sigma(a, t) = \frac{\mathrm{d}}{\mathrm{d}t} \int_a^b \rho v \, \mathrm{d}x, \tag{8.1}$$

where ρ is the initial density of the solid and v the velocity of a particle (in the x-direction). Equation (8.1) is of the form of eqn (6.3).

Therefore, if the derivatives of v exist, using eqn (6.4),

$$\frac{\partial\sigma}{\partial x}=\rho\frac{\partial v}{\partial t}=\rho\frac{\partial^2 u}{\partial t^2}, \tag{8.2}$$

and when v has a discontinuity at $x=x_s(t)$, using eqn (6.8),

$$[\sigma]+\rho V[v]=0, \tag{8.3}$$

where

$$V=\mathrm{d}x_s(t)/\mathrm{d}t. \tag{8.4}$$

The rate of work of the forces external to the above cylinder is

$$\dot{W}_{ab}=(\sigma v)_b-(\sigma v)_a,$$

and the rate of gain of heat of the cylinder from across its boundaries is

$$\dot{Q}_{ab}=-q_b+q_a,$$

where q is the x-component of the heat flux vector. Since the normal velocity is zero on the sides of the cylinder, the stresses σ_{yy} and σ_{zz} do no work and by symmetry the y and z components of the heat flux vector are zero.

The rates of gain of kinetic and internal energies of the mass within the cylinder are

$$\dot{K}_{ab}=\frac{\mathrm{d}}{\mathrm{d}t}\int_a^b \tfrac{1}{2}\rho v^2\,\mathrm{d}x \quad\text{and}\quad \dot{U}_{ab}=\frac{\mathrm{d}}{\mathrm{d}t}\int_a^b U\rho\,\mathrm{d}x,$$

where U is the internal energy per unit mass.

By the first law of thermodynamics,

$$\dot{W}_{ab}+\dot{Q}_{ab}=\dot{U}_{ab}+\dot{K}_{ab};$$

hence,

$$(\sigma v-q)_b-(\sigma v-q)_a=\frac{\mathrm{d}}{\mathrm{d}t}\int_a^b(\tfrac{1}{2}\rho v^2+\rho U)\,\mathrm{d}x. \tag{8.5}$$

When the first derivatives of v and U exist,

$$\frac{\partial}{\partial x}(\sigma v-q)=\frac{\partial}{\partial t}(\tfrac{1}{2}\rho v^2+\rho U); \tag{8.6}$$

when v and U have discontinuities across $x=x_s(t)$,

$$[\sigma v-q]+V[\tfrac{1}{2}\rho v^2+\rho U]=0. \tag{8.7}$$

The jump conditions (8.3), (8.4) and (8.7) will be needed later. We continue with eqns (8.2) and (8.6).

Now, using

$$m = \partial u/\partial x, \tag{8.8}$$

$$\frac{\partial}{\partial x}(\sigma v - q) = \sigma\frac{\partial v}{\partial x} + v\frac{\partial \sigma}{\partial x} - \frac{\partial q}{\partial x}$$

$$= \sigma\frac{\partial m}{\partial t} + \rho v\frac{\partial v}{\partial t} - \frac{\partial q}{\partial x},$$

since

$$\frac{\partial v}{\partial x} = \frac{\partial}{\partial x}\frac{\partial u}{\partial t} = \frac{\partial}{\partial t}\frac{\partial u}{\partial x} = \frac{\partial m}{\partial t} \quad \text{and} \quad \frac{\partial \sigma}{\partial x} = \rho\frac{\partial v}{\partial t},$$

eqn (8.2). Substitute for $\partial(\sigma v - q)/\partial x$ into eqn (8.6):

$$\rho\frac{\partial U}{\partial t} = \sigma\frac{\partial m}{\partial t} - \frac{\partial q}{\partial x}.$$

The rate of heat gain per unit mass is $\partial Q/\partial t$. Hence, see Fig. 8.2,

$$\delta x\,\rho\frac{\partial Q}{\partial t} = -q_{x+\delta x} + q_x = -\frac{\partial q}{\partial x}\,\delta x$$

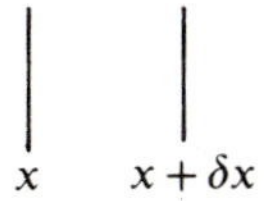

$x \qquad x+\delta x$

Fig. 8.2

or

$$\frac{\partial q}{\partial x} = -\rho\frac{\partial Q}{\partial t}, \tag{8.9}$$

the heat balance equation. Therefore

$$\rho\frac{\partial U}{\partial t} = \sigma\frac{\partial m}{\partial t} + \rho\frac{\partial Q}{\partial t}. \tag{8.10}$$

Since $\partial/\partial t$ denotes time differentiation at constant x, i.e. following a particle, eqn (8.10) can be written

$$\rho\,\mathrm{d}U = \sigma\,\mathrm{d}m + \rho\,\delta Q, \tag{8.11}$$

where the increments refer to particular particles or groups of particles.[†]

[†] The symbol d denotes an increment in a variable of state, δ an increment in a quantity which is not a variable of state.

By the second law of thermodynamics,

$$\frac{\delta Q}{T} \leqslant \mathrm{d}S \tag{8.12}$$

with equality holding for reversible processes; S is the entropy per unit mass or specific entropy and T is the absolute temperature. Restricting attention to reversible processes and substituting $\delta Q = T\,\mathrm{d}S$ in eqn (8.11),

$$\mathrm{d}U = \frac{\sigma\,\mathrm{d}m}{\rho} + T\,\mathrm{d}S. \tag{8.13}$$

An elastic solid is defined by the two hypotheses that (i) the components of strain and the entropy form a complete and independent set of variables of state, and (ii) a reversible path (which may not necessarily be followed) exists between any two states. For dilatational waves in the x-direction, only the strain component m is non-zero. Hence, $U = U(m, S)$ and

$$\mathrm{d}U = \frac{\partial U}{\partial m}\,\mathrm{d}m + \frac{\partial U}{\partial S}\,\mathrm{d}S. \tag{8.14}$$

Since the variables m and S are independent, we may first put $\mathrm{d}S = 0$ and $\mathrm{d}m \neq 0$ in eqns (8.13) and (8.14) and then put $\mathrm{d}S \neq 0$ and $\mathrm{d}m = 0$. Equating the two values of $\mathrm{d}U$ in the two cases, we obtain respectively

$$\sigma = \rho\frac{\partial U}{\partial m} \tag{8.15}$$

and

$$T = \frac{\partial U}{\partial S}. \tag{8.16}$$

For linear elasticity, it is sufficient to expand $U(m, S)$ in a generalized Taylor's series in m and S up to second-order terms. With the initial conditions that $U = 0$, $\sigma = 0$ and $T = T_0$ when $m = 0$ and $S = 0$, the equations for U, σ and T, on use of eqns (8.15) and (8.16), become

$$U = T_0 S + \frac{1}{2\rho}(\lambda + 2\mu)m^2 - \kappa m S + \tfrac{1}{2}\eta S^2, \tag{8.17}$$

$$\sigma = \rho\frac{\partial U}{\partial m} = (\lambda + 2\mu)m - \rho\kappa S \tag{8.18}$$

and

$$T=\frac{\partial U}{\partial S}=T_0-\kappa m+\eta S, \tag{8.19}$$

where $\lambda+2\mu$, κ and η are constants. Eliminate σ and m from eqns (8.2), (8.8) and (8.18) to give

$$\rho\frac{\partial^2 u}{\partial t^2}=(\lambda+2\mu)\frac{\partial^2 u}{\partial x^2}-\rho\kappa\frac{\partial S}{\partial x}. \tag{8.20}$$

Equation (8.20) is one equation between u and S. It is known as the mechanical equation because it reduces to the classical wave equation for u with velocity $(\lambda+2\mu)^{\frac{1}{2}}\rho^{-\frac{1}{2}}$ when $\kappa=0$.

The second equation between u and S is the heat balance equation (8.9). Use Newton's law of heat conduction for q, namely

$$q=-k\frac{\partial T}{\partial x}, \tag{8.21}$$

where k is the coefficient of conductivity, and the linearized form of equation (8.12)

$$\frac{\partial Q}{\partial t}=T_0\frac{\partial S}{\partial t}$$

in eqn (8.9) to give

$$k\frac{\partial^2 T}{\partial x^2}=\rho T_0\frac{\partial S}{\partial t}. \tag{8.22}$$

Eliminate T by eqn (8.19):

$$\rho T_0\frac{\partial S}{\partial t}=k\eta\frac{\partial^2 S}{\partial x^2}-k\kappa\frac{\partial^3 u}{\partial x^3}. \tag{8.23}$$

Equation (8.23) is the second equation between u and S. It is known as the thermal equation, because when $\kappa=0$ it reduces to the classical heat conduction equation for T, on substitution for S from eqn (8.19). Equations (8.20) and (8.23) can be written

$$c_1^2\frac{\partial^2 u}{\partial x^2}-\frac{\partial^2 u}{\partial t^2}=\kappa\frac{\partial S}{\partial x} \tag{8.24}$$

and

$$\beta\frac{\partial^2 S}{\partial x^2}-\frac{\partial S}{\partial t}=\frac{\kappa\beta}{\eta}\frac{\partial^3 u}{\partial x^3}. \tag{8.25}$$

Here, as in Chapter 7,

$$c_1^2=\frac{\lambda+2\mu}{\rho} \tag{8.26}$$

and

$$\beta=\frac{k\eta}{\rho T_0}. \tag{8.27}$$

When $\kappa=0$, there is no mechanical-thermal interaction and eqn (8.17) is of the form

$$U=U_1(m)+U_2(S).$$

κ is a measure of the mechanical-thermal interaction.

8.2. Sinusoidal thermoelastic waves

Look for a solution of eqns (8.20) and (8.23) of the form

$$u=A\mathrm{e}^{\mathrm{i}\omega(t-\alpha x)} \quad \text{and} \quad S=B\mathrm{e}^{\mathrm{i}\omega(t-\alpha x)}, \tag{8.28}$$

where ω is a real constant and A, B and α are constants, possibly complex. Substitution into eqns (8.24) and (8.25) yields

$$-\omega^2 A=-\omega^2\alpha^2 c_1^2 A+\mathrm{i}\omega\alpha\kappa B$$

and

$$\mathrm{i}\omega B=-\omega^2\alpha^2\beta B-\mathrm{i}\omega^3\alpha^3\beta\kappa A/\eta.$$

Therefore,

$$B/A=-\frac{\alpha^3\omega^2\kappa\beta}{\eta(1-\mathrm{i}\omega\alpha^2\beta)} \tag{8.29}$$

and

$$c_1^2\alpha^2-1+\frac{\mathrm{i}\alpha^4\omega\kappa^2\beta}{\eta(1-\mathrm{i}\omega\alpha^2\beta)}=0. \tag{8.30}$$

Equation (8.30) is the dispersion equation for thermoelastic waves.

When $\kappa=0$, $\alpha=\pm 1/c_1$ or $\alpha=\pm(\mathrm{i}\omega\beta)^{-\frac{1}{2}}$. The former represents the mechanical wave and the latter the thermal, derivable from eqns (8.24) and (8.25) respectively when $\kappa=0$. We shall be concerned in this book only with the mechanical thermoelastic wave. It is that wave for which $\alpha\to\pm 1/c_1$ as $\kappa\to 0$. To be specific, we treat the mechanical wave moving in the positive sense, i.e. the wave for which $\alpha\to +1/c_1$ as $\kappa\to 0$.

Equation (8.30) is a quadratic for α^2. To simplify the solution algebra, we introduce two non-dimensional variables and a non-dimensional constant given by

$$\hat{\omega}=\frac{\beta}{c_1^2}\,\omega \tag{8.31}$$

$$\hat{\alpha}=\hat{\alpha}(\hat{\omega})=c_1\alpha \tag{8.32}$$

and

$$\varepsilon=\frac{\kappa^2}{\eta c_1^2}. \tag{8.33}$$

Typical values of ε, c_1^2/β and $(2\pi\beta/c_1)(1-\varepsilon/4)$ are given in Table 8.1.

Equation (8.30), on use of eqns (8.31), (8.32) and (8.33) becomes

$$\hat{\alpha}^2-1+\frac{\mathrm{i}\varepsilon\hat{\omega}\hat{\alpha}^4}{1-\mathrm{i}\hat{\omega}\hat{\alpha}^2}=0,$$

or

$$\mathrm{i}\hat{\omega}(1-\varepsilon)\hat{\alpha}^4-(1+\mathrm{i}\hat{\omega})\hat{\alpha}^2+1=0.$$

The solution for $\hat{\alpha}$ that tends to $+1$ as $\varepsilon\to 0$ is given by

$$\hat{\alpha}=\left(\frac{1+\mathrm{i}\hat{\omega}-(1-\mathrm{i}\hat{\omega})[1+4\mathrm{i}\hat{\omega}\varepsilon(1-\mathrm{i}\hat{\omega})^{-2}]^{\frac{1}{2}}}{2\mathrm{i}\hat{\omega}(1-\varepsilon)}\right)^{\frac{1}{2}}. \tag{8.34}$$

Now

$$\left|\frac{4\mathrm{i}\hat{\omega}}{(1-\mathrm{i}\hat{\omega})^2}\right|=\frac{4\hat{\omega}}{1+\hat{\omega}^2}=\frac{4\hat{\omega}}{(1-\hat{\omega})^2+2\hat{\omega}}\leqslant 2 \quad \text{for all } \hat{\omega},\ 0\leqslant\hat{\omega}<\infty.$$

Table 8.1 shows that $\varepsilon\ll 1$ for most solids. We therefore expand the square roots in eqn (8.34) by the binomial theorem:

$$\begin{aligned}\hat{\alpha}&=\left(\frac{(1+\mathrm{i}\hat{\omega})-(1-\mathrm{i}\hat{\omega})-2\mathrm{i}\hat{\omega}\varepsilon(1-\mathrm{i}\hat{\omega})^{-1}+\mathrm{O}(\varepsilon^2)}{2\mathrm{i}\hat{\omega}(1-\varepsilon)}\right)^{\frac{1}{2}}\\&=[(1-\varepsilon(1-\mathrm{i}\hat{\omega})^{-1})(1-\varepsilon)^{-1}+\mathrm{O}(\varepsilon^2)]^{\frac{1}{2}}\\&=1+\tfrac{1}{2}\varepsilon[1-(1-\mathrm{i}\hat{\omega})^{-1}]+\mathrm{O}(\varepsilon^2),\end{aligned}$$

or

$$\hat{\alpha}=1+\tfrac{1}{2}\varepsilon\frac{\hat{\omega}^2-\mathrm{i}\hat{\omega}}{1+\hat{\omega}^2}+\mathrm{O}(\varepsilon^2). \tag{8.35}$$

Table 8.1

	Aluminium	Copper	Rubber	Lead	Steel	Silica
ε	0.027	0.019	0.036	0.053	0.0079	0.36×10^{-5}
$\omega_0=c_1^2/\beta(\mathrm{s}^{-1})$	4.06×10^{11}	2.06×10^{11}	2.48×10^{13}	1.79×10^{11}	3.16×10^{12}	4.24×10^{13}
$\lambda_0=\dfrac{2\pi\beta}{c_1}\left(1-\dfrac{\varepsilon}{4}\right)(\mathrm{cm})$	0.99×10^{-5}	1.46×10^{-5}	4.04×10^{-8}	0.77×10^{-5}	1.17×10^{-6}	8.90×10^{-8}
$\lambda_0/(\varepsilon\pi)(\mathrm{cm})$	1.17×10^{-4}	2.45×10^{-4}	3.57×10^{-7}	4.62×10^{-5}	4.71×10^{-5}	7.9×10^{-3}

The wave velocity $c(\omega)$ and the attenuation $a(\omega)$ are[†]

$$c(\omega)=(\mathrm{R}[\alpha])^{-1}=c_1(\mathrm{R}[\hat{\alpha}])^{-1}$$
$$=c_1\left(1-\frac{1}{2}\varepsilon\frac{\hat{\omega}^2}{1+\hat{\omega}^2}\right)+\mathrm{O}(\varepsilon^2) \tag{8.36}$$

and

$$a(\omega)=-\omega\mathrm{I}[\alpha]=-\frac{c_1}{\beta}\hat{\omega}\mathrm{I}[\hat{\alpha}]$$
$$=\frac{c_1\varepsilon}{2\beta}\frac{\hat{\omega}^2}{1+\hat{\omega}^2}+\mathrm{O}(\varepsilon^2). \tag{8.37}$$

The low- and high-frequency limits of velocity and attenuation are

$$c(\omega)\to c_1 \quad \text{and} \quad a(\omega)\to 0 \quad \text{as } \omega\to 0 \tag{8.38}$$

and

$$c(\omega)\to(1-\tfrac{1}{2}\varepsilon)c_1 \quad \text{and} \quad a(\omega)\to\frac{c_1\varepsilon}{2\beta} \quad \text{as } \omega\to\infty. \tag{8.39}$$

Exercise 8.1. Find the limits of $\hat{\alpha}$ as $\omega\to 0$ and as $\omega\to\infty$ directly from eqn (8.34). Show that these limits agree with eqns (8.38) and (8.39) for $\varepsilon \ll 1$.

The non-dimensional reduced wave velocity $\hat{c}(\hat{\omega})$ and attenuation $\hat{a}(\hat{\omega})$ are defined by and are equal to

$$\hat{c}(\hat{\omega})=\frac{c(\omega)}{c_1}=1-\tfrac{1}{2}\varepsilon\,\hat{\omega}^2(1+\omega^2)^{-1}+\mathrm{O}(\varepsilon^2) \tag{8.40}$$

and

$$\hat{a}(\hat{\omega})=\frac{\beta}{c_1}a(\omega)=\tfrac{1}{2}\varepsilon\,\hat{\omega}^2(1+\hat{\omega}^2)^{-1}+\mathrm{O}(\varepsilon^2). \tag{8.41}$$

Correct to order ε, $\hat{c}(\hat{\omega})$ is plotted against $\hat{\omega}$ in Fig. 8.3.

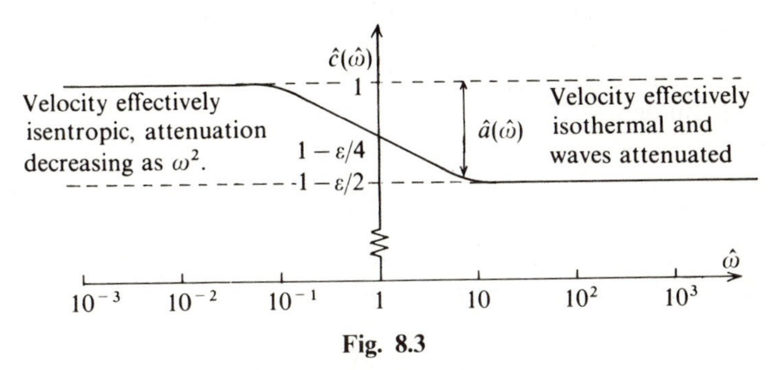

Fig. 8.3

[†] $\exp(\mathrm{i}\omega(t-\alpha x))=\exp(\mathrm{i}\omega(t-R[\alpha]x))\exp(\omega\mathrm{I}[\alpha]x)$.

$\hat{c}(\hat{\omega})$ decreases from 1 to $1-\varepsilon/2$ as $\hat{\omega}$ increases from 0 to $+\infty$. When $\hat{\omega}=1$, i.e. $\omega=c_1^2/\beta$, $\hat{c}(1)=1-\varepsilon/4$. Since $\hat{\omega}^2/(\hat{\omega}^2+1)$ equals 1/101 for $\hat{\omega}=10^{-1}$ and equals 100/101 for $\hat{\omega}=10$, all but 2 per cent of the total change in wave velocity from $\hat{\omega}=0$ to $\hat{\omega}=\infty$ occurs over the two decades about $\hat{\omega}=1$ or $\omega=\omega_0$, where

$$\omega_0=c_1^2/\beta. \tag{8.42}$$

Since the r.h.s. of eqns (8.40) and (8.41) have sum unity, probably only a mathematical curiosity, the reduced attenuation $\hat{a}(\hat{\omega})$ is represented in Fig. 8.3 by the distance between the curve for $\hat{c}(\hat{\omega})$ and the straight line $\hat{c}(\hat{\omega})=1$. Values of ω_0 and the corresponding wavelength

$$\lambda_0=\frac{2\pi\beta}{c_1}\left(\frac{1-\varepsilon}{4}\right),$$

are given in Table 8.1.

The limiting velocity as $\omega\to\infty$ is frequently called the 'isothermal' velocity, for the reasons that follow. If the approximation $T=T_0$ is introduced, eqn (8.19) gives $S=\kappa m/\eta$ and eqn (8.22) is left unsatisfied. Equation (8.24) becomes

$$\frac{\partial^2 u}{\partial t^2}=\left(c_1^2-\frac{\kappa^2}{\eta}\right)\frac{\partial^2 u}{\partial x^2}, \tag{8.43}$$

the classical wave equation with velocity

$$c_T=c_1\sqrt{1-\varepsilon}=c_1\left(\frac{1-\varepsilon}{2}+O(\varepsilon^2)\right) \tag{8.44}$$

in agreement with eqn (8.39). The fractional difference between c_1 and c_T is $(c_1-c_T)/c_1\simeq\varepsilon/2$, which is of the order of 1 per cent.

For $\omega<\omega_0/10$, the fractional error in the velocity of the isentropic approximation of Chapter 7 is less than $\frac{1}{2}\varepsilon(\omega^2/\omega_0^2)$. For a given maximum acceptable error θ, this requires

$$\frac{\omega}{\omega_0}<\sqrt{\frac{2\theta}{\varepsilon}}. \tag{8.45}$$

The attenuation $a(\omega)$ for $\omega>10\omega_0$ is within 1 per cent of its limiting value $\varepsilon c_1/2\beta\simeq\varepsilon\pi/\lambda_0$. For $\omega<\omega_0/10$,

$$a(\omega)\simeq\frac{\varepsilon\pi}{\lambda_0}\frac{\omega^2}{\omega_0^2}.$$

The reciprocal of $a(\omega)$ gives the length the wave travels for its amplitude to be reduced by a factor 1/e. For $\omega > 10\omega_0$, the length is $\lambda_0/\varepsilon\pi$ and for $\omega < \omega_0/10$ it is

$$\frac{\lambda_0}{\varepsilon\pi}\frac{\omega_0^2}{\omega^2}.$$

Values of $\lambda_0/\varepsilon\pi$ are given in Table 8.1. Whether the effect of attenuation can be neglected, as far as use of isentropic theory is concerned, depends on a typical dimension L of the body through which the wave propagates, the wave frequency ω, and the maximum acceptable fractional decrease ϕ in the amplitude. After a distance of travel L for $\omega < \omega_0/10$, the amplitude has decreased by a factor $\exp(-\varepsilon\pi\omega^2/\lambda_0\omega_0^2 L)$. Hence, one requires

$$1-\exp\left(-\frac{\varepsilon\pi\omega^2}{\lambda_0\omega_0^2 L}\right) < \phi.$$

If $\phi \ll 1$, then

$$\frac{\omega}{\omega_0} < \sqrt{\frac{\lambda_0\phi}{\varepsilon\pi L}}. \qquad (8.46)$$

For both velocity and attenuation criteria for the use of isentropic theory to be satisfied, ω/ω_0 must be less than the minimum of $\sqrt{2\theta/\varepsilon}$ and $\sqrt{\lambda_0\phi/\varepsilon\pi L}$.

It is a fundamental hypothesis of all continuum mechanics that the intrinsic properties of the smallest element considered are the same as those of the medium as a whole. This hypothesis breaks down for waves when the wavelength becomes sufficiently small for it to be comparable to the distance between atoms. The Debye cut-off length, about 5×10^{-8} cm for metals, is a measure of the lower limit of wavelength to which continuum theory can be applied; below this wavelength, waves are rapidly attenuated. The values of $\lambda_0/\varepsilon\pi$ in Table 8.1 show that, as wavelength decreases or frequency increases, thermoelastic damping effectively attenuates sinusoidal elastic waves about two decades before the Debye cut-off becomes operational. For non-sinusoidal waves or signals, the length Λ over which there is a significant change in amplitude must be large compared to λ_0 for use of isentropic theory. If $\Lambda < \lambda_0/10$, the attenuation is $\varepsilon\pi/\lambda_0$; in particular this is true for shocks.

The non-dimensionalization of ω and α and of the constants has so simplified the algebra of this section that we now consider how to non-dimensionalize the variables x and t and hence the governing differential equations (8.24) and (8.25). We start by requiring that the non-dimensional coordinate ξ and time τ should leave the form of $\omega(t-\alpha x)$ invariant, i.e. that $\omega(t-\alpha x)=\hat{\omega}(\tau-\hat{\alpha}\xi)$. Use of eqns (8.31) and (8.32) gives

$$\tau=\frac{c_1^2}{\beta}t \quad \text{and} \quad \xi=\frac{c_1}{\beta}x. \tag{8.47}$$

To leave the already non-dimensional strain $\partial u/\partial x$ unaltered, u must be non-dimensionalized to $\hat{u}$ by

$$\hat{u}=\frac{c_1}{\beta}u. \tag{8.48}$$

To make the mechanical-thermal interaction terms, on the r.h.s. of eqns (8.24) and (8.25), have the same coefficients, non-dimensionalize S by

$$\hat{S}=\frac{\eta^{\frac{1}{2}}}{c_1}S. \tag{8.49}$$

Use of eqns (8.33), (8.47), (8.48) and (8.49) in eqns (8.24) and (8.25) yields

$$\frac{\partial^2\hat{u}}{\partial\xi^2}-\frac{\partial^2\hat{u}}{\partial\tau^2}=\varepsilon^{\frac{1}{2}}\frac{\partial\hat{S}}{\partial\xi} \tag{8.50}$$

and

$$\frac{\partial^2\hat{S}}{\partial\xi^2}-\frac{\partial\hat{S}}{\partial\tau}=\varepsilon^{\frac{1}{2}}\frac{\partial^3\hat{u}}{\partial\xi^3}. \tag{8.51}$$

We have already seen that ε is of the order of 10^{-2} for real materials. Therefore, solutions that depend upon ε being small compared to unity will have real applications. Since elimination of $\hat{S}$ from eqns (8.50) and (8.51) will produce a fourth-order equation for $\hat{u}$, in the next section we study a mathematically simpler equation before returning to thermoelastic solids in Section 8.4.

8.3. Voigt solid; half-space problem

The Voigt solid is one of the two simplest examples of a viscoelastic solid. Its mechanical model consists of a spring and a dashpot in parallel (Fig. 8.4). The spring has stress–strain equation $\sigma_s=Mm_s$

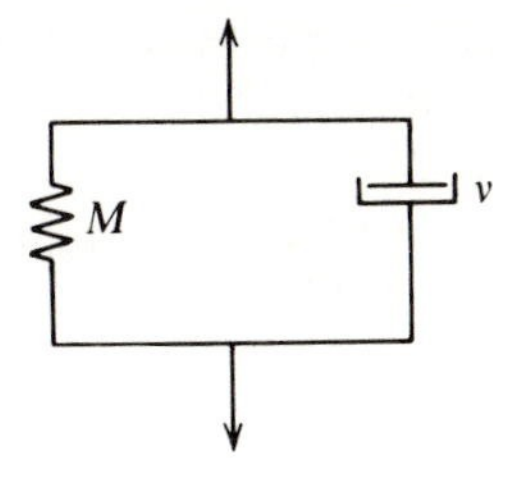

Fig. 8.4

and the dashpot $\sigma_d = v\,\partial m_d/\partial t$, M and v constants. Then, as the stress σ in the solid is the sum of the stresses in the two elements and the strain in the solid is the same as the strain in each of the elements,

$$\sigma = \sigma_s + \sigma_d = Mm + v\frac{\partial m}{\partial t}, \tag{8.52}$$

where $m = m_s = m_d$. Equations (8.2) and (8.8) are the other two governing equations for σ, u and m. Elimination of u and m gives

$$\rho\frac{\partial^2 \sigma}{\partial t^2} = M\frac{\partial^2 \sigma}{\partial x^2} + v\frac{\partial^3 \sigma}{\partial x^2 \partial t} \tag{8.53}$$

as the equation for stress waves in a Voigt solid. The equation is third order with characteristics $t = \text{constant}$ (double) and $x = \text{constant}$.

To see how sinusoidal waves propagate through the Voigt solid, put

$$\sigma = A\mathrm{e}^{\mathrm{i}\omega(t-\alpha x)}, \quad \omega \text{ real},$$

in eqn (8.53) to give

$$\alpha = \sqrt{\frac{\rho}{M + \mathrm{i}\omega v}}.$$

Hence,

$$\alpha = \sqrt{\frac{\rho}{M}}\left(1 + \frac{\mathrm{i}\omega v}{M}\right)^{-\frac{1}{2}} \underset{\omega \to 0}{\sim} \sqrt{\frac{\rho}{M}}\left(1 - \frac{\mathrm{i}\omega v}{2M} + \mathrm{O}(\omega^2)\right)$$

and

$$\alpha = \sqrt{\frac{\rho}{\mathrm{i}\omega v}}\left(1 + \frac{M}{\mathrm{i}\omega v}\right)^{-\frac{1}{2}} \underset{\omega \to \infty}{\sim} \sqrt{\frac{\rho}{\mathrm{i}\omega v}}\left(1 - \frac{M}{2\mathrm{i}\omega v} + \mathrm{O}\left(\frac{1}{\omega^2}\right)\right).$$

By equations (8.36) and (8.37), the low- and high-frequency limits of wave velocity and attenuation are

$$c(\omega) \underset{\omega\to 0}{\to} \sqrt{\frac{M}{\rho}} \quad \text{with} \quad a(\omega) \underset{\omega\to 0}{\to} 0 \tag{8.54}$$

and, using $\mathrm{i}^{-\frac{1}{2}} = \mathrm{e}^{-\mathrm{i}\pi/4} = (1-\mathrm{i})/\sqrt{2}$,

$$c(\omega) \underset{\omega\to\infty}{\sim} \sqrt{\frac{2\nu\omega}{\rho}} \quad \text{with} \quad a(\omega) \underset{\omega\to\infty}{\sim} \sqrt{\frac{\rho\omega}{2\nu}}. \tag{8.55}$$

At the low-frequency limit, the Voigt solid behaves like an elastic solid with waves travelling undamped with velocity $\sqrt{M/\rho}$; as the upper frequency limit is approached, both the velocity and attenuation tend to $+\infty$. This latter behaviour is analogous to that of temperature waves governed by the heat conduction equation, which also has a double characteristic $t=$constant (see Exercise 8.3 below).

Exercise 8.2. The Maxwell material has a mechanical model consisting of a spring and dashpot in series (Fig. 8.5). Show that its stress–strain relation is

$$\frac{1}{M}\frac{\partial\sigma}{\sigma t} + \frac{1}{\nu}\sigma = \frac{\partial m}{\partial t}.$$

Fig. 8.5

Hint: The stress is the same in the two elements, but the strain in the material is the sum of the strains in the two elements.

Exercise 8.3. By looking for solutions of the form $T = Ae^{\mathrm{i}\omega(t-\alpha x)}$, ω real, to the equation $\kappa\,\partial^2 T/\partial x^2 = \partial T/\partial t$, show that $c(\omega) = \sqrt{2\omega\kappa}$ and $a(\omega) = \sqrt{\omega/2\kappa}$.

Consider a semi-infinite Voigt solid $x \geqslant 0$ to which a stress $\sigma = \sigma_0 H(t)$, σ_0 constant, is applied on the boundary $x=0$. We treat the case in which ν is small; the precise definition of 'small' will follow. If ν is zero, eqn (8.53) reduces to the classical wave equation with the solution to the stated problem being

$$\sigma = \sigma_0 H\left(t - \sqrt{\frac{\rho}{M}}\,x\right). \tag{8.56}$$

For $x<\sqrt{(M/\rho)}t$ and $x>\sqrt{(M/\rho)}t$ the stress is constant and equal to σ_0 and zero respectively. At $x=\sqrt{(M/\rho)}t$, the stress jumps from 0 to σ_0. The lengths over which changes occur are effectively infinity, zero, and infinity for the three regions $x<$, $=$ and $>\sqrt{(M/\rho)}t$.

For ν small but non-zero, the wavelength $\lambda(\omega)$, equal to $2\pi c(\omega)/\omega$, has limits, using eqns (8.54) and (8.55), given by

$$\lim_{\omega\to 0} \lambda(\omega)\to\frac{2\pi}{\omega}\sqrt{\frac{M}{\rho}}\underset{\omega\to 0}{\to}\infty$$

and

$$\lim_{\omega\to\infty} \lambda(\omega)\to\frac{2\pi}{\omega}\sqrt{\frac{2\nu\omega}{\rho}}\underset{\omega\to\infty}{\to}0.$$

The velocity corresponding to infinite wavelength is the wave velocity as $\omega\to 0$, i.e. $\sqrt{M/\rho}$, but the velocity corresponding to zero wavelength is the wave velocity as $\omega\to\infty$, i.e. $\sqrt{2\nu\omega/\rho\nu}\underset{\omega\to\infty}{\to}\infty$. The solution for ν zero is satisfactory for ν small except in the region containing $x=\sqrt{(M/\rho)}t$.

How can we 'improve' the solution (8.56) for small ν? We first non-dimensionalize the variables x, t and σ to ξ, τ and ϕ by the substitutions

$$x=L\xi,\quad t=L\sqrt{\frac{\rho}{M}}\tau\quad\text{and}\quad\sigma=\sigma_0\phi, \tag{8.57}$$

where L is a constant length, so that eqn (8.53) becomes

$$\frac{\partial^2\phi}{\partial\tau^2}=\frac{\partial^2\phi}{\partial\xi^2}+\varepsilon\frac{\partial^3\phi}{\partial\xi^2\partial\tau}, \tag{8.58}$$

where the non-dimensional constant ε is

$$\varepsilon=\frac{\nu}{L\sqrt{M\rho}}. \tag{8.59}$$

ν is taken to be sufficiently small for $\varepsilon\ll 1$. Note that in this particular problem there is apparently no restriction on ν because L is arbitrary. If the body were finite and if L were chosen as a typical dimension of the body, then the requirement $\varepsilon\ll 1$ imposes a restriction on the magnitude of ν. We return to this discussion in the penultimate paragraph of this section.

Look for a solution of eqn (8.58) in the form of a power series in ε, i.e.

$$\phi(\xi, \tau, \varepsilon)=\phi_0(\xi, r)+\varepsilon\phi_1(\xi, \tau)+\varepsilon^2\phi_2(\xi, \tau)+\ldots, \tag{8.60}$$

where $\phi_0, \phi_1, \phi_2, \ldots$ are independent of ε. Substitute eqn (8.60) into eqn (8.58) and equate successive powers of ε to zero:

$$\frac{\partial^2\phi_0}{\partial\tau^2}=\frac{\partial^2\phi_0}{\partial\xi^2}, \tag{8.61}$$

$$\frac{\partial^2\phi_1}{\partial\tau^2}=\frac{\partial^2\phi_1}{\partial\xi^2}+\frac{\partial^3\phi_0}{\partial\xi^2\partial\tau},$$

etc.

The solution of the first of eqns (8.60) representing a forward moving wave is

$$\phi_0=f(\tau-\xi). \tag{8.62}$$

The boundary condition on $\xi=0$ is $\phi=H(\tau)$. If eqn (8.62) is to be a solution for all ε including $\varepsilon=0$, then

$$\phi_0=H(\tau-\xi). \tag{8.63}$$

Substitute for ϕ_0 into the second of eqns (8.61):

$$\frac{\partial^2\phi_1}{\partial\tau^2}-\frac{\partial^2\phi_1}{\partial\xi^2}=H'''(\tau-\xi)=\delta''(\tau-\xi), \tag{8.64}$$

where n primes represent n derivatives of a function with respect to its argument.

The solution of eqn (8.64) is

$$\phi_1=F(\tau-\xi)+G(\tau+\xi)+\tfrac{1}{4}(\tau+\xi)\,\delta'(\tau-\xi)$$
$$=F(\tau-\xi)+G(\tau+\xi)+\tfrac{1}{2}\xi\,\delta'(\tau-\xi);$$

since $\delta'(\tau-\xi)=0$ for $\tau\neq\xi$. F and G are arbitrary functions of their arguments. The boundary conditions are not generalized functions and so the term $\frac{1}{2}\xi\,\delta'(\tau-\xi)$ remains in ϕ_1. On substitution back into eqn (8.60),

$$\phi(\xi, \tau, \varepsilon)=H(\tau-\xi)+\varepsilon\tfrac{1}{2}\xi\,\delta'(\tau-\xi)+\text{other terms}. \tag{8.65}$$

In formulating eqn (8.60), it is assumed that the terms in the series are decreasing for ε sufficiently small but non-zero. This assumption is clearly violated by eqn (8.65) in the neighbourhood of $\tau=\xi$. Also, real physical quantities do not equal derivatives of δ-functions. Hence eqn (8.65) cannot be used in the neighbourhood of $\tau-\xi=0$. Away from $\tau-\xi=0$, $\delta'(\tau-\xi)=0$ and $\phi(\xi, \tau, \varepsilon)=H(\tau-\xi)$. We have failed to find a solution of the form of eqn (8.60) valid throughout the whole ξ, τ-plane.

The solution $\phi_0=H(\tau-\xi)$ is illustrated in Fig. 8.6. The tangential derivatives of ϕ_0 on either side of $\tau=\xi$ are both zero but the normal derivative of ϕ_0 across $\tau=\xi$ is infinite. The solution procedure adopted in the last two paragraphs assumes implicitly that all partial derivatives are of the same order. This assumption is clearly false in the neighbourhood of $\tau=\xi$.

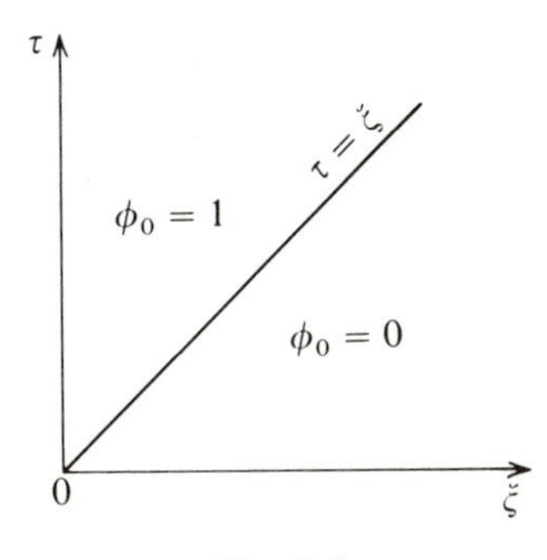

Fig. 8.6

Since the function $\phi_0=H(\tau-\xi)$ appears to be a satisfactory solution away from $\tau=\xi$, we now divide the τ, ξ-plane into three regions: a region about $\tau=\xi$ called an inner region and two outer regions in which $\phi_0=0$ or $\phi_0=1$ as illustrated in Fig. 8.7. In the inner region, the partial derivative of ϕ normal to $\tau=\xi$ is much larger than the tangential derivative. Since grad ϕ is a vector, this means that the partial derivative of ϕ in any direction making finite angle with the tangential direction is much larger than the tangential derivative. The change in ϕ has been spread out across the inner region.

Introduce new coordinates η and ζ, known as inner region coordinates (Fig. 8.8), such that

$$\eta=\varepsilon^n(\xi-\tau) \tag{8.66}$$

and

$$\zeta=\varepsilon^p\tau. \tag{8.67}$$

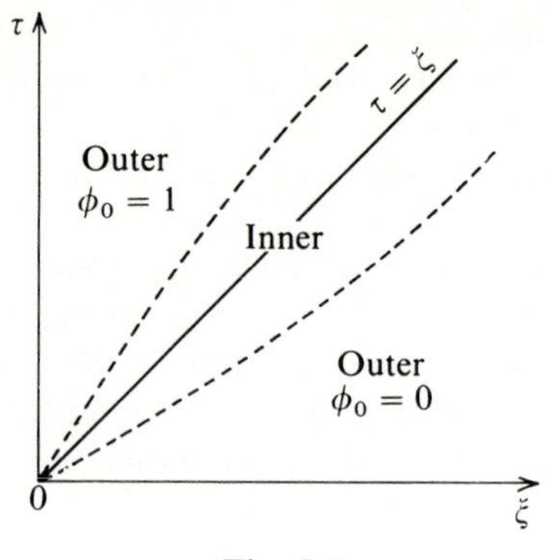

Fig. 8.7

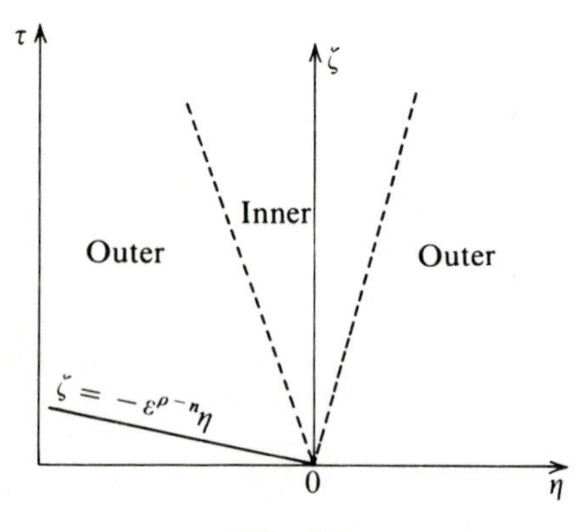

Fig. 8.8

Then η is constant when $\xi-\tau$ is constant and ζ is constant when τ is constant; $\eta=0$ when $\xi=\tau$. The left upper quartile from $\zeta=-\varepsilon^{p-n}\eta$ to the ζ-axis of the η, ζ-plane corresponds to $\tau>\xi$, the right upper quartile to $\tau<\xi$. Since a partial derivative with respect to $\xi-\tau$ at constant τ is much larger than a partial derivative with respect to τ at constant $\xi-\tau$, partial derivatives with respect to η and ζ will only be of the same order of magnitude if a large change in η and a small change in ζ correspond to the same change in $\xi-\tau$ and τ.† This requires $\varepsilon^n \gg \varepsilon^p$ and therefore, since $0<\varepsilon \ll 1$,

$$n<p. \tag{8.68}$$

From eqns (8.66) and (8.67),

$$\frac{\partial}{\partial \xi}=\frac{\partial \eta}{\partial \xi}\frac{\partial}{\partial \eta}+\frac{\partial \zeta}{\partial \xi}\frac{\partial}{\partial \zeta}=\varepsilon^n \frac{\partial}{\partial \eta} \tag{8.69}$$

and

$$\frac{\partial}{\partial \tau}=\frac{\partial \eta}{\partial \tau}\frac{\partial}{\partial \eta}+\frac{\partial \zeta}{\partial \tau}\frac{\partial}{\partial \zeta}=-\varepsilon^n \frac{\partial}{\partial \eta}+\varepsilon^p \frac{\partial}{\partial \zeta}. \tag{8.70}$$

Equation (8.58) in η, ζ coordinates becomes

$$-2\varepsilon^{n+p}\frac{\partial^2 \phi}{\partial \eta\, \partial \zeta}+\varepsilon^{2p}\frac{\partial^2 \phi}{\partial \zeta^2}=-\varepsilon^{3n+1}\frac{\partial^3 \phi}{\partial \eta^3}+\varepsilon^{2n+p+1}\frac{\partial^3 \phi}{\partial \zeta\, \partial \eta^2}; \tag{8.71}$$

n and p must be chosen so that at least two larger terms balance, i.e. the same smaller power of ε, and all other terms are smaller, i.e. higher powers of ε. Subject to the inequality (8.68), this requires

$$p=2n+1 \tag{8.72}$$

and eqn (8.71) becomes

$$\frac{\partial^3 \phi}{\partial \eta^3}-2\frac{\partial^2 \phi}{\partial \eta\, \partial \zeta}+\varepsilon^{n+1}\left(\frac{\partial^2 \phi}{\partial \zeta^2}-\frac{\partial^3 \phi}{\partial \zeta\, \partial \eta^2}\right)=0. \tag{8.73}$$

The strain–displacement equation $m=\partial u/\partial x$ becomes in inner region coordinates, using eqns (8.57) and (8.69),

$$m=C\varepsilon^n \frac{\partial u}{\partial \eta}, \tag{8.74}$$

where C is a finite non-zero constant. If both u and m are to remain both finite and non-zero in general as $\varepsilon \to 0$, then

$$n=0 \quad \text{and} \quad p=1, \tag{8.75}$$

† If $x=\alpha y$, α constant, then

$$\frac{\mathrm{d}f}{\mathrm{d}x}=\frac{1}{\alpha}\frac{\mathrm{d}f}{\mathrm{d}y}.$$

If $\mathrm{d}f/\mathrm{d}y \gg 1$, then $\mathrm{d}f/\mathrm{d}x$ can be made of order unity by a suitable choice of $\alpha \gg 1$.

on use of eqn (8.72). Equations (8.66), (8.67) and (8.73) become

$$\eta=\xi-\tau, \tag{8.76}$$

$$\zeta=\varepsilon\tau \tag{8.77}$$

and

$$\frac{\partial^3\phi}{\partial\eta^3}-2\frac{\partial^2\phi}{\partial\eta\,\partial\zeta}+\varepsilon\left(\frac{\partial^2\phi}{\partial\zeta^2}-\frac{\partial^3\phi}{\partial\zeta\,\partial\eta^2}\right)=0. \tag{8.78}$$

Look for a solution of eqn (8.78) of the form

$$\phi(\eta,\zeta,\varepsilon)=\Phi_0(\eta,\zeta)+\varepsilon\Phi_1(\eta,\zeta)+\ldots. \tag{8.79}$$

Equation (8.79) is referred to as an inner region expansion and eqn (8.60) an outer region expansion. Substitute for ϕ from eqn (8.79) into eqn (8.78) and equate successive powers of ε to zero:

$$\begin{aligned}&\frac{\partial^3\Phi_0}{\partial\eta^3}-2\frac{\partial^2\Phi_0}{\partial\eta\,\partial\zeta}=0,\\&\frac{\partial^3\Phi_1}{\partial\eta^3}-2\frac{\partial^2\Phi_1}{\partial\eta\,\partial\zeta}+\frac{\partial^2\Phi_0}{\partial\zeta^2}-\frac{\partial^3\Phi_0}{\partial\zeta\,\partial\eta^2}=0,\\&\text{etc.}\end{aligned} \tag{8.80}$$

The equation for Φ_0 can be integrated to give

$$\frac{\partial^2\Phi_0}{\partial\eta^2}-2\frac{\partial\Phi_0}{\partial\zeta}=g(\zeta).$$

Since the inner region merges on either side into an outer region in which ϕ is constant for all ε, Φ_0 is constant, $g(\zeta)=0$, and hence

$$\frac{\partial^2\Phi_0}{\partial\eta^2}-2\frac{\partial\Phi_0}{\partial\zeta}=0. \tag{8.81}$$

Equation (8.81) has a solution

$$\Phi_0=A\ \operatorname{erfc}\frac{\eta}{\sqrt{2\zeta}}+B. \tag{8.82}$$

As $\eta\to+\infty$, the inner region merges into outer region in which $\phi=0$ for all ε; and as $\eta\to-\infty$, the same occurs except that $\phi=1$ for all ε. Hence $B=0$ and $A=\frac{1}{2}$. Therefore,

$$\Phi_0=\frac{1}{2}\operatorname{erfc}\frac{\eta}{\sqrt{2\zeta}}. \tag{8.83}$$

Note, however, that the inner region intersects the τ axis in the neighbourhood of the origin and that therefore the above solution may break down in this neighbourhood. Further terms in the inner region expansion can be obtained by successive substitution of the Φ_i into eqns (8.80). The boundary conditions on the Φ_i from eqn (8.79) and from the facts that $\phi \to 0$ as $\eta \to +\infty$ and $\phi \to 1$ as $\eta \to -\infty$ for all ε, are

$$\Phi_i \to 0 \quad \text{as} \quad \eta \to +\infty \quad \text{and} \quad \Phi_i \to 0 \quad \text{as} \quad \eta \to -\infty, \quad i \neq 0.$$

The solution as $\varepsilon \to 0$ is given by ϕ_0 and Φ_0. We use the symbol† $\sim$, pronounced 'squiggles', to denote this limit, i.e.

$$\phi \sim 0 \text{ in the outer region ahead of } \tau = \zeta;$$

$$\phi \sim \frac{1}{2} \operatorname{erfc} \frac{\eta}{\sqrt{2\zeta}} \text{ in the inner region;}$$

$$\phi \sim 1 \text{ in the outer region behind } \tau = \zeta.$$

Since $\frac{1}{2} \operatorname{erfc}(\eta/\sqrt{2\zeta}) \to 1$ and 0 as $\eta \to -\infty$ and $+\infty$ respectively, the expression $\frac{1}{2} \operatorname{erfc}(\eta/\sqrt{2\zeta})$ can be used at both limits.

Substitute back for η and ζ in terms of ξ and η and then in terms of x and t, eqns (8.76), (8.77) and (8.57), for ε from eqn (8.59) and for ϕ from eqn (8.57), to give the solution for small v as

$$\sigma \sim \frac{\sigma_0}{2} \operatorname{erfc} \frac{x - \sqrt{(M/\rho)}\, t}{\sqrt{2vt/\rho}}. \tag{8.84}$$

At fixed t, the change in x between two fixed values of σ is proportional to $\sqrt{t}$. The effect of the viscous term in eqn (8.53) is to spread out the discontinuous change of the elastic solution. In the limit as $v \to 0$ for x and t both positive, the r.h.s. of eqn (8.84) tends to zero if $x > \sqrt{(M/\rho)}t$ and to σ_0 if $x < \sqrt{(M/\rho)}t$, i.e. the r.h.s. of eqn (8.56), the solution for the elastic solid.

As $t \to +0$, the r.h.s. of eqn (8.84) tends to zero for all $x > 0$ and so satisfies the given initial condition. When $x = 0$ and $t \neq 0$,

$$\sigma_{x=0} \sim \frac{\sigma_0}{2} \operatorname{erfc}\left(-\sqrt{\frac{Mt}{2v}}\right).$$

† $\sim$ is defined as 'is asymptotically equal to'. A notation alternative to $f(x) \sim g(x)$ is $f(x) = g(x)(1 + o(1))$ as a given limit is approached.

As $t \to +0$, $\sigma_{x=0} \to \frac{1}{2}\sigma_0$; and as $t \to -\infty$, $\sigma_{x=0} \to \sigma_0$. Since $\mathrm{erfc}(-2) = 1.995$, $\sigma_{x=0}$ is within $\frac{1}{4}$ per cent of σ_0 for $Mt/v > 8$ or $t > 8v/M$. For $t \gg v/M$, $\sigma_{x=0}$ is nearly equal to the given boundary condition $\sigma_{x=0} = \sigma_0$. Since the problem has a typical velocity $\sqrt{M/\rho}$, $t \gg v/M$ corresponds to $x \gg (v/M)\sqrt{M/\rho} = v/\sqrt{M\rho}$. If we now choose $L \gg v/\sqrt{M\rho}$, then by eqn (8.59) $\varepsilon \ll 1$ as required. Both the stated boundary condition and the asymptotic series (8.60) and (8.79) are only valid for $t \gg v/M$ and $x \gg v/\sqrt{M\rho}$. It follows that the solution (8.84) is only valid in the x, t-plane for $x > 0$ and $t > 0$, provided a small region about the origin is excluded.

The procedure adopted in this section of joining together different asymptotic expansions in inner and outer regions is known as 'matched asymptotic expansions'. The technical difficulty increases exponentially with the number of terms required. That is why we have stopped at one term in each expansion for the problem of this section. It is not uncommon to arrive at solutions in asymptotics valid only in part of the region of interest.[†] The final check on the solution frequently can be made only against experiment. For linear problems it is often possible to check results partially or totally by Laplace transform, but this method is not available for non-linear problems. It will not be pursued here except in the next exercise, where the solution of the problem of this section is summarized for small time, the details being left to the reader.

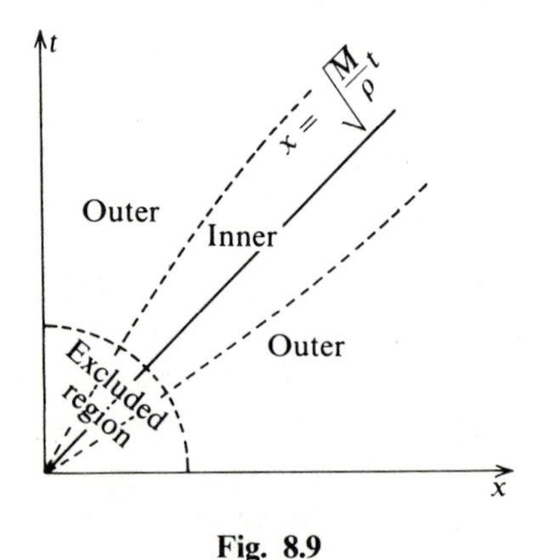

Fig. 8.9

[†] For example, the Stokes phenomenon for asymptotic expansions in the complex plane.

Exercise 8.4. Introduce non-dimensional variables

$$\xi=\frac{\sqrt{\rho M}}{v}x, \quad \tau=\frac{Mt}{v} \quad \text{and} \quad S=\frac{\sigma}{\sigma_0}$$

so that eqn (8.53) becomes

$$S_{\tau\tau}=S_{\xi\xi}+S_{\xi\xi\tau} \quad \text{in} \quad \tau>0,\ \xi>0$$

with

$$S=0 \quad \text{on} \quad \tau=0 \quad \text{for} \quad \xi>0$$

and

$$S=1 \quad \text{on} \quad \xi=0 \quad \text{for} \quad \tau>0.$$

Application of the Laplace transform

$$\bar{S}(\xi, p)=\int_0^\infty S(\xi, \tau)\mathrm{e}^{-p\tau}\,\mathrm{d}\tau$$

gives

$$p^2\bar{S}=(1+p)\bar{S}_{\xi\xi} \quad \text{and} \quad \bar{S}(0, p)=p^{-1},$$

provided $S_\tau=0$ on $\tau=0$ for $\xi>0$. This must be checked from the solution below.

The solution for $\bar{S}$ which tends to zero for large ξ is

$$S(\xi, p)=p^{-1}\exp\left(-\frac{p\xi}{\sqrt{p+1}}\right).$$

The behaviour of $S(\xi, \tau)$ as $\tau\to 0$ depends on the behaviour of $\bar{S}(\xi, p)$ as $|p|\to\infty$. This limit gives the solution in the excluded region,

$$\bar{S}(\xi, p)=p^{-1}\exp\left(-\sqrt{p}\,\xi\left(1+\frac{1}{p}\right)^{-\frac{1}{2}}\right).$$

Therefore,

$$\bar{S}(\xi, p) \underset{|p|\to\infty}{\sim} p^{-1}\exp(-\sqrt{p}\,\xi).$$

Therefore,

$$S(\xi, \tau) \underset{\tau\to 0}{\sim} \operatorname{erfc}\frac{\xi}{2\sqrt{\tau}}, \quad \xi>0,$$

or

$$S(x, t) \underset{t\to 0}{\sim} \operatorname{erfc}\left(\frac{x}{2}\sqrt{\frac{p}{vt}}\right), \quad x>0.$$

M does not appear in the above equation. When $M=0$, eqn (8.53) after one integration reduces to the heat conduction or diffusion equation. The solution in the excluded region is diffusive and not wave-like as is the rest of the solution. Note that the scaling for the exact Laplace transform solution removes the explicit dependence on a small parameter, whereas the asymptotic solution depends on the presence of a small parameter which therefore must be retained in any scaling.

8.4. The thermoelastic half-space problem

The half-space $x>0$ is at rest for $t<0$ and a stress is applied to the boundary $x>0$ for $t>0$ so that the strain m is given by

$$m=\varepsilon_{xx}=\frac{\partial u}{\partial x}=\frac{\partial \hat{u}}{\partial \xi}=A\,H(t), \tag{8.85}$$

A constant, on $x=0$. When $\varepsilon=0$, eqn (8.50) for $\hat{u}$ becomes

$$\frac{\partial^2 \hat{u}}{\partial \xi^2}-\frac{\partial^2 \hat{u}}{\partial \tau^2}=0. \tag{8.86}$$

ξ, τ and $\hat{u}$ are defined by eqns (8.47) and (8.48).

$$\frac{\partial^2 m}{\partial \xi^2}-\frac{\partial^2 m}{\partial \tau^2}=0,$$

with solution to satisfy eqn (8.85),

$$m=A\,H(\tau-\xi). \tag{8.87}$$

For ε small but non-zero, the isentropic solution given by $\varepsilon=0$ and shown in Fig. 8.10 is only a good approximation provided the typical length of the part of the wave under consideration is large compared to λ_0, see page 239. It follows that ahead of and behind $\xi=\tau$ eqn (8.87) is a good approximation, but in the region about $\xi=\tau$ the approximation is invalid and the full eqns (8.50) and (8.51) must be used.

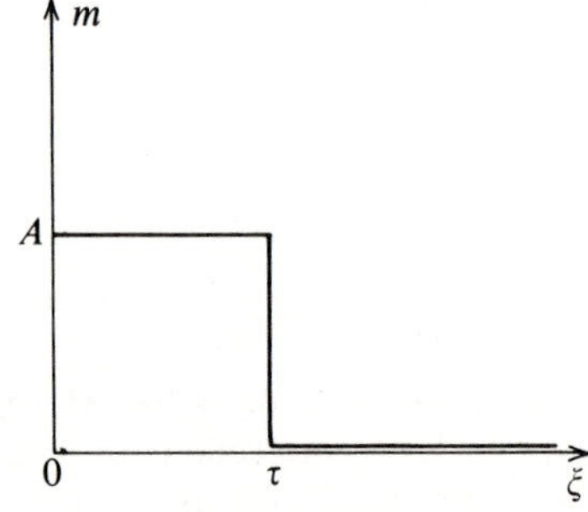

Fig. 8.10

If $\hat{S}$ is eliminated from eqns (8.50) and (8.51),

$$\left(\frac{\partial^2}{\partial\xi^2}-\frac{\partial}{\partial\tau}\right)\left(\frac{\partial^2\hat{u}}{\partial\xi^2}-\frac{\partial^2\hat{u}}{\partial\tau^2}\right)=\varepsilon\frac{\partial^4\hat{u}}{\partial\xi^4} \tag{8.88}$$

or

$$(1-\varepsilon)\frac{\partial^4\hat{u}}{\partial\xi^4}-\frac{\partial^4\hat{u}}{\partial\xi^2\partial\tau^2}=\frac{\partial}{\partial\tau}\left(\frac{\partial^2\hat{u}}{\partial\xi^2}-\frac{\partial^2\hat{u}}{\partial\tau^2}\right), \tag{8.89}$$

a fourth-order equation with characteristics $d\xi/d\tau=\pm\sqrt{1-\varepsilon}$ and $d\tau=0$ (twice), or

$$\xi=\pm\sqrt{1-\varepsilon}\,\tau+\text{constant and } \tau=\text{constant (twice)}. \tag{8.90}$$

The characteristics of eqn (8.86) are

$$\xi=\pm\tau+\text{constant}. \tag{8.91}$$

Since $\sqrt{1-\varepsilon}<1<+\infty$, the characteristic velocity of the approximate equation (8.86) lies between those of the more exact eqn (8.89) as illustrated in Fig. 8.11. The same result arose for eqns (7.133) and (7.142) in Section 7.6, where stability required $\gamma>c_0$, eqn (7.145), and hence $\gamma>c_0>c_s$.

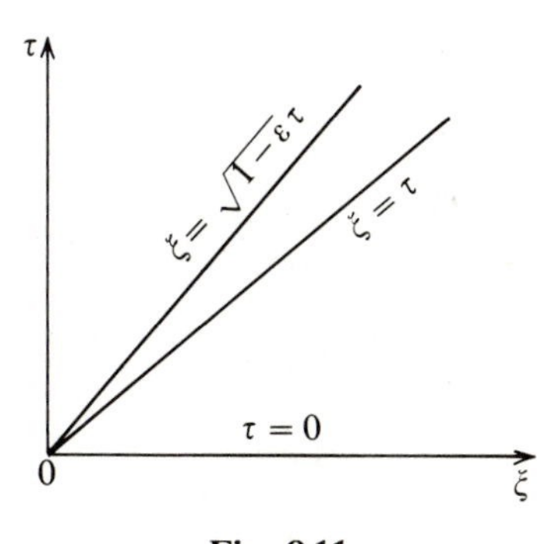

Fig. 8.11

If we look for a solution of eqn (8.88), $m=\partial\hat{u}/\partial\xi$ satisfies the same equation, of the form

$$m(\zeta,\tau,\varepsilon)=m_0(\zeta,\tau)+\varepsilon m_1(\zeta,\tau)+\ldots,$$

then m_0 satisfies the classical wave equation with $m_0=A\,H(\tau-\zeta)$ as before. The alternative equation for m_0, namely

$$\frac{\partial^2 m_0}{\partial \xi^2}-\frac{\partial m_0}{\partial \tau}=0$$

is the thermal equation.

Since the solution $m_0=A\,H(\tau-\xi)$ is inadmissible about $\xi=\tau$ and the spatial derivative normal to $\xi=\tau$ is much larger than the tangential derivative, try the same inner coordinates as in the last section, namely

$$\eta=\xi-\tau \quad \text{and} \quad \zeta=\varepsilon\tau, \tag{8.92}$$

so that

$$\frac{\partial}{\partial \xi}=\frac{\partial}{\partial \eta} \quad \text{and} \quad \frac{\partial}{\partial \tau}=-\frac{\partial}{\partial \eta}+\varepsilon\frac{\partial}{\partial \zeta}. \tag{8.93}$$

Equation (8.88) becomes

$$\left(\frac{\partial^2}{\partial \eta^2}+\frac{\partial}{\partial \eta}-\varepsilon\frac{\partial}{\partial \zeta}\right)\left(2\varepsilon\frac{\partial^2 \hat{u}}{\partial \eta\,\partial \zeta}-\varepsilon^2\frac{\partial^2 \hat{u}}{\partial \zeta^2}\right)=\varepsilon\frac{\partial^4 \hat{u}}{\partial \eta^4}. \tag{8.94}$$

Divide through by ε and put

$$\hat{u}(\eta,\zeta,\varepsilon)=F_0(\eta,\zeta)+\varepsilon F_1(\eta,\zeta)+\ldots.$$

The term independent of ε is

$$2\frac{\partial^4 F_0}{\partial \eta^3\,\partial \zeta}+2\frac{\partial^3 F_0}{\partial \eta^2\,\partial \zeta}-\frac{\partial^4 F_0}{\partial \eta^4}=0.$$

Put $m(\eta,\zeta,\varepsilon)=M_0(\eta,\zeta)+\varepsilon M_1(\eta,\zeta)+\ldots$. Since $m=\partial\hat{u}/\partial\xi=\partial\hat{u}/\partial\eta$,

$$M_0=\frac{\partial F_0}{\partial \eta} \tag{8.95}$$

and

$$2\frac{\partial^3 M_0}{\partial \eta^2\,\partial \zeta}+2\frac{\partial^2 M_0}{\partial \eta\,\partial \zeta}-\frac{\partial^3 M_0}{\partial \eta^3}=0.$$

On integration,

$$2\frac{\partial^2 M_0}{\partial \eta\,\partial \zeta}+2\frac{\partial M_0}{\partial \zeta}-\frac{\partial^2 M_0}{\partial \eta^2}=0, \tag{8.96}$$

provided $M_0\to 0$ as $\eta\to+\infty$ for all ζ; this proviso will be verified later.

There is a discontinuity in m at the origin, since $m=A$ on $x=0$ and $m=0$ on $t=0$. Unlike the problem of Section 8.3, the full governing equation of this problem, eqn (8.89) possess a characteristic with finite non-zero reduced velocity $\mathrm{d}\xi/\partial\tau=\sqrt{1-\varepsilon}=1-\varepsilon/2$, correct to first order in ε. Since eqn (8.89) is linear, shock velocity is equal to the characteristic velocity and the attenuation of the shock is equal to the limiting attenuation of sinusoidal waves as $\lambda\to 0$ or $\omega\to\infty$, i.e. $c_1\varepsilon/(2\beta)$ by eqn (8.39). These results will be checked in Chapter 10. Since $\xi=(c_1/\beta)x$, the attenuation in ξ, τ-space is $\varepsilon/2$. The attenuation is the reciprocal of the distance x_0 or ξ_0 travelled by the wave for its amplitude to decrease by $1/\mathrm{e}$. Hence

$$x_0 c_1\frac{\varepsilon}{2\beta}=\frac{\beta}{c_1}\xi_0 c_1\frac{\varepsilon}{2\beta}=\frac{\varepsilon}{2}\xi_0 .$$

The last row in Table 8.1, apart from a factor $1-\varepsilon/4$, gives the distance in centimetres travelled by the shock in some real materials for it to decrease in amplitude by a factor $1/\mathrm{e}$.

The forward moving shock passing through the origin has equation

$$\xi=\left(1-\frac{\varepsilon}{2}\right)\tau, \tag{8.97}$$

correct to first order in ε. Since $[m]=A-0=A$ at the origin, on the shock

$$[m]=A\,\mathrm{e}^{-\frac{1}{2}\varepsilon\xi}=A\,\mathrm{e}^{-\frac{1}{2}\varepsilon\tau}, \tag{8.98}$$

correct to first order ε in the exponential index. Substitute from eqns (8.92) into eqns (8.97) and (8.98):

$$[m]=A\,\mathrm{e}^{-\frac{1}{2}\zeta} \tag{8.99}$$

on

$$\eta+\tfrac{1}{2}\zeta=0. \tag{8.100}$$

Consistent with the outer region solution $m_0=A\,H(\tau-\xi)$, we take across the shock $m_-=A$ and $m_+=M_- -[m]=A(1-\mathrm{e}^{-\frac{1}{2}\zeta})$. Now $m=0$ for $\zeta=\varepsilon\tau=0$ with $\eta\geqslant 0$ and $m\to 0$ as $\eta=\xi-\tau\to\infty$ at fixed τ, i.e. at fixed ζ, both limits for all ε. The condition as $\eta\to\infty$ follows from the requirement that the inner region solution must merge into the outer region solution $m_0=A\,H(\tau-\zeta)=A\,H(-\eta)$ as $\eta\to\infty$. Substitute from eqn (8.95) into $m_+=A(1-\mathrm{e}^{-\frac{1}{2}\zeta})$, $m=0$ on $\zeta=0$ and $m\to 0$ as $\eta\to\infty$ and let $\varepsilon\to 0$. This gives

$$M_0 = A(1-e^{-\frac{1}{2}\zeta}) \quad \text{on} \quad \eta+\tfrac{1}{2}\zeta=0, \tag{8.101}$$

$$M_0 = 0 \quad \text{on} \quad \zeta=0, \quad \eta>0, \tag{8.102}$$

$$M_0 \to 0 \quad \text{as} \quad \eta\to\infty, \quad \text{all} \quad \zeta>0, \tag{8.103}$$

as the boundary conditions for eqn (8.96). The relevant section of the η, ζ plane is on and to the right of the line $\eta+\frac{1}{2}\zeta=0$ in Fig. 8.12.

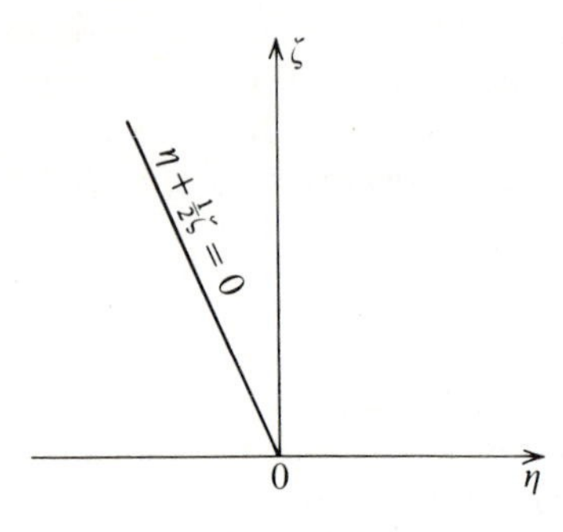

Fig. 8.12

The characteristics of eqn (8.96) are given by eqn (5.7) as

$$2\,d\eta\,d\zeta + (d\zeta)^2 = 0,$$

or

$$\zeta = \text{constant} \quad \text{and} \quad \eta+\tfrac{1}{2}\zeta = \text{constant}. \tag{8.104}$$

Since $\zeta=0$ and $\eta+\frac{1}{2}\zeta=0$ are both characteristics of and boundaries for the partial differential equations, it is convenient to introduce what are known as characteristic coordinates ζ, ϕ, one of which is constant on each characteristic, given by

$$\phi = \eta+\tfrac{1}{2}\zeta \quad \text{and} \quad \zeta=\zeta. \tag{8.105}$$

Then

$$\frac{\partial}{\partial\eta} = \frac{\partial}{\partial\phi}, \quad \left(\frac{\partial}{\partial\zeta}\right)_\eta = \frac{1}{2}\frac{\partial}{\partial\phi} + \left(\frac{\partial}{\partial\zeta}\right)_\phi \tag{8.106}$$

and eqn (8.96) becomes

$$2\frac{\partial^2 M_0}{\partial\phi\,\partial\zeta} + \frac{\partial M_0}{\partial\phi} + 2\frac{\partial M_0}{\partial\zeta} = 0, \tag{8.107}$$

with boundary conditions

$$M_0 = A(1 - e^{-\frac{1}{2}\zeta}) \quad \text{on} \quad \phi = 0, \tag{8.108}$$

$$M_0 = 0 \quad \text{on} \quad \zeta = 0, \quad \phi \geqslant 0, \tag{8.109}$$

$$M_0 \to 0 \quad \text{as} \quad \phi \to \infty, \quad \text{all} \quad \zeta \geqslant 0. \tag{8.110}$$

The solution is over the quarter-plane $\phi > 0$ and $\zeta > 0$. The lines in Fig. 8.13 are characteristics. Note that a second-order hyperbolic p.d.e. referred to characteristic coordinates contains only a mixed second-order partial derivative.

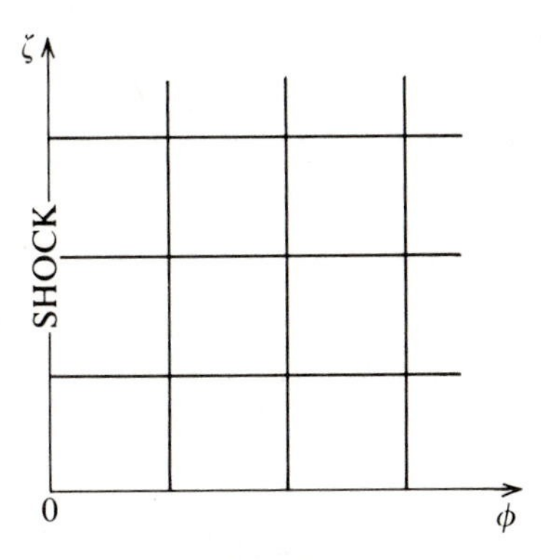

Fig. 8.13

Apply the Laplace transform to the ζ variable in eqn (8.107):

$$2\left(p\frac{d\bar{M}_0}{d\phi} - \left(\frac{\partial M_0}{\partial \phi}\right)_{\zeta=0}\right) + \frac{d\bar{M}_0}{d\phi} + 2(p\bar{M}_0 - (M_0)_{\zeta=0}) = 0.$$

Use of boundary condition (8.109) gives

$$(2p+1)\frac{d\bar{M}_0}{d\phi} + 2p\bar{M}_0 = 0;$$

on integration,

$$\bar{M}_0(\phi, p) = B(p)\exp\left(-\frac{p}{p+\frac{1}{2}}\phi\right).$$

From eqn (8.108),

$$\bar{M}_0(0, p) = A\left(\frac{1}{p} - \frac{1}{p+\frac{1}{2}}\right).$$

Hence,

$$B(p)=A\left(\frac{1}{p}-\frac{1}{p+\frac{1}{2}}\right),$$

$$\bar{M}_0(\phi,p)=\frac{1}{2}A\frac{1}{p(p+\frac{1}{2})}\exp\left(-\frac{p}{p+\frac{1}{2}}\phi\right),$$

and

$$\frac{\mathrm{d}\bar{M}_0}{\mathrm{d}\phi}=-\frac{1}{2}A\,\mathrm{e}^{-\phi}\frac{1}{(p+\frac{1}{2})^2}\exp\left(\frac{1}{p+\frac{1}{2}}\frac{\phi}{2}\right).$$

The inverse Laplace transform† of $p^{-2}\exp(\frac{1}{2}\phi/p)$ is $(2\zeta/\phi)^{\frac{1}{2}}$ $I_1(\sqrt{2\phi\zeta})$. Hence,

$$\frac{\partial M_0}{\partial\phi}=-\frac{1}{2}A\,\mathrm{e}^{-\phi}\,\mathrm{e}^{-\frac{1}{2}\zeta}\left(\frac{2\zeta}{\phi}\right)^{\frac{1}{2}}I_1(\sqrt{2\phi\zeta}),$$

and, using eqn (8.109),

$$M_0(\phi,\zeta)=A(1-\mathrm{e}^{-\frac{1}{2}\zeta})-A\mathrm{e}^{-\frac{1}{2}\zeta}\sqrt{\frac{\zeta}{2}}\int_0^{\phi}\psi^{-\frac{1}{2}}\,\mathrm{e}^{-\psi}I_1(\sqrt{2\psi\zeta})\,\mathrm{d}\psi\,.$$

Now,‡

$$\int_0^{\infty}\psi^{-\frac{1}{2}}\,\mathrm{e}^{-\psi}I_1(\sqrt{2\psi\zeta})\,\mathrm{d}\psi=\sqrt{\frac{2}{\zeta}}(\mathrm{e}^{\frac{1}{2}\zeta}-1)\,.$$

Hence

$$M_0(\phi,\zeta)=A\,\mathrm{e}^{-\frac{1}{2}\zeta}\sqrt{\frac{\zeta}{2}}\int_{\phi}^{\infty}\psi^{-\frac{1}{2}}\,\mathrm{e}^{-\psi}I_1(\sqrt{2\psi\zeta})\,\mathrm{d}\psi\,. \quad (8.111)$$

The boundary condition (8.110) is satisfied.

In Fig. 8.14, M_0/A is plotted against ϕ/ζ for particular values of ζ. Using eqns (8.105) and (8.92), $\phi/\zeta=0$ corresponds to the thermoelastic shock, $\phi/\zeta=\frac{1}{2}$ to the approximate isentropic shock. For $\zeta=40$, M_0/A has decreased from almost unity to almost zero as ϕ/ζ increases from $0+$ to 1. This suggests that a simpler asymptotic form for M_0/A may exist for large ζ for non-zero ϕ.

Now $I_1(x)\sim\mathrm{e}^x(2\pi x)^{-\frac{1}{2}}$ as $x\to+\infty$. Therefore from equation (8.111),

---

† See tables of Laplace transforms.

‡ See tables of definite integrals.

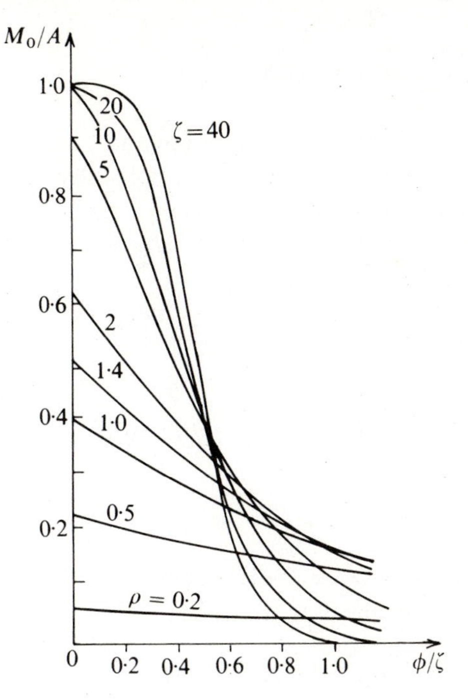

Fig. 8.14

$$M_0 \underset{\zeta\to\infty}{\sim} A\,2^{-\frac{5}{4}}\zeta^{\frac{1}{4}}\pi^{-\frac{1}{2}}\int_{\phi}^{\infty}\psi^{-\frac{3}{4}}\exp(-(\tfrac{1}{2}\zeta+\psi-\sqrt{2\psi\zeta}))\,\mathrm{d}\psi$$

$$\underset{\psi=\zeta\theta}{\sim} A\,2^{-\frac{5}{4}}\zeta^{\frac{1}{2}}\pi^{-\frac{1}{2}}\int_{\phi/\zeta}^{\infty}\theta^{-\frac{3}{4}}\exp-\zeta\left(\left(\sqrt{\theta}-\frac{1}{\sqrt{2}}\right)^{2}\right)\mathrm{d}\theta$$

$$\underset{\sqrt{\theta}-1/\sqrt{2}=x}{\sim} A\,2^{-\frac{1}{4}}\zeta^{\frac{1}{2}}\pi^{-\frac{1}{2}}\int_{\sqrt{\phi/\zeta}-1/\sqrt{2}}^{\infty}\left(x+\frac{1}{\sqrt{2}}\right)^{-\frac{1}{2}}\exp(-\zeta x^2)\mathrm{d}x\,.$$

As $\zeta\to+\infty$, the main contribution to the integral comes from the neighbourhood of the smallest value of x^2; the term $(x+1/\sqrt{2})^{-\frac{1}{2}}$ is replaced by its value at this minimum. When $\sqrt{\phi/\zeta}<1/\sqrt{2}$, the minimum is at $x=0$; when $\sqrt{\phi/\zeta}>1/\sqrt{2}$, the minimum is at the lower limit. Two simple calculations give

$$M_0 \underset{\zeta\to\infty}{\sim} \frac{A}{2}\operatorname{erfc}\left(\sqrt{\phi}-\sqrt{\frac{\zeta}{2}}\right) \qquad \text{for } \sqrt{\frac{\phi}{\zeta}}<\frac{1}{\sqrt{2}} \qquad (8.112)$$

and

$$M_0 \underset{\zeta\to\infty}{\sim} \frac{A}{2}\left(\frac{\zeta}{2\phi}\right)^{\frac{1}{4}} \operatorname{erfc}\left(\sqrt{\phi}-\sqrt{\frac{\zeta}{2}}\right) \quad \text{for } \sqrt{\frac{\phi}{\zeta}} \geqslant \frac{1}{\sqrt{2}}. \tag{8.113}$$

Since ϕ and ζ are both very large in eqn (8.113) and erfc $x\to 0$ as $x\to\infty$ (erfc $2\simeq 0.005$), the factor $(\frac{1}{2}\zeta/\phi)^{\frac{1}{4}}$ differs appreciably from unity only when $\operatorname{erfc}(\sqrt{\phi}-\sqrt{\zeta/2})$ is exponentially small. Hence, eqn (8.112) may be used for all $\phi>0$ for large ζ.

When $\phi=\zeta/2$ or $\xi=\tau$, $\operatorname{erfc}(\sqrt{\phi}-\sqrt{\zeta/2})=1$. For $|\sqrt{\phi}-\sqrt{\zeta/2}|$ of order unity with ζ large, ϕ is also large and $\phi-\zeta/2$ is small compared to $\zeta/2$. Hence

$$\sqrt{\phi}-\sqrt{\zeta/2}=\sqrt{\frac{\zeta}{2}+\left(\phi-\frac{\zeta}{2}\right)}-\sqrt{\frac{\zeta}{2}}=\sqrt{\frac{\zeta}{2}}\left(1+\frac{1}{2}\frac{\phi-\zeta/2}{\zeta/2}-1\right)$$
$$=\frac{\phi-\zeta/2}{\sqrt{2\zeta}},$$

neglecting terms of $O(2\phi/\zeta-1)^2$ compared to unity. Substitute for $\sqrt{\phi}-\sqrt{\zeta/2}$ into eqns (8.112) to give

$$M_0 \underset{\zeta\to\infty}{\sim} \frac{A}{2}\operatorname{erfc}\frac{\phi-\zeta/2}{\sqrt{2\zeta}} \quad \text{or} \quad M_0 \underset{\zeta\to\infty}{\sim} \frac{A}{2}\operatorname{erfc}\frac{\eta}{\sqrt{2\zeta}}, \tag{8.114}$$

i.e.

$$M_0 \underset{\tau\to\infty}{\sim} \frac{A}{2}\operatorname{erfc}\frac{\xi-\tau}{\sqrt{2\varepsilon\tau}}. \tag{8.115}$$

For $|\sqrt{\phi}-\sqrt{\zeta/2}|\gg 1$, eqn (8.115) still gives the same result as eqn (8.112), because for $\sqrt{\phi}-\sqrt{\zeta/2}\gg 1$ both erfcs are effectively zero and for $\sqrt{\phi}-\sqrt{\zeta/2}\ll -1$ both erfcs are effectively $+2$.

Equations (8.83) and (8.114) are formally identical. After the appropriate scalings, the roles of viscoelasticity and thermoelasticity are the same at large values of the time—the zero-thickness shock of isentropic elasticity is spread out about $\xi=\tau$, the width of spread being proportional to $t^{\frac{1}{2}}$. However, at very small times the thermoelastic material supports a rapidly decaying shock. The corresponding solution for the viscoelastic solid has not been found but it cannot contain a shock, because there are no finite non-zero characteristic velocities.

To sum up, the solution of this section is obtained from the full shock equations, from the first term of the asymptotic expansions in the inner region about the position of the isentropic shock, and from the constant first terms in the two outer regions. It is

$$m = A, \quad 0 \leqslant \xi < (1-\varepsilon/2)\tau \tag{8.87}$$

and

$$m = A\,\mathrm{e}^{-\frac{1}{2}\zeta}\sqrt{\zeta/2}\int_{\phi}^{\infty}\psi^{-\frac{1}{2}}\mathrm{e}^{-\psi}I_1(\sqrt{2\psi\zeta})\,\mathrm{d}\psi, \quad \zeta \geqslant (1-\varepsilon/2)\tau, \tag{8.111}$$

where the equations between coordinates are (8.47), (8.92) and (8.105). The asymptotic form of the solution, valid for times after the shock has decayed away is

$$m \sim \frac{1}{2}A\,\mathrm{erfc}\,\frac{\xi-\tau}{\sqrt{2\varepsilon\tau}}. \tag{8.115}$$

The last line of Table 8.1 shows that for typical lengths greater than 1 mm, eqn (8.115) will be satisfactory. However, in such applications as coatings, where thicknesses are measured in micrometres, eqns (8.87) and (8.111) will be needed.

Answers to exercises

8.1. As $\omega \to 0$, $\hat{\alpha} \to 1 + \mathrm{O}(\hat{\omega})$;

as $\omega \to \infty$, $\hat{\alpha} \to \dfrac{1}{\sqrt{1-\varepsilon}} - \dfrac{\mathrm{i}\varepsilon}{2\sqrt{1-\varepsilon}\,\hat{\omega}}$.

9 Unidirectional traffic flow

9.1. The governing partial differential equation

The simplest wave equation describing motion in the positive x-direction only is

$$\frac{\partial \phi}{\partial t}+c\frac{\partial \phi}{\partial x}=0, \quad c \text{ constant,} \tag{9.1}$$

with solution $\phi=\phi(x-ct)$. More complicated first-order equations of the form

$$P(t,x,\phi)\frac{\partial \phi}{\partial t}+Q(t,x,\phi)\frac{\partial \phi}{\partial x}=R(t,x,\phi) \tag{9.2}$$

can still be solved by the methods of Chapter 5 unless either characteristics intersect or there is a region in which there are apparently no characteristics. How is the solution then continued? It is often helpful when faced with a new mathematical problem to study simultaneously a physical problem to which the mathematics applies. This may suggest new means of approach. It will now be shown that unidirectional traffic flow is governed by a non-linear first-order p.d.e., which is a particular case of eqn (9.2).

Consider a stretch of one-lane motorway between two junctions. Flow is in one direction only and no overtaking occurs. The basic assumption of the theory is that many essential features of the flow will be reproduced if the flow is treated as a continuum instead of a succession of individual vehicles, all called cars for simplicity. The density $\rho(x,t)$ is equal to the number of cars per unit length at position x at time t and $v(x,t)$ is the velocity of a car. The car flux $q(x,t)$, the number of cars passing a fixed point in unit time, is equal to ρv. By conservation of cars, the net number of cars entering the segment of motorway between x and $x+\mathrm{d}x$ in time interval $\mathrm{d}t$ must equal the increase in the number of cars in this segment in time $\mathrm{d}t$, i.e.

$$q(x)\,\mathrm{d}t-q(x+\mathrm{d}x)\,\mathrm{d}t=\frac{\partial \rho}{\partial t}\,\mathrm{d}x\,\mathrm{d}t$$

or

$$\frac{\partial \rho}{\partial t}+\frac{\partial q}{\partial x}=0. \tag{9.3}$$

We now postulate that each driver adjusts his speed to the traffic density in his immediate vicinity, i.e. $v=v(\rho)$, and hence

$$q=\rho v=Q(\rho). \tag{9.4}$$

Eliminate q between eqns (9.3) and (9.4):

$$\frac{\partial \rho}{\partial t}+Q'(\rho)\frac{\partial \rho}{\partial x}=0, \tag{9.5}$$

a first-order non-linear p.d.e. for $\rho(x, t)$.

$Q(\rho)=q=\rho v$ is zero when $\rho=0$ or when $v=0$. In the latter case the cars are stationary, nose to tail, and $\rho=\rho_j$, j for jam. $0\leqslant\rho\leqslant\rho_j$. We shall assume that $Q(\rho)$ has the form shown in Fig. 9.1 with $Q''(\rho)<0$, $0\leqslant\rho\leqslant\rho_j$. Near $\rho=0$, the continuum theory is not a good approximation, but clearly $q=0$ when $\rho=0$, since there are no cars on the road!

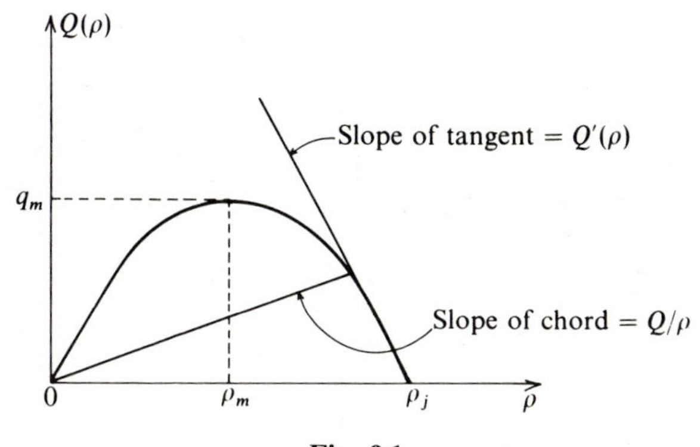

Fig. 9.1

The associated equations for the first-order p.d.e., eqn (9.5), are

$$\frac{\mathrm{d}t}{1}=\frac{\mathrm{d}x}{Q'(\rho)}=\frac{\mathrm{d}\rho}{0}, \tag{9.6}$$

with solution

$$\rho=\text{constant on } \frac{\mathrm{d}x}{\mathrm{d}t}=Q'(\rho). \tag{9.7}$$

Since $Q'(\rho)$ is constant when ρ is constant, the characteristics are straight lines in the x, t-plane. The velocity of a car, $v=q/\rho$, is the slope of the chord from the origin to the point (ρ, q) in Fig. 9.1 and the characteristic velocity $Q'(\rho)$ is the slope of the tangent. Since $Q'(\rho)=dq/d\rho=v+\rho\,\partial v/\partial\rho$ and $\partial v/\partial\rho<0$, i.e. drivers decrease speed as density increases, $Q'(\rho)<v$. The characteristic velocity is less than the car velocity and waves propagate backwards down the stream of cars. $Q'(\rho)$ decreases as ρ increases.

Consider at time t_0 a density distribution $\rho(x, t_0)$ in the form of a hump which has a maximum ρ_m at some point as shown in the diagram below. Let $\rho=\rho_0$ either side of the hump and let $\rho_a=(\rho_m+\rho_0)/2$. Since $\rho_m>\rho_a>\rho_0$, $Q'(\rho_m)<Q'(\rho_a)<Q'(\rho_0)$. Hence, using eqn (8.7), the point on the 'hump' with density ρ_m travels slower than the point with density ρ_a, which travels slower than the point with density ρ_0. The shape of the 'hump' at three later times t_1, t_2 and t_3, $t_0<t_1<t_2<t_3$, can be sketched—the point with density ρ_m has travelled a distance $(t_1-t_0)Q'(\rho_m)$ by time t_1, etc. The result is shown in Fig. 9.2.

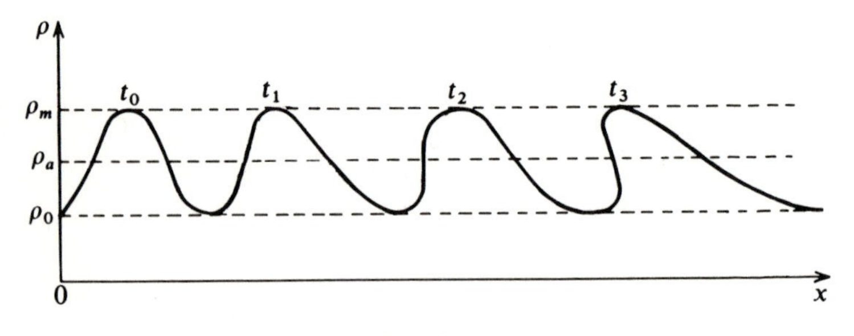

Fig. 9.2

The pulse steepens at the back and flattens at the front (time t_1). Eventually a vertical segment appears at the back (time t_2). The vertical segment, however, corresponds to a point in x, t-space at which ρ is discontinuous and therefore at which at least one of the partial derivatives $\partial\rho/\partial x$ and $\partial\rho/\partial t$ no longer exists; the governing p.d.e., eqn (8.5), is then meaningless. However, the physical principle, which eqn (8.5) expresses when $\partial\rho/\partial x$ and $\partial\rho/\partial t$ exist, namely conservation of cars, still applies.

9.2. Strong discontinuities or shocks

Let the discontinuity in ρ move with velocity U. Let quantities just on that side of the discontinuity with greater value of x be denoted by the suffix $+$, and those on that side with lesser value of x with suffix $-$, as shown in Fig. 9.3; U may not be positive. Denote by $[f]$ the change in any variable across the discontinuity from the positive to the negative side, i.e.

$$[f]=f_{-}-f_{+}. \tag{9.8}$$

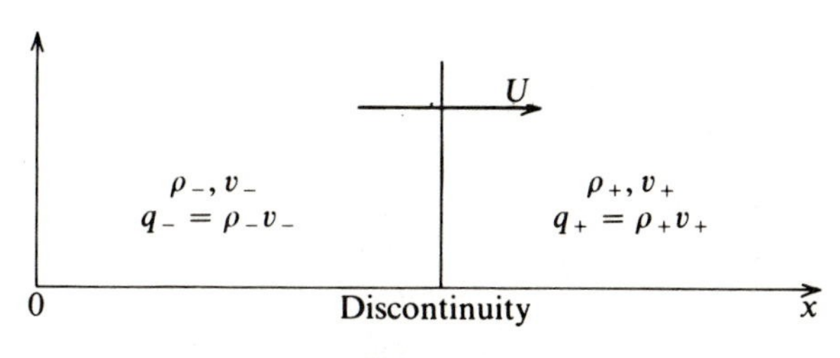

Fig. 9.3

Since the discontinuity is of zero thickness, the car flux into the discontinuity from the negative side must equal the car flux out on the positive side, i.e.

$$\rho_{-}(v_{-}-U)=\rho_{+}(v_{+}-U)$$

or

$$\rho_{-}v_{-}-\rho_{+}v_{+}=U\rho_{-}-U\rho_{+}$$

or

$$[q]=U[\rho]. \tag{9.9}$$

Equation (9.9) is the mathematical form of the principle of conservation of cars when ρ is discontinuous. When $\partial\rho/\partial x$ and $\partial\rho/\partial t$ exist, it is eqn (9.5).

Since the highest-order derivatives of ρ which occur in the partial differential equation (8.5) are first order, discontinuities in ρ itself are strong. Following the nomenclature of continuum mechanics, henceforth we shall call a strong discontinuity a 'shock' and eqn (8.9) will be called the shock equation for unidirectional traffic flow.

The shock velocity $U=(q_{-}-q_{+})/(\rho_{-}-\rho_{+})$ is the slope of the chord joining the two states on either side of the shock (Fig. 9.4). It is not equal to the characteristic velocity on either side, the slope of the

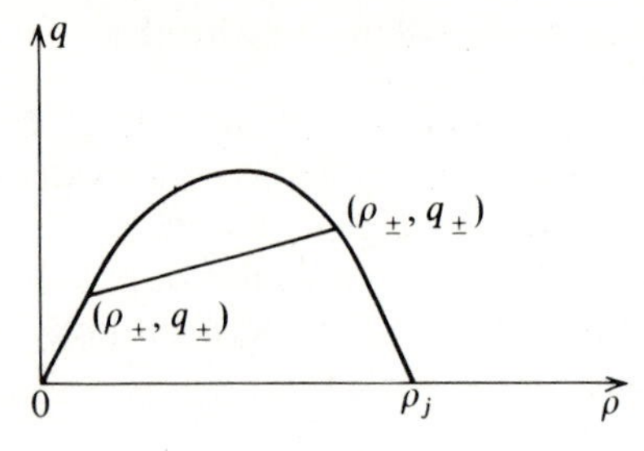

Fig. 9.4

tangents at (ρ_-, q_-) and (ρ_+, q_+). We have yet to determine whether ρ_- is greater than or less than ρ_+.

Next look at the initial value problem

$$\rho(x,0)=\begin{cases}\rho_1, & x<0,\\ \rho_2, & x>0,\end{cases} \tag{9.10}$$

where ρ_1 and ρ_2 are constants. There are two cases: $\rho_1<\rho_2$ and $\rho_1>\rho_2$.

Case (i). $\rho_1<\rho_2$; hence $v_1>v_2$ and $Q'(\rho_1)>Q'(\rho_2)$. A provisional sketch of the characteristic field using the initial conditions (9.10) on the x-axis is given in Fig. 9.5: $\rho=\rho_1$ on each characteristic intersecting the negative x-axis; $\rho=\rho_2$ on each characteristic intersecting the positive x-axis. The sketch cannot represent the solution, because at any point where two characteristics intersect, $\rho=\rho_1$ and $\rho=\rho_2$, which is physically impossible. A jump in ρ can occur across a curve, i.e. as a shock, but two values of ρ cannot co-exist over a region of finite area.

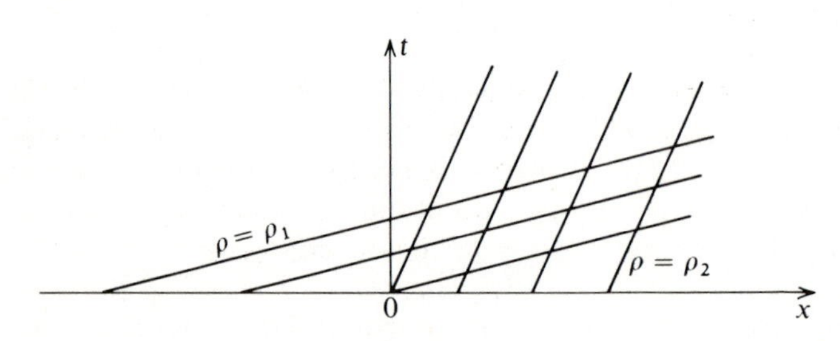

Fig. 9.5

The behaviour of the 'hump' in ρ in x, t-space just treated suggests that a shock should be inserted between two regions of constant ρ. Since the intersections start at the origin, look for a solution of the form illustrated in Fig. 9.6, i.e. two regions of constant ρ separated by a constant velocity shock. It must be shown that eqn (9.5) and boundary conditions (9.10) are satisfied on both sides of the shock and that eqn (9.9) is satisfied on the shock. Clearly $\rho=\rho_1$, constant, satisfies eqn (9.5) and eqn (9.10) for $x<0$ at $t=0$; similarly, $\rho=\rho_2$ for $x>0$ at $t=0$.

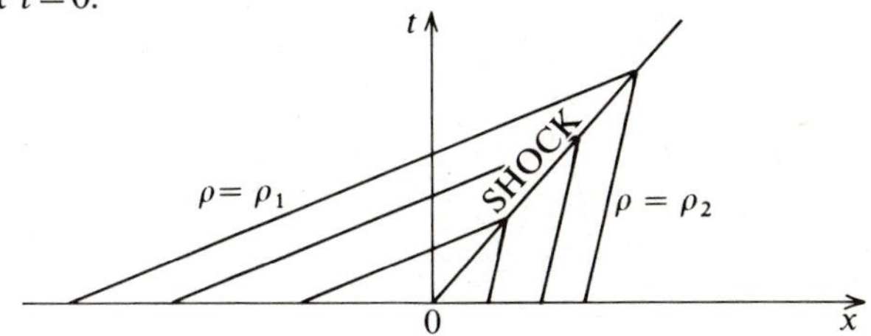

Fig. 9.6

The shock condition (9.9) is satisfied if

$$U=\frac{q_1-q_2}{\rho_1-\rho_2}, \tag{9.11}$$

i.e. if the shock travels at the constant velocity given by eqn (9.11). The shock path is a straight line through the origin in the x, t-plane. Not only are all equations and boundary conditions satisfied, but ρ is single-valued for all x and t except across the shock.

Case (ii). $\rho_1>\rho_2$; hence $v_1<v_2$ and $Q'(\rho_1)<Q'(\rho_2)$. A provisional sketch for this case using initial conditions (9.10) is given in Fig. 9.7.

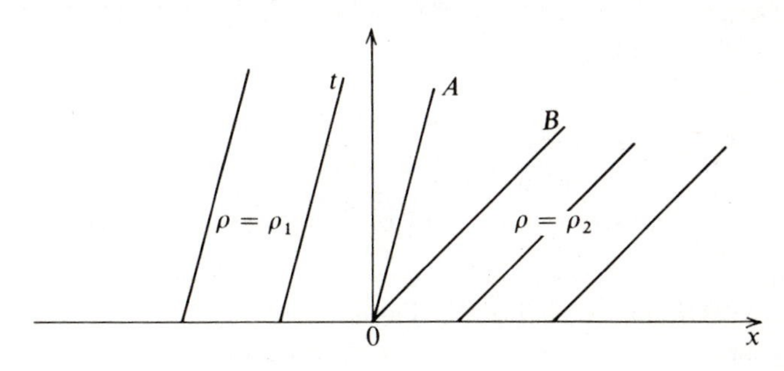

Fig. 9.7

Within the infinite sector A0B there are no characteristics and hence no solution for $\rho(x, t)$. In reality, changes rarely take place instantaneously, provided one goes down to a large enough scale. This physical argument can be expressed in mathematical form. Suppose that the initial conditions at t_2 change from ρ_1 to ρ_2 over the space interval $(-\varepsilon, \varepsilon)$ and then let $\varepsilon \to +0$. These two steps are illustrated in Figs. 9.8(*a*) and (*b*).

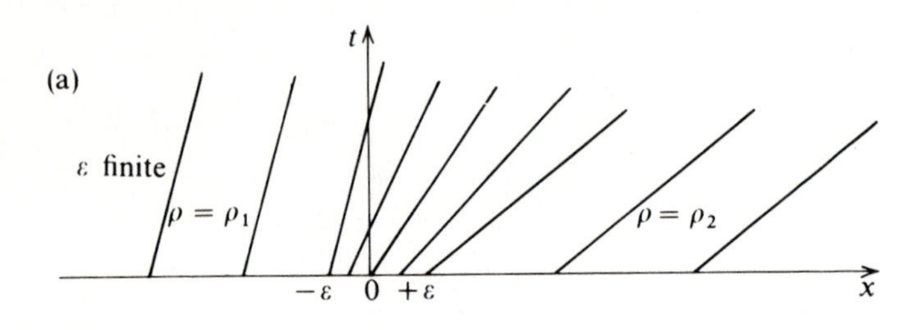

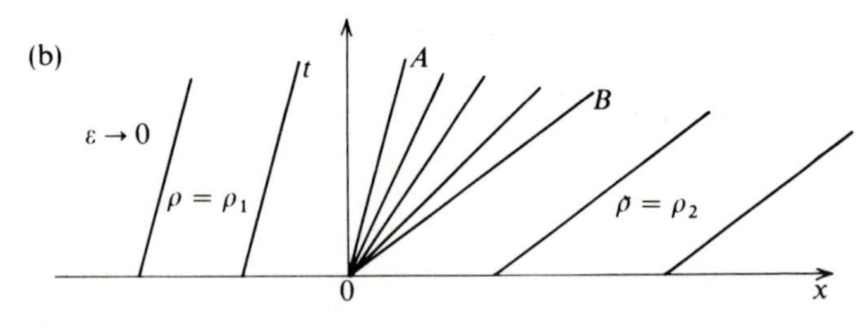

Fig. 9.8

We have arrived at two regions, each of constant ρ, separated by a fan centred at the origin. It must now be shown that the solution represented by the latter diagram satisfies all the governing equations and boundary conditions.

The solution is

$$\rho = \begin{cases} \rho_2 & \text{to the right of 0B} \\ \rho_1 & \text{to the left of 0A} \\ k, & \text{constant, } \rho_2 < k < \rho_1 \text{ on } x = Q'(k)t \text{ in the sector A0B.} \end{cases}$$

Again, eqn (9.5) and boundary condition (9.10) are satisfied to the right of 0B and to the left of 0A. The solution in the sector A0B is the characteristic form, eqn (9.7), of the solution of eqn (9.5). Across 0A and 0B, ρ is continuous but $\partial\rho/\partial x$ and $\partial\rho/\partial t$ are discontinuous

($\partial\rho/\partial x$ and $\partial\rho/\partial t$ are non-zero in the sector A0B but zero elsewhere). The discontinuities across 0A and 0B are weak. Since ρ is continuous, by Hadamard's lemma across 0A or 0B, it is required that

$$\left[\frac{\partial \rho}{\partial t}\right]+Q'(\rho)\left[\frac{\partial \rho}{\partial x}\right]=0. \tag{9.12}$$

Equation (9.5) holds on both sides of 0A, and subtraction of eqn (9.5) on one side of 0A from eqn (9.5) on the other produces eqn (9.12); similarly for 0B.

Solutions have now been obtained for both cases, (i) $\rho_1 < \rho_2$ and (ii) $\rho_1 > \rho_2$. However, it is possible to construct another solution for case (ii) which satisfies all equations and boundary conditions. It consists of a shock dividing two regions of constant ρ (Fig. 9.9). All characteristics on the left of the shock are parallel and so are those to the right. For case (ii) the fan solution and the shock solution both satisfy all mathematical requirements. If the solution is to be unique, a further physical requirement must be introduced to eliminate at least one of the two solutions so far obtained.

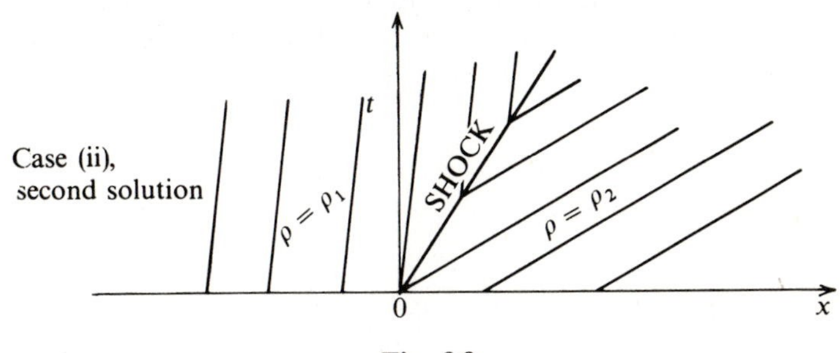

Fig. 9.9

The physical requirement is stability. The reader will be familiar with stability of equilibrium. In nature any object in equilibrium is subject to small chance disturbances. After the disturbance is removed, if the object returns to its former position of equilibrium, the equilibrium is said to be 'stable'; if the object departs further from the equilibrium position, the equilibrium is 'unstable'; and if the object remains in the new displaced position caused by the small disturbance, the equilibrium is 'neutral'. A simple example of these three types of equilibrium is provided by a cone under gravity resting on a horizontal plane, illustrated in Fig. 9.10.

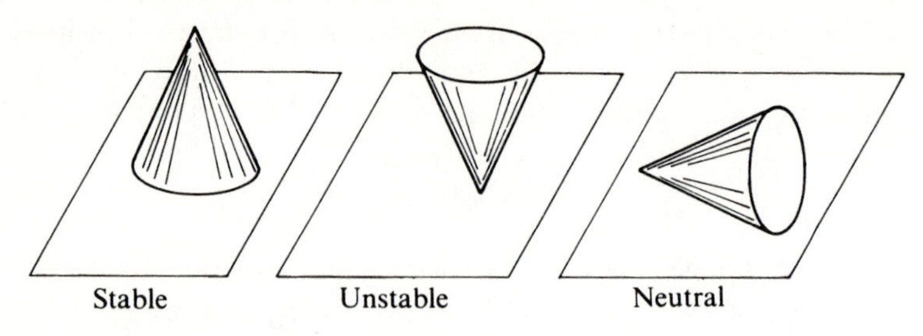

Fig. 9.10

The physical concept of stability is now applied to a shock. For case (i) in which $\rho_1 < \rho_2$, Fig. 9.11(*a*) illustrates the solution previously obtained; Figs. 9.11(*b*) and (*c*) illustrate what happens if a small disturbance changes the profile either just behind or just ahead of the shock; Fig. 9.11(*d*) is the relevant q–ρ curve. In (*b*) the small section detached just behind the shock will travel at approximately the characteristic velocity $Q'(\rho_1)$, which is greater than the shock velocity U, and therefore the detached section rejoins the shock. In (*c*) the detached section travels at approximately velocity $Q'(\rho_2) < U$ and therefore the detached section is overtaken by the shock and the original shock reforms. The shock is stable.

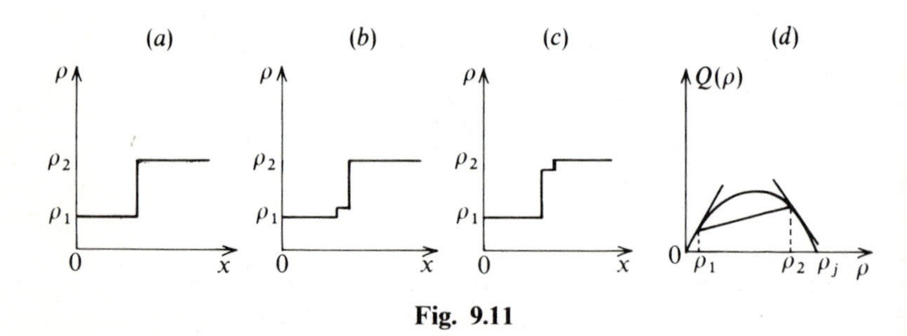

Fig. 9.11

The diagrams for case (ii), $\rho_1 > \rho_2$ are given in Fig. 9.12. In (*b*) the detached section travels at velocity $Q'(\rho_1) < U$ and falls further behind the shock; in (*c*) the detached section travels at velocity $Q'(\rho_2) > U$ and draws further ahead of the shock. In neither case does the shock reform and so the shock is unstable.

In the traffic flow problem, small disturbances are always present. Different drivers take different times to react and there is the 'sneeze'

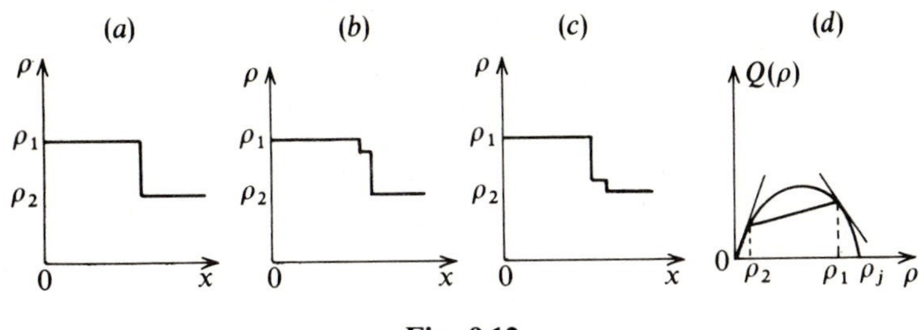

Fig. 9.12

factor considered by automobile engineers. The $v(\rho)$ relationship is an average valid over times greater than about 1/10 second but not over smaller time intervals. Consequently, shocks will only arise when the density to the right of the shock is greater than that to the left and not vice versa; right and left are defined as seen on a ρ–x diagram, as above.

Since $U < Q'(\rho_1) < v_1$, shocks travel backwards down the line of cars. In the particular case of a line of cars travelling at velocity v_1 being brought to a halt, $\rho_2 = \rho_j$, $v_2 = 0$ and $U = (q_2 - q_1)/(\rho_2 - \rho_1) = -q_1/(\rho_j - \rho_1)$. The stationary jam travels backwards down the motorway at velocity $|\rho_1 v_1/(\rho_j - \rho_1)|$.

The reader will have noted that uniqueness in this first-order non-linear problem no longer follows from governing equations and boundary conditions alone, as in some parts of linear theory. It is necessary to introduce a further physical principle. This feature is common to non-linear problems. In some examples in continuum mechanics, the second law of thermodynamics is needed to decide between two possible solutions or to decide in which sense a particular wave can travel; see for example Chapter 10.

Returning to the pulse propagation problem, discussion was broken off at the time t_2 at which a shock formed at the back of the pulse. If the construction of Fig. 9.2 is continued to a time $t_3 > t_2$, then a $\rho(x)$ curve like that illustrated in Fig. 9.2 and repeated in Fig. 9.13 will be obtained. This curve represents an inadmissible solution, because ρ must be single-valued in x, except across a shock. Try inserting a shock at some value of x. Since both equations (9.5) and (9.9) represent the law of conservation of cars and the total number of cars in any interval (a, b), see Fig. 9.13, is $\int_a^b \rho \, dx$, the shock must be inserted at that value of x which leaves the total area under the curve unaltered, i.e. at the point S such that area ADB = area

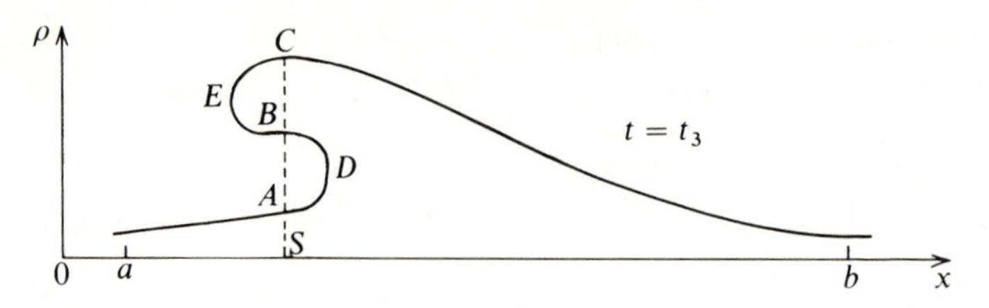

Fig. 9.13

BEC. This simple result that the two areas cut off by the line through S are equal is known as Whitham's rule.

What is the physical interpretation of the profile containing the shock? Since $v_- > U$, $v_+ > U$ by the equation $\rho_-(v_- - U) = \rho_+(v_+ - U)$; cars enter the shock from behind, instantly decelerating, come out at the front, and gradually accelerate back to their original speed. This agrees with the common experience of driving on a motorway except that the deceleration, although more rapid than the acceleration, takes place over a small time interval and not instantaneously as in Fig. 9.14. In the next section, the model will be modified to predict this observation.

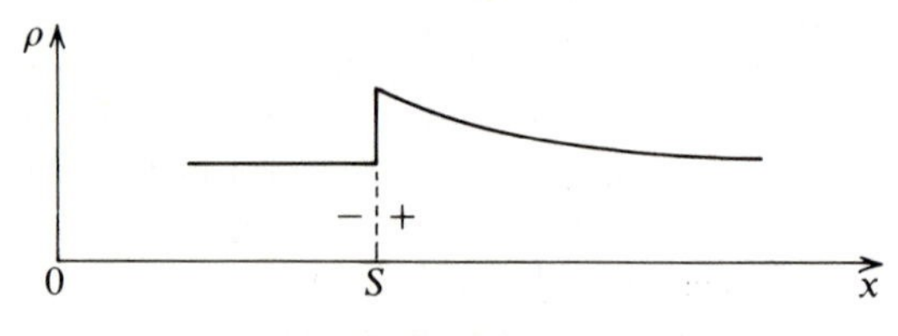

Fig. 9.14

Exercise 9.1. A simplified form for the equation of propagation of flood waves is

$$\frac{\partial A}{\partial t} + c(A)\frac{\partial A}{\partial x} = 0,$$

where $c'(A) > 0$.

If at time $t = t_0$, the A–x profile is as shown in Fig. 9.15, sketch the development of the profile at later times.

Exercise 9.2. The integral form of the equation for conservation of cars is

$$q(x_1, t) - q(x_2, t) = \frac{\mathrm{d}}{\mathrm{d}t}\int_{x_2}^{x_1} \rho\,(x, t)\,\mathrm{d}x.$$

Use eqns (6.4) and (6.8) to verify eqns (9.3) and (9.9).

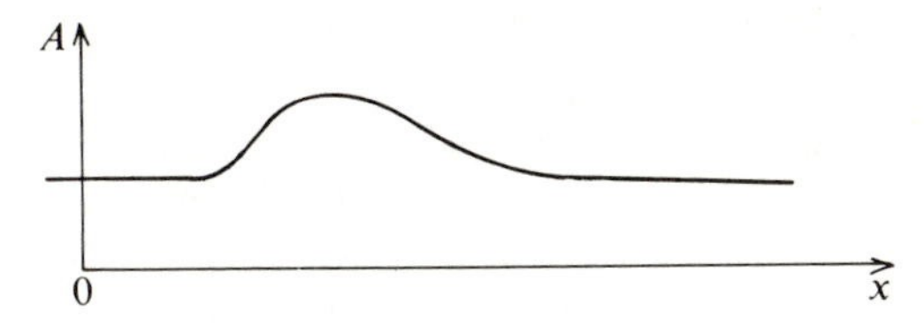

Fig. 9.15

9.3. Refined speed adjustment

The basic postulate $q = \rho v = Q(\rho)$ assumes that drivers adjust their speed according to the traffic density in their immediate neighbourhood. However, most drivers also keep at least half an eye on the traffic density ahead in the middle distance and if this is denser than in their neighbourhood they go slower and vice versa. This behaviour can be represented by the equation

$$v = \frac{1}{\rho} Q(\rho) - \nu \frac{\partial \ln \rho}{\partial x}, \quad \nu > 0, \tag{9.13}$$

so that

$$q = \rho v = Q(\rho) - \nu \frac{\partial \rho}{\partial x}. \tag{9.14}$$

Substitution in eqn (9.3) gives

$$\frac{\partial \rho}{\partial t} + Q'(\rho) \frac{\partial \rho}{\partial x} = \nu \frac{\partial^2 \rho}{\partial x^2}. \tag{9.15}$$

The precise form of eqn (9.13) has been chosen so that manipulation of eqn (9.15) is as easy as possible.

Consider the same initial conditions as before, equations (9.10) with $\rho_1 < \rho_2$, for eqn (9.15). The problem consists of finding ρ at later times for all x. If eqn (9.15) had r.h.s. zero, the solution would contain a shock and would be illustrated by Fig. 9.6. If eqn (9.15) did not contain the second term on the l.h.s., it would become $\partial \rho / \partial t = \partial^2 \rho / \partial x^2$ with solution

$$\rho = \frac{1}{2}(\rho_1 + \rho_2) - \frac{1}{2}(\rho_1 - \rho_2) \operatorname{erf} \frac{x}{2\sqrt{t}}. \tag{9.16}$$

The first solution contains a shock and other solutions of the same equation build up a shock, as shown in the last section; eqn (9.16) represents a solution in which the disturbance spreads out as $t^{\frac{1}{2}}$. Is it possible that for the full equation (9.15), the steepening and dispersive effects balance? If such a balance is reached, then the profile of a pulse remains constant as it travels. If the pulse travels with velocity V, V a constant to be determined, then ρ will depend on x and t through $x-Vt$. We look for a solution of eqn (9.15) of the form

$$\rho=\rho(\xi), \tag{9.17}$$

where

$$\xi=x-Vt, \quad V \text{ constant.} \tag{9.18}$$

When $\rho=\rho(\xi)$,

$$\frac{\partial\rho}{\partial x}=\frac{\mathrm{d}\rho}{\mathrm{d}\xi}\frac{\partial\xi}{\mathrm{d}x}=\frac{\mathrm{d}\rho}{\mathrm{d}\xi}, \quad \frac{\partial^2\rho}{\partial x^2}=\frac{\partial}{\partial x}\frac{\partial\rho}{\partial x}=\frac{\partial\xi}{\partial x}\frac{\mathrm{d}}{\mathrm{d}\xi}\frac{\mathrm{d}\rho}{\mathrm{d}\xi}=\frac{\mathrm{d}^2\rho}{\mathrm{d}\xi^2}$$

and

$$\frac{\partial\rho}{\partial t}=\frac{\mathrm{d}\rho}{\mathrm{d}\xi}\frac{\partial\xi}{\partial t}=-V\frac{\mathrm{d}\rho}{\mathrm{d}\xi}.$$

Equation (9.15) becomes

$$-V\frac{\mathrm{d}\rho}{\mathrm{d}\xi}+Q'(\rho)\frac{\mathrm{d}\rho}{\mathrm{d}\xi}=v\frac{\mathrm{d}^2\rho}{\mathrm{d}\xi^2}. \tag{9.19}$$

On integration,

$$-V\rho+Q(\rho)=v\frac{\mathrm{d}\rho}{\mathrm{d}\xi}+C, \quad C \text{ constant.} \tag{9.20}$$

If the constant profile solution is to connect two constant states of ρ, say ρ_1 and ρ_2 with $\rho_2>\rho_1$, ξ extends to both $\pm\infty$, ρ tends to ρ_1 at one limit and to ρ_2 at the other with $\mathrm{d}\rho/\mathrm{d}\xi$ zero at both limits; at this stage it is not specified which value of ρ at which limit of ξ. Apply the conditions $\rho=\rho_1$ and $\mathrm{d}\rho/\mathrm{d}\xi=0$ to eqn (9.20) to give

$$-V\rho_1+Q(\rho_1)=C; \tag{9.21}$$

and then use $\rho=\rho_2$ and $\mathrm{d}\rho/\mathrm{d}\xi=0$ in eqn (9.20) to give

$$-V\rho_2+Q(\rho_2)=C. \tag{9.22}$$

On subtraction,

$$V=\frac{Q(\rho_2)-Q(\rho_1)}{\rho_2-\rho_1}. \tag{9.23}$$

The equation for the velocity is exactly the same as that previously obtained for the shock connecting the two states ρ_1 and ρ_2, eqn (9.9). Also

$$C = \frac{\rho_2 Q(\rho_1) - \rho_1 Q(\rho_2)}{\rho_2 - \rho_1}. \tag{9.24}$$

The relation between ρ and ξ can be obtained by integration of eqn (9.20):

$$\xi = v \int^{\rho} \frac{d\sigma}{Q(\sigma) - V\sigma - C}. \tag{9.25}$$

We now show that the denominator of the integrand in eqn (9.25),

$$f(\rho) = Q(\rho) - V\rho - C,$$

is positive for $\rho_1 < \rho < \rho_2$. It is zero at $\rho = \rho_1$ and at $\rho = \rho_2$ by eqns (9.21) and (9.22). Since $f''(\rho) = Q''(\rho) < 0$, $f'(\rho)$ is steadily decreasing as ρ increases from ρ_1 to ρ_2. Since $Q'(\rho_1) > V > Q'(\rho_2)$, $f'(\rho_1) > 0$ and $f'(\rho_2) < 0$. As ρ increases from ρ_1, $f(\rho)$ at first increases and then decreases to become zero again at ρ_2. $f(\rho)$ cannot be zero for $\rho_1 < \rho < \rho_2$; for suppose $f(\rho_3) = 0$, see Fig. 9.16, where ρ_3 is the first zero of $f(\rho)$ greater than ρ, but less than ρ_2, then $f'(\rho_3) < 0$, $f(\rho)$ continues to decrease and cannot be zero at $\rho = \rho_2$.

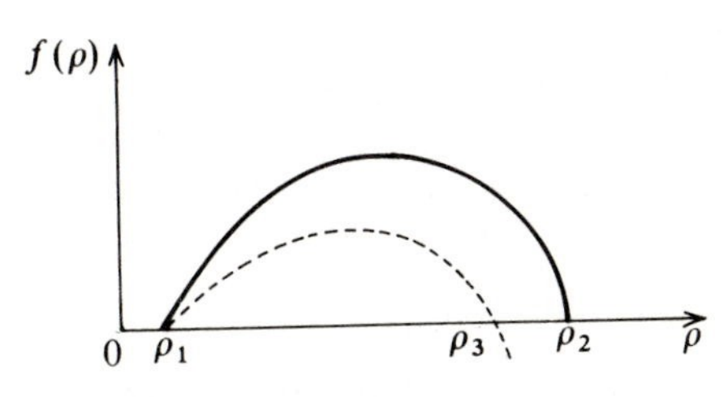

Fig. 9.16

Since

$$Q(\rho) - V\rho - C > 0 \quad \text{for } \rho_1 < \rho < \rho_2,$$

$$\frac{\mathrm{d}\rho}{\mathrm{d}\xi}>0 \quad \text{for} \quad \rho_1<\rho<\rho_2.$$

Hence, $\rho \to \rho_2$ as $\xi \to +\infty$ and $\rho \to \rho_1$ as $\xi \to -\infty$, i.e. ρ is smaller at the back of the profile than at the front. This result agrees with the previous analysis in the absence of the term $\nu(\mathrm{d}^2\rho/\mathrm{d}\xi^2)$. The effect of including the term $\nu(\mathrm{d}^2\rho/\mathrm{d}\xi^2)$ is to spread out the shock which is what we hoped to achieve by introducing that term (the effect is illustrated in Fig. 9.17).

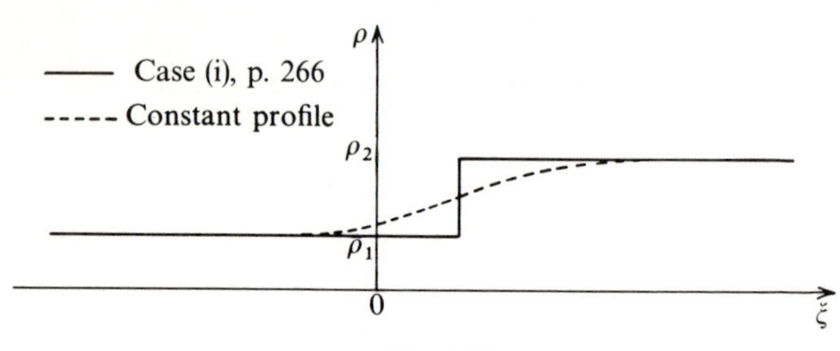

Fig. 9.17

To proceed further, an analytic form for $Q(\rho)$ is required. Choose the quadratic form which is zero at $\rho=0$ and at $\rho=\rho_j$ and has maximum q_m:

$$Q(\rho)=\frac{4q_m}{\rho_j^2}\rho(\rho_j-\rho). \tag{9.26}$$

Clearly $Q''(\rho)<0$. From eqn (9.26),

$$\rho=\frac{\rho_j}{2}-\frac{\rho_j^2}{8q_m}Q'(\rho). \tag{9.27}$$

Substitute for ρ into eqn (9.15):

$$\frac{\partial Q'(\rho)}{\partial t}+Q'(\rho)\frac{\partial Q'(\rho)}{\partial x}=\nu\frac{\partial^2 Q'(\rho)}{\partial x}. \tag{9.28}$$

Equation (9.28) is known as Burgers' equation. It is of frequent occurrence in applied mathematics and will now be treated separately before applying the results to traffic flow.

9.4. Burgers' equation

On replacing $Q'(\rho)$ by ϕ, eqn (9.28) becomes

$$\frac{\partial\phi}{\partial t}+\phi\frac{\partial\phi}{\partial x}=v\frac{\partial^2\phi}{\partial x^2}. \tag{9.29}$$

Burgers' equation (8.29) is the simplest wave equation which contains non-linear and diffusion terms. It is also the only non-linear wave equation for which a reduction to linear form has been found and partly in consequence of this it has been studied in detail.

The initial part of the constant profile solution of Burger's equation follows that of the last section and is set as an exercise.

Exercise 9.3

(i) If $\phi=\phi(\xi)$ where $\xi=x-Vt$, V constant, show that eqn (9.29) becomes, after substitution for $\partial/\partial x$ and $\partial/\partial t$ and one integration with respect to ξ,

$$-V\phi+\tfrac{1}{2}\phi^2=v\frac{\mathrm{d}\phi}{\mathrm{d}\xi}+C,\quad C \text{ constant.} \tag{9.30}$$

(ii) If $\phi\to\phi_1$, $\mathrm{d}\phi/\mathrm{d}\xi\to 0$, and $\phi\to\phi_2$, $\mathrm{d}\phi/\mathrm{d}\xi\to 0$ as $\xi\to\pm\infty$ (which way round not specified) and† $\phi_1>\phi_2$, show that

$$V=\tfrac{1}{2}(\phi_1+\phi_2)\quad\text{and}\quad C=-\tfrac{1}{2}\phi_1\phi_2. \tag{9.31}$$

(iii) Show that eqn (9.30) with eqns (9.31) integrates to

$$\frac{1}{2v}\xi=\int^{\phi}\frac{\mathrm{d}\Psi}{(\Psi-\phi_1)(\Psi-\phi_2)}. \tag{9.32}$$

Look for a solution for eqn (9.32) in which $\phi_2<\phi<\phi_1$ with equality only at the limits $\xi\to\pm\infty$. Hence,

$$\frac{1}{2v}\xi=\frac{1}{\phi_1-\phi_2}\int^{\phi}\left(-\frac{1}{\Psi-\phi_2}+\frac{1}{\Psi-\phi_1}\right)\mathrm{d}\Psi$$

$$=\frac{1}{\phi_1-\phi_2}(-\ln(\phi-\phi_2)+\ln|\phi-\phi_1|)+K,\quad K \text{ constant.}$$

---

† From eqn (9.28) to eqn (9.29), $Q'(\rho)$ has been replaced by ϕ. Hence by eqn (9.27) $\rho_2-\rho_1$ is replaced by $-(\rho_j^2/8q_m)(\phi_2-\phi_1)$ and $\rho_1<\rho_2$ corresponds to $\phi_1>\phi_2$.

Put $K = +\alpha/2v$, then

$$\frac{\phi_1-\phi_2}{2v}(\xi-\alpha) = -\ln(\phi-\phi_2)+\ln(\phi_1-\phi) = \ln\frac{\phi_1-\phi}{\phi-\phi_2},$$

$$\frac{\phi_1-\phi}{\phi-\phi_2} = \exp\left(\frac{\phi_1-\phi_2}{2v}(\xi-\alpha)\right)$$

$$\phi = \left[\phi_1+\phi_2\exp\left(\frac{\phi_1-\phi_2}{2v}(\xi-\alpha)\right)\right]\left[1+\exp\left(\frac{\phi_1-\phi_2}{2v}(\xi-\alpha)\right)\right]^{-1}. \tag{9.33}$$

Two further algebraic forms of this solution are useful:

$$\phi = \phi_2 + \frac{\phi_1-\phi_2}{1+\exp\left(\frac{\phi_1-\phi_2}{2v}(\xi-\alpha)\right)} \tag{9.34}$$

and

$$\begin{aligned}\phi &= \frac{\phi_2+\phi_1}{2} - \frac{\phi_1-\phi_2}{2}\left[\exp\left(\frac{\phi_1-\phi_2}{2v}(\xi-\alpha)\right)-1\right] \\ &\quad\times\left[\exp\left(\frac{\phi_1-\phi_2}{2v}(\xi-\alpha)\right)+1\right]^{-1} \\ &= \frac{\phi_2+\phi_1}{2} - \frac{\phi_1-\phi_2}{2}\left[\tanh\left(\frac{\phi_1-\phi_2}{4v}(\xi-\alpha)\right)\right].\end{aligned} \tag{9.35}$$

on multiplying through both terms containing exponentials by

$$\exp\left(-\frac{\phi_1-\phi_2}{4}(\xi-\alpha)\right).$$

When $\xi-\alpha=0$, $\phi=\frac{1}{2}(\phi_1+\phi_2)$. When $\xi\to+\infty$, $\phi\to\phi_2$; when $\xi\to-\infty$, $\phi\to\phi_1$. Since $\tanh(-\xi) = -\tanh\,\xi$, $\phi-\frac{1}{2}(\phi_1+\phi_2)$ is antisymmetric about $\xi-\alpha$. The constant α determines the position of the centre of the profile, i.e. the value of ξ for which $\phi=\frac{1}{2}(\phi_1+\phi_2)$. As $\phi_1>\phi_2$, ϕ decreases monotonically with ξ from ϕ_1 at $\xi\to-\infty$ to ϕ_2 as $\xi\to+\infty$ (Fig. 9.18).

The width w of the profile is defined as the distance between the two values of ξ at which ϕ has changed 5 per cent and 95 per cent from its initial to its final values. Since $\tanh 1.5 = 0.905$ ($\tanh x$

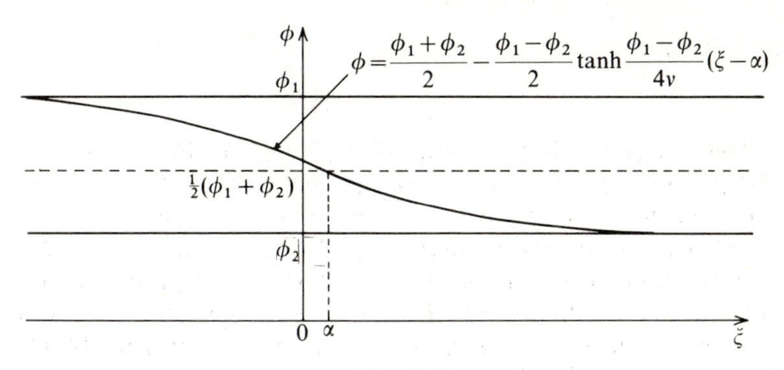

Fig. 9.18

changes from -1 to $+1$ in $-\infty<x<\infty$),

$$\frac{\phi_1-\phi_2}{4\nu}\left(\frac{1}{2}w\right)\simeq 1.5=\frac{3}{2},$$

$$w\simeq\frac{12\nu}{\phi_1-\phi_2}. \tag{9.36}$$

Hence, the width of the profile is proportional to ν, the coefficient of the diffusion term $\partial^2\phi/\partial x^2$, and inversely proportional to the change in ϕ. Substitution of $\xi=x-Vt$ gives the final solution for the profile:

$$\phi=\frac{\phi_1+\phi_2}{2}-\frac{\phi_1-\phi_2}{2}\tanh\left(\frac{\phi_1-\phi_2}{4\nu}(x-Vt-\alpha)\right). \tag{9.37}$$

Return to the problem of traffic flow. Since eqns (9.28) and (9.29) differ only in that $Q'(\rho)$ has been replaced by ϕ and since, by eqn (9.27),

$$Q'(\rho)=-\frac{8q_m}{\rho_j^2}\rho+\frac{4q_m}{\rho_j},$$

eqns (9.36) and (9.37) can be taken over to traffic flow if ϕ is replaced by

$$\phi=-\frac{8q_m}{\rho_j^2}\rho+\frac{4q_m}{\rho_j}. \tag{9.38}$$

With reference back to Fig. 9.17, the shock of zero thickness has been replaced by a continuous change with $\rho(x, t)$, equation

$$\rho(x,t)=\frac{1}{2}(\rho_1+\rho_2)+\frac{1}{2}(\rho_2-\rho_1)\tanh\left(\frac{2q_m(\rho_2-\rho_1)}{\rho_j^2\nu}(x-Vt-\alpha)\right) \tag{9.39}$$

and 'width'

$$w=\frac{3\rho_j^2 \nu}{2q_m(\rho_2-\rho_1)}. \tag{9.40}$$

The problem of instantaneous deceleration has been overcome, but at a price. The time for change of speed in the real world is not zero, but neither is it infinite. The definition of 'width' of profile adopted above attempts to meet this problem. The price now paid is that the beginning and end of the period of deceleration is not correctly modelled by eqn (9.15). It has been implicitly assumed in this discussion that the shock is replaced by the constant profile when the term $\nu(\mathrm{d}^2\rho/\mathrm{d}x^2)$ is introduced. It will be shown that this does occur after some time has elapsed.

Exercise 9.4. The Korteweg–de Vries equation is

$$\frac{\partial\eta}{\partial t}+C_0\left(1+\frac{3}{2}\frac{\eta}{h_0}\right)\frac{\partial\eta}{\partial x}+\gamma\frac{\partial^3\eta}{\partial x^3}=0.$$

Look for solutions of the form $\eta=h_0 f(\xi)$, where $\xi=x-Ut$, and where $f(\xi)$ and $f'(\xi)$ tend to zero as $\xi\to\pm\infty$. In this case the constant profile solution represents a pulse.

Does the velocity of the pulse depend on its amplitude?

Cole and Hopf independently discovered a transformation that reduces Burgers' equation to a linear p.d.e. First write eqn (9.29) as

$$\frac{\partial}{\partial x}\left(\nu\frac{\partial\phi}{\partial x}-\frac{1}{2}\phi^2\right)=\frac{\partial\phi}{\partial t},$$

which is the compatibility condition for a function Ψ to exist such that

$$\phi=\frac{\partial\Psi}{\partial x} \quad\text{and}\quad \nu\frac{\partial\phi}{\partial x}-\frac{1}{2}\phi^2=\frac{\partial\Psi}{\partial t}.$$

Elimination of ϕ between the last two equations gives

$$\nu\frac{\partial^2\Psi}{\partial x^2}-\frac{1}{2}\left(\frac{\partial\Psi}{\partial x}\right)^2=\frac{\partial\Psi}{\partial t}.$$

Put $\Psi=-2\nu\ln\theta$, then

$$\frac{\partial\Psi}{\partial x}=-\frac{2\nu}{\theta}\frac{\partial\theta}{\partial x},\quad \frac{\partial^2\Psi}{\partial x^2}=+\frac{2\nu}{\theta^2}\left(\frac{\partial\theta}{\partial x}\right)^2-\frac{2\nu}{\theta}\frac{\partial^2\theta}{\partial x^2}$$

and

$$\frac{\partial\Psi}{\partial t}=-\frac{2\nu}{\theta}\frac{\partial\theta}{\partial t}.$$

Hence,

$$+\frac{2v^2}{\theta^2}\left(\frac{\partial\theta}{\partial x}\right)^2-\frac{2v^2}{\theta}\frac{\partial^2\theta}{\partial x^2}-\frac{2v^2}{\theta^2}\left(\frac{\partial\theta}{\partial x}\right)^2=-\frac{2v}{\theta}\frac{\partial\theta}{\partial t}$$

or

$$\frac{\partial\theta}{\partial t}=v\frac{\partial^2\theta}{\partial x^2}, \tag{9.41}$$

the heat conduction or diffusion equation. Eliminating Ψ between $\phi=\partial\Psi/\partial x$ and $\Psi=-2v\ln\theta$, the transformation is

$$\phi=-\frac{2v}{\theta}\frac{\partial\theta}{\partial x}. \tag{9.42}$$

Many solutions of eqn (9.41) are known. The solution can then be substituted into eqn (9.42) for ϕ to obtain solutions to Burgers' equation. The existence of this method has meant that the behaviour of the solutions of Burgers' equation are better understood than those of any other non-linear wave equation. This knowledge also suggests what to look for in the solution of other non-linear equations even when the solution has to be carried out numerically. Note that θ can be multiplied by an arbitrary constant without affecting either equation (9.41) or (9.42) in which θ appears.

The above method will be used to solve Burgers' equation

$$\frac{\partial\phi}{\partial t}+\phi\frac{\partial\phi}{\partial x}=v\frac{\partial^2\phi}{\partial x^2}$$

when

$$\phi(x,0)=\begin{cases}\phi_1 & \text{for } x<0,\\ \phi_2 & \text{for } x>0,\end{cases}\quad \phi_1>\phi_2. \tag{9.43}$$

Since $\partial\ln\theta/\partial x=-\phi/2v$,

$$\ln\theta=\begin{cases}-\dfrac{\phi_1}{2v}x & \text{for } x<0,\ t=0,\\[2ex] -\dfrac{\phi_2}{2v}x & \text{for } x>0,\ t=0,\end{cases}$$

if the arbitrary multiplicative constants are chosen so that $\theta=1$ at $x=0$ when $t=0$. Therefore,

$$\theta(x,0)=\begin{cases}\exp(-\phi_1 x/(2v)), & x<0,\\ \exp(-\phi_2 x/(2v)), & x>0.\end{cases} \tag{9.44}$$

The solution of the heat conduction equation for given $\theta(x,0)$ was found by Laplace. It is[†]

$$\theta(x,t)=(4\pi\nu t)^{-\frac{1}{2}}\int_{-\infty}^{\infty}\theta(\eta,0)\exp\left(-\frac{(x-\eta)^2}{4\nu t}\right)\mathrm{d}\eta. \tag{9.45}$$

$\theta(\eta,0)$ must be such that the integral converges at both limits. Since the other factor in the integrand behaves like $\mathrm{e}^{-\eta^2}$, the convergence condition is not usually restrictive. In particular, it is satisfied by the $\theta(x,0)$ of eqn (9.44). $\phi(x,t)$ is given by

$$\phi=-\frac{2\nu}{\theta}\frac{\partial\theta}{\partial x}=\frac{\displaystyle\int_{-\infty}^{\infty}\frac{x-\eta}{t}\theta(\eta,0)\exp\left(-\frac{(x-\eta)^2}{4\nu t}\right)\mathrm{d}\eta}{\displaystyle\int_{-\infty}^{\infty}\theta(\eta,0)\exp\left(-\frac{(x-\eta)^2}{4\nu t}\right)\mathrm{d}\eta}. \tag{9.46}$$

To evaluate the integrals, split them both up into $\int_{-\infty}^{0}+\int_{0}^{\infty}$ so that

$$\phi=\frac{I_4+I_3}{I_2+I_1},$$

$$I_1=\int_0^{\infty}\exp\left(\frac{-\phi_2\eta}{2\nu}\right)\exp\left(-\frac{(x-\eta)^2}{4\nu t}\right)\mathrm{d}\eta$$

$$=\int_0^{\infty}\exp\left(-\frac{(\eta-x+\phi_2 t)^2}{4\nu t}\right)\exp\left(\frac{\phi_2^2 t}{4\nu}-\frac{x\phi_2}{2\nu}\right)\mathrm{d}\eta,$$

on completing the square in η in the exponential index.
Put $\eta-x+\phi_2 t=2\sqrt{\nu t}\,\xi$, so that

$$I_1=\int_{-(x-\phi_2 t)/2\sqrt{\nu t}}^{\infty}\mathrm{e}^{-\xi^2}\exp\left(\frac{\phi_2^2 t}{4\nu}-\frac{x\phi_2}{2\nu}\right)2\sqrt{\nu t}\,\mathrm{d}\xi$$

$$=\exp\left(\frac{\phi_2^2 t}{4\nu}-\frac{x\phi_2}{2\nu}\right)\sqrt{\pi\nu t}\,\operatorname{erfc}\left(-\frac{x-\phi_2 t}{2\sqrt{\nu t}}\right). \tag{9.47}$$

$$I_2=\int_{-\infty}^{0}\exp\left(\frac{-\phi_1\eta}{2\nu}\right)\exp\left(-\frac{(x-\eta)^2}{4\nu t}\right)\mathrm{d}\eta$$

$$=\int_0^{\infty}\exp\left(\frac{\phi_1\eta}{2\nu}\right)\exp\left(-\frac{(x+\eta)^2}{4\nu t}\right)\mathrm{d}\eta$$

[†] See e.g. Carslaw, H. S. and Jaeger, J. C. (1959). *Conduction of heat in solids*, (2nd edn.). Section 2.2. Oxford University Press.

after replacing η by $-\eta$, which is the same as I_1 with x replaced by $-x$ and ϕ_2 by $-\phi_1$. Hence,

$$I_2=\exp\left(\frac{\phi_1^2 t}{4v}-\frac{x\phi_1}{2v}\right)\sqrt{\pi vt}\,\operatorname{erfc}\frac{x-\phi_1 t}{2\sqrt{vt}}. \tag{9.48}$$

Now

$$\frac{dI_1}{dx}=-\frac{1}{2v}\int_0^\infty \frac{x-\eta}{t}\exp\left(-\frac{\phi_2\eta}{2v}\right)\exp\left(-\frac{(x-\eta)^2}{4vt}\right)d\eta$$
$$=-\frac{1}{2v}I_3.$$

Therefore,

$$I_3=-2v\frac{dI_1}{dx}$$
$$=\phi_2 I_1-2v\exp\left(\frac{\phi_2^2 t}{4v}-\frac{x\phi_2}{2v}\right)\sqrt{\pi vt}\,\frac{2}{\sqrt{\pi}}\frac{1}{2\sqrt{vt}}\exp\left(-\frac{(x-\phi_2 t)^2}{4vt}\right)$$
$$=\phi_2 I_1-2v\,\exp\left(-\frac{x^2}{4vt}\right). \tag{9.49}$$

$$I_4=\int_{-\infty}^0 \frac{x-\eta}{t}\exp\left(-\frac{\phi_1\eta}{2v}\right)\exp\left(-\frac{(x-\eta)^2}{4vt}\right)d\eta$$
$$=\int_0^\infty \frac{x+\eta}{t}\exp\left(\frac{\phi_1\eta}{2v}\right)\exp\left(-\frac{(x+\eta)^2}{4vt}\right)d\eta,$$

which is the same as $-I_3$ except that x is replaced by $-x$ and ϕ_2 by $-\phi_1$. Therefore,

$$I_4=\phi_1 I_2+2v\,\exp\left(-\frac{x^2}{4vt}\right). \tag{9.50}$$

On substitution for I_1, I_2, I_3 and I_4,

$$\phi=\frac{I_3+I_4}{I_1+I_2}=\frac{\phi_2 I_1+\phi_1 I_2}{I_1+I_2}=\phi_2+\frac{\phi_1-\phi_2}{1+I_1/I_2},$$

where

$$\frac{I_1}{I_2}=\frac{\operatorname{erfc}\left(-\dfrac{x-\phi_2 t}{2\sqrt{vt}}\right)}{\operatorname{erfc}\left(\dfrac{x-\phi_1 t}{2\sqrt{vt}}\right)}\exp\left(\frac{\phi_1-\phi_2}{2v}(x-Ut)\right)$$

and

$$U=\tfrac{1}{2}(\phi_1+\phi_2). \tag{9.51}$$

Hence,

$$\phi=\phi_2+(\phi_1-\phi_2)\left[1+\frac{\operatorname{erfc}\left(-\dfrac{x-\phi_2 t}{2\sqrt{\nu t}}\right)}{\operatorname{erfc}\left(\dfrac{x-\phi_1 t}{2\sqrt{\nu t}}\right)}\exp\left(\frac{\phi_1-\phi_2}{2\nu}(x-Ut)\right)\right]^{-1}. \tag{9.52}$$

Exercise 9.5. Check that eqn (9.52) as $t\to+0$ satisfies the initial conditions (9.43).

Change the variables from x and t to $\xi=x-Ut$ and t[†], so that

$$\phi=\phi_2+(\phi_1-\phi_2)\left[1+\frac{\operatorname{erfc}\left(-\dfrac{\xi+(U-\phi_2)t}{2\sqrt{\nu t}}\right)}{\operatorname{erfc}\left(\dfrac{\xi-(\phi_1-U)t}{2\sqrt{\nu t}}\right)}\exp\left(\frac{\phi_1-\phi_2}{2\nu}\xi\right)\right]^{-1}. \tag{9.53}$$

Now, holding ξ fixed, let $t\to+\infty$. Since $\phi_1>U>\phi_2$, the arguments of both erfcs tend to $-\infty$ and $\operatorname{erfc}(-\infty)=2$. Hence

$$\lim_{\substack{t\to\infty\\ \xi\ \text{fixed}}}\phi=\phi_2+(\phi_1-\phi_2)\left[1+\exp\left(\frac{\phi_1-\phi_2}{2\nu}\xi\right)\right]^{-1}, \tag{9.54}$$

the constant profile solution, eqn (9.34). We have shown that the initial step function profile leads to the constant profile after sufficient time has elapsed.

The ratio of the two erfcs in eqn (9.53) is

$$\operatorname{erfc}\left(\frac{-\frac{1}{2}(\phi_1-\phi_2)t-\xi}{2\sqrt{\nu t}}\right)\Big/\operatorname{erfc}\left(\frac{-\frac{1}{2}(\phi_1-\phi_2)t+\xi}{2\sqrt{\nu t}}\right), \tag{9.55}$$

and from this ratio it is possible to estimate the time at which the profile differs from the constant profile by a prescribed amount.

† In effect substitute $x=\xi+Ut$.

Answers to exercises

9.1. The pulse steepens at the front and eventually forms a shock as shown in Fig. 9.19.

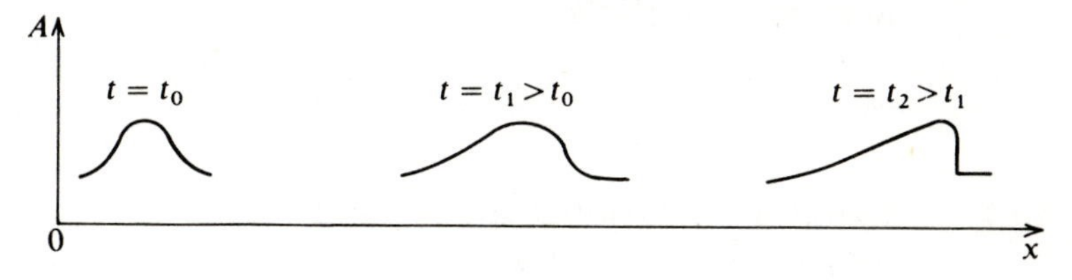

Fig. 9.19

9.4. $\phi = 3V \operatorname{sech}^2 \sqrt{(V/\gamma)}\,(x - Vt + B)$, where V and B are constants. The amplitude A is given by $A = 3V$.

10 Non-linear plane dilatational waves in solids

10.1. Adiabatic dilatational shocks and waves in non-linear elasticity

In Chapter 8 it was seen that the inclusion of thermal terms modified the solutions for isentropic linear elastodynamics. In this section, the equations of linear elastodynamics will be modified by the inclusion of a third-order term in the strain m in the expression for the internal energy U, eqn (8.17), which now becomes

$$U = T_0 S + \frac{1}{2\rho}(\lambda + 2\mu)\left(m^2 + \frac{\alpha}{3}m^3\right) - \kappa m S + \frac{1}{2}\eta S^2, \qquad (10.1)$$

where α is a non-dimensional constant of order of magnitude unity. The stress σ and temperature T become

$$\sigma = \rho\frac{\partial U}{\partial m} = (\lambda + 2\mu)\left(m + \frac{\alpha}{2}m^2\right) - \rho\kappa S \qquad (10.2)$$

and

$$T = \frac{\partial U}{\partial S} = T_0 - \kappa m + \eta S. \qquad (10.3)$$

Compared to the linear theory of Chapter 8, eqn (10.2) contains a non-linear term and eqn (10.3) in unaltered. The theory of eqns (10.1) to (10.3) is intermediate between linear theory and the full non-linear theory; m is sufficiently large to allow us retain m but to neglect m^2, both compared to unity.

The shock conditions already derived are eqns (8.3) and (8.7), namely

$$[\sigma] + \rho V[v] = 0, \qquad (10.4)$$

where V is the shock velocity, and

$$[\sigma v - q] + \rho V[\tfrac{1}{2}v^2 + U] = 0. \qquad (10.5)$$

Since u is continuous everywhere, by Hadamard's lemma,

$$\left[\frac{\partial u}{\partial x}\right]+V\left[\frac{\partial u}{\partial x}\right]=0$$

or

$$[v]+V[m]=0. \tag{10.6}$$

The second law of thermodynamics, eqn (8.12), states

$$\mathrm{d}S \geqslant \frac{\delta Q}{T},$$

where S and Q are specific quantities, i.e. per unit mass. In the adiabatic approximation, heat transfer within the solid is neglected, so that $Q \equiv 0$ and the above equation becomes

$$\mathrm{d}S \geqslant 0. \tag{10.7}$$

Across a shock, the change is finite and eqn (10.7) becomes

$$[S] \geqslant 0. \tag{10.8}$$

This is the form of the second law of thermodynamics for shocks in the adiabatic approximation, frequently called the entropy condition. If in a particular case $[S] \neq 0$, then the shock represents an irreversible change between two states A and B and the inequality (10.8) determines in which sense, i.e. either A to B or B to A, the shock passes. It cannot pass in the opposite sense. The form of the second law for shocks in an elastic solid when heat conduction is included is found later, eqn (10.79).

For reversible changes, equality holds in eqn (10.7), which is true for each material element. Hence $S=S(x)$. If S is identically zero for all x initially, it remains identically zero at later times, giving the isentropic approximation. Since increase of entropy can occur across a shock, the adiabatic approximation is more general than the isentropic. We next investigate adiabatic dilatational shock propagation into a media at rest in the undeformed state.

Let the shock move in the positive sense; the suffix $-$ denotes conditions behind the shock; $+$ denotes those ahead. Then $m_+ = v_+ = S_+ = U_+ = \sigma_+ = 0$ and $T_+ = T_0$. Since $k=0$, $q=0$. Equations (10.4), (10.5) and (10.6) become

$$\sigma+\rho V v=0, \quad \sigma v+\tfrac{1}{2}\rho V v^2+\rho V U=0 \quad \text{and} \quad v+Vm=0, \tag{10.9}$$

where the suffix $-$ has been dropped since all quantities are evaluated behind the shock. Eliminate v and use $\sigma=\rho(\partial U/\partial m)$, eqn (10.2):

$$\frac{\partial U}{\partial m}=V^2 m \quad \text{and} \quad U=\tfrac{1}{2}V^2 m^2; \tag{10.10}$$

two equations for V and S in terms of $m=[m]$, the shock strength. Elimination of V^2 gives

$$U=\frac{1}{2}m\frac{\partial U}{\partial m} \tag{10.11}$$

as the relation between S and m. As shock amplitude $m\to 0$, $S\to 0$. For shocks of small amplitude, considered in this intermediate theory, retain the lowest powers of m and S on substitution from eqn (10.1) into eqn (10.11), to give

$$S=\frac{\lambda+2\mu}{12\rho T_0}\alpha m^3. \tag{10.12}$$

The inequality (10.8), using eqn (10.12), requires $\alpha m>0$. If $\alpha>0$, $m>0$ and the shock is tensile; if $\alpha<0$, $m<0$ and the shock is compressive.† In both cases the entropy change is of the order of the cube of the strain. Although the entropy change is small, condition (10.8) determines in which direction the shock can go.

The velocity of the shock is determined by substitution for S and U from eqns (10.1) and (10.12) into the second of eqns (10.10):

$$V^2=\frac{2U}{m^2}=\frac{\lambda+2\mu}{6\rho}\alpha m+\frac{\lambda+2\mu}{\rho}\left(1+\frac{\alpha}{3}m\right)+\mathrm{O}(m^2),$$

$$V=\sqrt{\frac{\lambda+2\mu}{\rho}}\left(1+\frac{1}{4}\alpha m\right), \tag{10.13}$$

correct to order m. The velocity is dependent upon the shock amplitude m. As $m\to 0$, $V\to\sqrt{(\lambda+2\mu)/\rho}=c_1$, the linear isentropic dilatational wave velocity.

In Section 9.2, shock stability was used to determine in which direction a shock between two states would travel. The conclusion was that the shock was stable if $c_{(+)}<V<c_{(-)}$ where $c_{(+)}$ and $c_{(-)}$ were the characteristic velocities just ahead of and just behind the shock respectively. We now investigate whether the stability and entropy criteria and lead to the same results.

† α can be positive in elasticity. However, in the corresponding equation for compressible gases, α is negative.

Using the isentropic non-linear equation for stress in terms of strain, eqn (10.2) with $S=0$, and substituting in eqn (8.2) yields

$$(\lambda+2\mu)(1+\alpha m)\frac{\partial^2 u}{\partial x^2}=\rho\frac{\partial^2 u}{\partial t^2}. \tag{10.14}$$

Equation (10.14) has positive characteristic velocity correct to order m, using eqn (5.7),

$$\frac{\mathrm{d}x}{\mathrm{d}t}=+\sqrt{\frac{\lambda+2\mu}{\rho}(1+\alpha m)}=+\sqrt{\frac{\lambda+2\mu}{\rho}}(1+\tfrac{1}{2}\alpha m). \tag{10.15}$$

Since $m_+=0$, $c_{(+)}=\sqrt{(\lambda+2\mu)/\rho}$; and since $m_-=m$, $c_{(-)}=\sqrt{[(\lambda+2\mu)/\rho]}(1+\frac{1}{2}\alpha m)$. Clearly $c_{(+)}<V<c_{(-)}$ for $\alpha m>0$; hence the shock considered is stable for $\alpha m>0$. The second law of thermodynamics and stability lead to the same conclusion.

We now investigate the half-space problem for a non-linear elastic solid. We have already seen that, provided no shocks have previously passed through a material element, then $S=0$ in the adiabatic approximation. The governing p.d.e. is eqn (10.14), valid in the quarter-plane $x>0$, $t>0$. At time $t=0$, $u=\partial u/\partial t=0$ for all $x>0$. $m(=\partial u/\partial x)$ on $x=0$ for $0<t\leq t_1$ is given as $m(0, t)=M(t)$, where $M(t)$ is continuous, and until further notice $M(0)=0$.

Before solving this problem, the equations and boundary conditions will be non-dimensionalized and scaled. All variables and derivatives occurring explicitly in the differential equation will be made order unity, if necessary by the introduction of a small constant. The scaling is then said to be explicit, as the order of magnitude of all terms is immediately obvious. The method is not foolproof. When a solution is obtained, the variables must be checked to be of order of magnitude unity over most of their range. If they are not, then the scaling is incorrect and a new scaling, possibly suggested by the failed solution, must be attempted.

$$t=\sqrt{\frac{\rho}{\lambda+2\mu}}\,t', \tag{10.16}$$

then

$$(1+\alpha m)\frac{\partial^2 u}{\partial x^2}=\frac{\partial^2 u}{\partial t'^2}; \tag{10.17}$$

u, x, t' all have the dimensions of length, α and m are non-dimensional. There is no typical length or time in the equation, but there may be in the boundary condition. Let L be a typical length; if instead there were a typical time T then take $L=\sqrt{[(\lambda+2\mu)/\rho]}\,T$. Put $u=L\bar{u}$, $x=L\xi$, $t'=L\tau$. Equation (10.17) becomes

$$(1+\alpha m)\frac{\partial^2\bar{u}}{\partial\xi^2}=\frac{\partial^2\bar{u}}{\partial\tau^2},$$

where $m=\partial u/\partial x=\partial\bar{u}/\partial\xi$.

The form adopted for $U(m, S)$, eqn (10.1), assumes that m is sufficiently small to neglect its square compared to unity and that α is of order of magnitude unity. For any solution of the governing p.d.e., boundary and initial conditions must be given. Let N be the greatest value of $|m|$ either explicit or implicit in these conditions. Put

$$\varepsilon=|\alpha|N, \tag{10.18}$$

then ε is positive and sufficiently small to neglect its square compared to unity. Put

$$\bar{u}=\frac{\varepsilon}{\alpha}\eta, \tag{10.19}$$

so that the governing equation becomes

$$\left(1+\varepsilon\frac{\partial\eta}{\partial\xi}\right)\frac{\partial^2\eta}{\partial\xi^2}=\frac{\partial^2\eta}{\partial\tau^2}. \tag{10.20}$$

$$m=\frac{\partial\bar{u}}{\partial\xi}=\frac{\varepsilon}{\alpha}\frac{\partial\eta}{\partial\xi}, \tag{10.21}$$

so that $\partial\eta/\partial\xi$ is of order of magnitude unity. We work henceforth in this chapter correct to order ε.

The net transformation from x, t, u to ξ, τ, η is

$$x=L\xi,\quad t=\sqrt{\frac{\rho}{\lambda+2\mu}}\,L\tau \quad\text{and}\quad u=\varepsilon\frac{L}{\alpha}\eta. \tag{10.22}$$

Note that the initial and boundary conditions determine two of the constants, L and ε, used in the scaling.

The boundary conditions must be scaled. The boundary condition on $x=0$ is $m=M(t)$ or, using eqns (10.21) and (10.22),

$$\frac{\partial\eta}{\partial\xi}=\frac{\alpha}{\varepsilon}M\left(\sqrt{\frac{\rho}{\lambda+2\mu}}\,L\tau\right)\quad\text{on}\quad\xi=0.$$

If $|m|$ has its boundary maximum N at a point P on $\xi=0$, then it follows from the above equation and eqn (10.18) that $\partial\eta/\partial\xi=\pm 1$ at P. Put

$$\frac{\partial\eta}{\partial\xi}=\mu(\tau) \quad \text{on} \quad \xi=0, \tag{10.23}$$

so that

$$\mu(\tau)=\frac{\alpha}{\varepsilon}M\left(\sqrt{\frac{\rho}{\lambda+2\mu}}L\tau\right), \quad M \text{ a given function.} \tag{10.24}$$

The initial condition $u=\partial u/\partial t=0$ on $t=0$ becomes

$$\eta=\frac{\partial\eta}{\partial\tau}=0 \quad \text{on} \quad \tau=0. \tag{10.25}$$

The characteristics of eqn (10.20) are

$$\frac{\mathrm{d}\xi}{\mathrm{d}\tau}=\pm\left(1+\varepsilon\frac{\partial\eta}{\partial\xi}\right)^{\frac{1}{2}}=\pm\left(1+\frac{1}{2}\varepsilon\frac{\partial\eta}{\partial\xi}\right), \tag{10.26}$$

correct to order ε. The equations holding along the characteristics, found by the same method as eqns (5.11), are

$$\mathrm{d}\frac{\partial\eta}{\partial\tau}\mathrm{d}\tau-\mathrm{d}\frac{\partial\eta}{\partial\xi}\mathrm{d}\xi=0 \quad \text{on} \quad \frac{\mathrm{d}\xi}{\mathrm{d}\tau}=\pm\left(1+\frac{1}{2}\varepsilon\frac{\partial\eta}{\partial\xi}\right) \quad \text{on } C_{\pm}, \tag{10.27}$$

or

$$\mathrm{d}\frac{\partial\eta}{\partial\tau}\mp\left(1+\frac{1}{2}\varepsilon\frac{\partial\eta}{\partial\xi}\right)\mathrm{d}\frac{\partial\eta}{\partial\xi}=0 \quad \text{on} \quad C_{\pm}.$$

On integration,

$$\frac{\partial\eta}{\partial\tau}\mp\left[\frac{\partial\eta}{\partial\xi}+\frac{1}{4}\varepsilon\left(\frac{\partial\eta}{\partial\xi}\right)^2\right]=\text{constant on } C_{\pm}. \tag{10.28}$$

Since η, and hence $\partial\eta/\partial\xi$, and $\partial\eta/\partial\tau$ are all zero on $\tau=0$, the constant carried by any characteristic cutting the ξ-axis is zero. If C_{+0} denotes the C_+ characteristic passing through the origin in Fig. 10.1, then $u\equiv 0$ at all points to the right of C_{+0}, provided that no shock formed to the left of C_{+0} intersects C_{+0}. The equation of C_{+0} is $\xi=\tau$ and all characteristics to its right have slope ± 1.

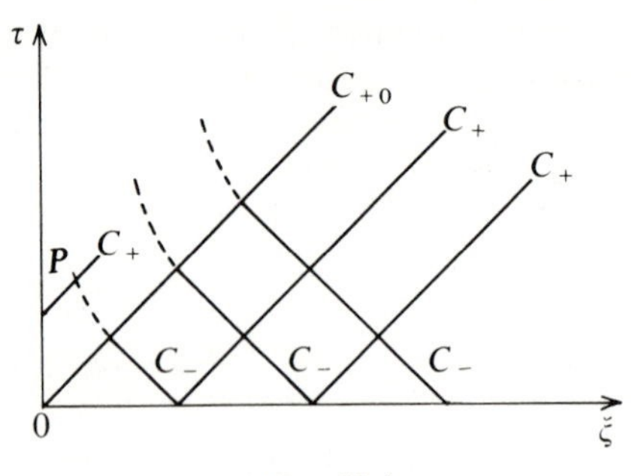

Fig. 10.1

The C_- characteristics continue to the left of C_{+0}. Since each C_- characteristic carries constant zero,

$$\frac{\partial \eta}{\partial \tau}+\frac{\partial \eta}{\partial \xi}+\frac{1}{4}\varepsilon\left(\frac{\partial \eta}{\partial \xi}\right)^2=0 \tag{10.29}$$

to the left of C_{+0}. On the C_+ characteristic passing through P, any point to the left of C_{+0}, see Fig. 10.1,

$$\frac{\partial \eta}{\partial \tau}-\left[\frac{\partial \eta}{\partial \xi}+\frac{1}{2}\left(\frac{\partial \eta}{\partial \xi}\right)^2\right]=-2k, \quad \text{constant on each } C_+.$$

Eliminating $\partial\eta/\partial\tau$ using eqn (10.29):

$$\frac{\partial \eta}{\partial \xi}+\frac{1}{2}\left(\frac{\partial \eta}{\partial \xi}\right)^2=k, \text{ constant on each } C_+. \tag{10.30}$$

Let the C_+ characteristic intersect the τ axis at $(0, \theta)$. Then, by eqn (10.23),

$$k=\mu(\theta)+\tfrac{1}{2}[\mu(\theta)]^2.$$

Substituting back for k:

$$\frac{\partial \eta}{\partial \xi}=\mu(\theta), \text{ constant on each } C_+. \tag{10.31}$$

Equation (10.26) can now be integrated to give

$$\xi=[1+\tfrac{1}{2}\varepsilon\mu(\theta)](\tau-\theta) \tag{10.32}$$

as the equation of each C_+. All C_+ characteristics are straight lines.

Case (i). When $\mu(\tau)$ decreases with θ, the slopes $d\tau/d\xi$ of the C_+ characteristics increase as the ordinate θ of their intercept on the τ axis increases, eqn (10.32), Fig. 10.2. Consequently the C_+ characteristics do not intersect one another and $\partial\eta/\partial\xi$ and $\partial\eta/\partial\tau$ have been found at all points in the ξ, τ-plane below $C_{+\phi}$ where ϕ is the maximum value of θ for which μ is given.

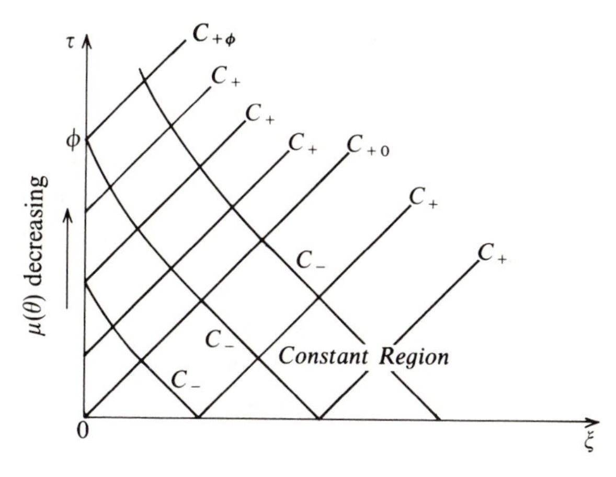

Fig. 10.2

L, ϕ and ε must now be found for case (i). The typical time T is taken as the time at which $\tau=\phi$, i.e. $T=t_1$. Hence,

$$L=\sqrt{\frac{\lambda+2\mu}{\rho}}\,t_1$$

and by substitution for L and of $t=t_1$ into the second of eqns (10.22), $\phi=1$. The maximum absolute value of $m=\partial u/\partial x$ on any boundary occurs at $(0, t_1)$ where $m=M(t_1)$. $N=|M(t_1)|$ and by eqn (10.18) $\varepsilon=|\alpha M(t_1)|$. The solution has now been obtained in particular everywhere to the left of the C_- characteristic through (0, 1). This intersects the ξ-axis at a point with abscissa $1+O(\varepsilon)$. The independent variables ξ and τ each have range whose magnitude is of order unity. By eqn (10.31), $\partial\eta/\partial\xi$ lies between 0 and -1 to the left of C_{0+}. Since

$$\eta=\int^{\xi}\frac{\partial\eta}{\partial\xi}\,d\xi$$

and the range of ξ is of order unity, η is O(1), apart from an arbitrary constant which represents a rigid body displacement. By eqn (10.29), $\partial\eta/\partial\tau$ is O(1). Conversion of the solution back to physical variables is achieved by substituting

$$\xi=\sqrt{\frac{\rho}{\lambda+\mu}}\frac{1}{T}x, \quad \tau=\frac{1}{T}t \quad \text{and} \quad \eta=\frac{\alpha}{|\alpha M(T)|\,T}\sqrt{\frac{\rho}{\lambda+2\mu}}\,u,$$

so that

$$\frac{\partial\eta}{\partial\xi}=\frac{\alpha}{|\alpha M(T)|}\frac{\partial u}{\partial x}=\frac{\alpha}{|\alpha M(T)|}m. \tag{10.33}$$

The solution is also valid to the right of the C_- characteristic through (0, 1) and below $C_{+\phi}$. In this region ξ, τ and η become large compared to unity. However ξ, τ and η do not occur explicitly in eqn (10.20), only derivatives of η with respect to ξ and T and these derivatives are order unity.

In case (i), $\partial\eta/\partial\xi$ is negative behind $\xi=\tau$; hence by eqn (10.33) $\partial\eta/\partial\xi=-1$ at $\phi=1$ and the sign of m depends on α. If $\alpha>0$, $m<0$, which represents a compressive wave. If $\alpha<0$, $m>0$, which represents a tensile wave.

Case (ii). When $\mu(\theta)$ increases with θ, the slopes $d\tau/d\xi$ of the C_+ characteristics decrease with θ, eqn (10.32), and C_+ characteristics will eventually intersect. Let the least reduced time at which intersection occurs be τ_0. Since each characteristic carries a different value of $\partial\eta/\partial\xi$, $\partial\eta/\partial\xi$ becomes discontinuous at the point of intersection and a shock will form. The analysis for times $\tau>\tau_0$ becomes quite complicated, in general requiring both the shock and continuous equations, but a qualitative sketch of the consequent characteristic field can be drawn since the shock velocity U satisfies $(d\xi/d\tau)_+<U<(d\xi/d\tau)_-$, where $d\xi/d\tau$ is the characteristic velocity, the suffices + and − denoting conditions just ahead of and just behind the shock. The shock may eventually intersect C_{+0} so that C_{+0} terminates on the shock (Fig. 10.3).

Cases (iii) and (iv) no longer have $\mu(\theta)$ continuous at the origin.

Case (iii). $\mu(\theta)=+1$. In this case $\partial\eta/\partial\xi$ has different limits at the origin when approached along each axis in turn. This suggests we look for a solution which contains a shock starting at the origin. As the shock velocity is greater than the characteristic velocity ahead, the shock bounds the region of rest ahead of the shock; as shown in Fig. 10.4, the C_+ characteristic ahead run into the shock. If the shock is of constant strength, then, since conditions ahead of the shock are

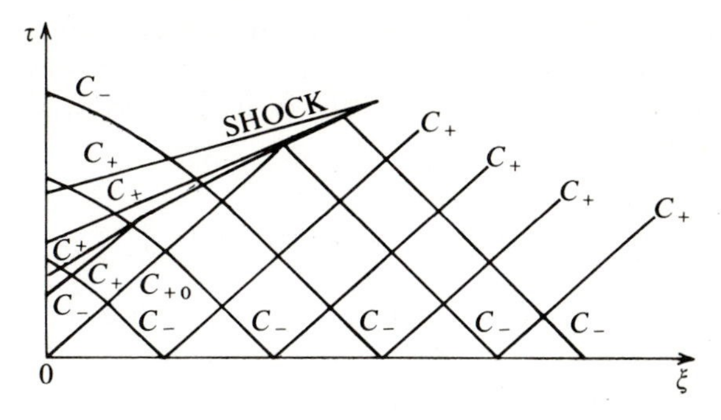

Fig. 10.3

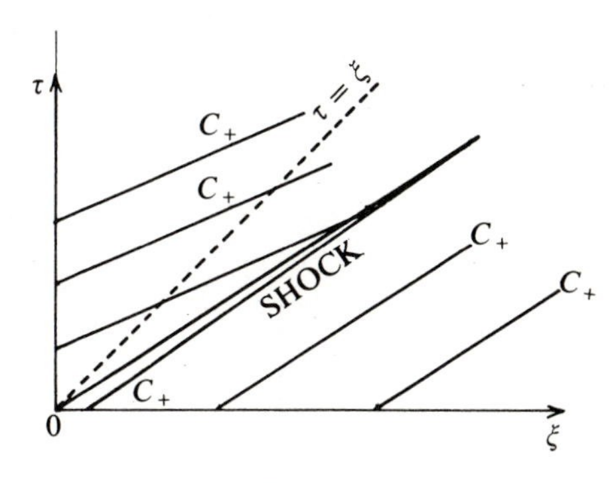

Fig. 10.4

also constant, so are conditions just behind the shock. Hence the C_- characteristics behind the shock all carry the same constant and, by eqn (10.28) for the C_- characteristic, $\partial\eta/\partial\tau$ can be expressed in terms of $\partial\eta/\partial\xi$ behind the shock to within an additive constant, the same everywhere. Then, by eqn (10.28) for the C_+, $\partial\eta/\partial\xi$ is constant on all C_+ behind the shock. But such C_+ intersect the τ axis on which $\partial\eta/\partial\xi$ is given to be constant. Therefore $\partial\eta/\partial\xi$ and hence $\partial\eta/\partial\tau$ are constant behind the shock. The solution consists of two constant regions, the region in front at rest, divided by a constant shock, provided conditions across the shock can be satisfied. But the first part of this section treated a shock moving into a medium at rest and determined the conditions just behind the shock. These conditions now apply at all points behind the shock. The shock strength m, a parameter in the previous analysis, is determined by the given constant value on $x=0$. Since m and v are discontinuous across the shock, the slope of the C_- characteristics is discontinuous across the shock. The reduced shock velocity is $1+\frac{1}{4}\varepsilon$, corresponding to a physical velocity

$$V=\sqrt{\frac{\lambda+2\mu}{\rho}}\,(1+\tfrac{1}{4}\,\alpha M),$$

where M is the value of m behind the shock. Since $\partial\eta/\partial\xi=+1$, by eqn (10.26) $\mathrm{d}\xi/\mathrm{d}\tau=1+\frac{1}{2}\varepsilon$ on C_+ characteristics behind the shock. The shock is stable, since $1+\frac{1}{2}\varepsilon>1+\frac{1}{4}\varepsilon>1$.

The shock profile at constant time is a step function (Fig. 10.5). In the following sections we ask whether this profile is altered by the viscoelastic or head conduction properties generally present in real solids. The resulting study is known as 'shock structure'.

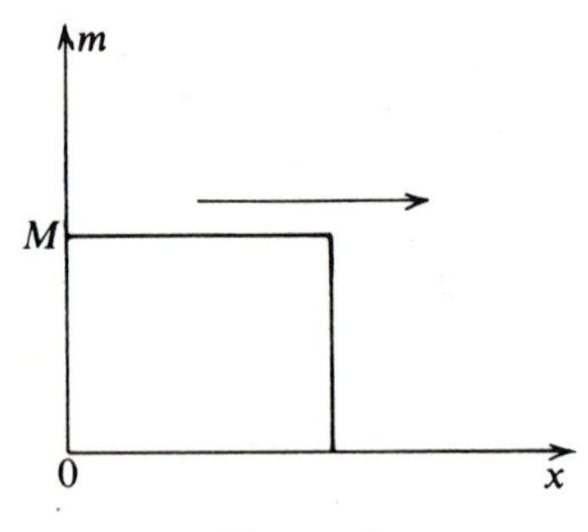

Fig. 10.5

Case (iv). $\mu(\theta)=-1$. Similar arguments to those used above and in Section 9.1 suggest that we look for a solution consisting of two constant regions divided by a fan (Fig. 10.6). Equation (10.30) remains valid on all C_+. In the region behind the fan, all C_+ intersect the τ axis on which $\partial\eta/\partial\xi=-1$. Hence $\partial\eta/\partial\xi=-1$ and by eqn (10.29), $\partial\eta/\partial\tau=1-\frac{1}{4}\varepsilon$ behind the fan. Across the fan, $\partial\eta/\partial\xi$ varies from 0 on C_{+0} to -1 at the back of the fan. The characteristic velocities behind the fan are $\mathrm{d}\xi/\mathrm{d}\tau=\pm(1-\frac{1}{2}\varepsilon)$. The value of $\partial\eta/\partial\xi$ on a C_+ characteristic in the fan on which $\mathrm{d}\xi/\mathrm{d}\tau=w$ is given by eqn (10.26) as

$$\frac{\partial\eta}{\partial\xi}=\frac{2(w-1)}{\varepsilon},\quad 1-\tfrac{1}{2}\varepsilon<w<1.$$

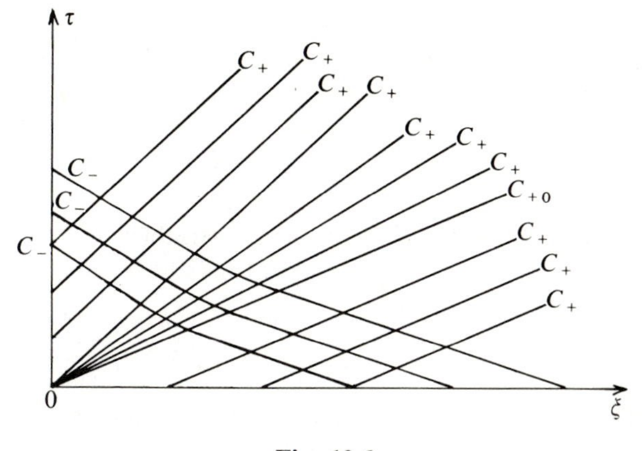

Fig. 10.6

10.2. Non-linear Voigt solid

The governing equation (8.97) for dilatational waves in a linear Voigt solid can be obtained formally from that for a linear elastic solid by replacing the modulus M by the operator $M+\nu(\partial/\partial t)$. If the same procedure is applied to the governing equation (10.14) for dilatational waves in a non-linear elastic solid, there results

$$\left((\lambda+2\mu)\left(1+\alpha\frac{\partial u}{\partial x}\right)+\nu\frac{\partial}{\partial t}\right)\frac{\partial^2 u}{\partial x^2}=\rho\frac{\partial^2 u}{\partial t^2}. \tag{10.34}$$

The mechanical model for this equation consists of a non-linear spring in series with a linear dashpot $(M=\lambda+2\mu)$ (Fig. 10.7).

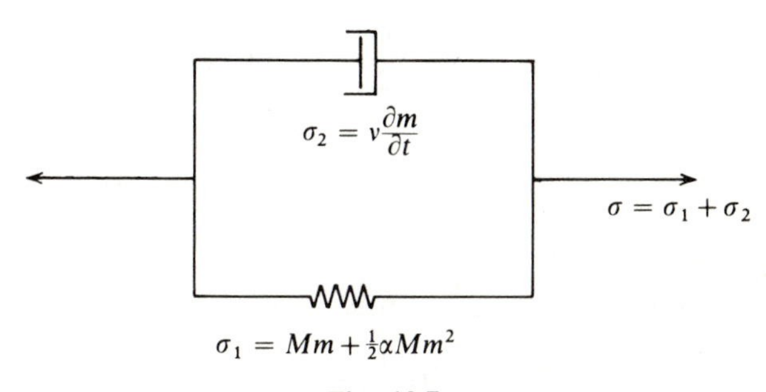

Fig. 10.7

Check:

$$\rho\frac{\partial^2 u}{\partial t^2}=\frac{\partial\sigma}{\partial x}=\frac{\partial\sigma_1}{\partial x}+\frac{\partial\sigma_2}{\partial x}=\frac{\partial}{\partial x}\left(M\frac{\partial u}{\partial x}+\frac{1}{2}\alpha M\left(\frac{\partial u}{\partial x}\right)^2\right)$$
$$+\frac{\partial}{\partial x}\left(\nu\frac{\partial}{\partial t}\left(\frac{\partial u}{\partial x}\right)\right).$$

Introduce the same dimensionless variables η, ξ, τ as in the previous section, defined by eqns (10.18), (10.19) and (10.20), to yield

$$\left(1+\varepsilon\frac{\partial\eta}{\partial\xi}\right)\frac{\partial^2\eta}{\partial\xi^2}+\frac{\nu}{L\sqrt{(\lambda+2\mu)\rho}}\frac{\partial^3\eta}{\partial\tau\partial\xi^2}=\frac{\partial^2\eta}{\partial\tau^2}. \tag{10.35}$$

The non-linear elastic solid had no typical length or time and this quantity had to be introduced from the boundary conditions. Both linear and non-linear Voigt solids have a typical time $=\nu/(\lambda+2\mu)$. We shall investigate Voigt solids for which $\nu/(\lambda+2\mu)$ is small compared to the transit time of a wave through the solid and waves for which $\partial u/\partial x$ is small enough for us to neglect its square compared to unity. We therefore choose

$$T=\frac{\nu}{\varepsilon(\lambda+2\mu)},$$

so that

$$L=\frac{\nu}{\varepsilon\sqrt{\rho(\lambda+2\mu)}}. \tag{10.36}$$

Substitute for L into eqn (10.35):

$$\left(1+\varepsilon\frac{\partial\eta}{\partial\xi}\right)\frac{\partial^2\eta}{\partial\xi^2}+\varepsilon\frac{\partial^3\eta}{\partial\tau\partial\xi^2}=\frac{\partial^2\eta}{\partial\tau^2}. \tag{10.37}$$

The transformations (10.22) become

$$x=\frac{\nu/\varepsilon}{\sqrt{(\lambda+2\mu)\rho}}\xi,\quad t=\frac{\nu/\varepsilon}{\lambda+2\mu}\tau\quad\text{and}\quad u=\frac{\nu}{\alpha\sqrt{(\lambda+2\mu)\rho}}\eta, \tag{10.38}$$

where, as before, $\varepsilon=|\alpha m_{\max}|$ where $m_{\max}$ is the greatest absolute value of $m(=\partial u/\partial x)$ occurring on the time or spatial boundaries; by hypothesis, $m_{\max}\ll 1$, α is order unity, so that $0<\varepsilon\ll 1$.

The solution for case (iii) can be written

$$\frac{\partial\eta}{\partial\xi}=H(-\xi+(1+\tfrac{1}{4}\varepsilon)\tau), \tag{10.39}$$

where H is the Heaviside unit function corresponding in physical variables to

$$m = \frac{\partial u}{\partial x} = m_0 H\left(-x + (1 + \tfrac{1}{4}\alpha m_0)\sqrt{\frac{\lambda + 2\mu}{\rho}}\, t\right), \tag{10.40}$$

where m_0 is the strain behind the shock and $\alpha m_0 > 0$. Equation (10.37) differs from eqn (10.20) by the presence of $\varepsilon(\partial^3\eta/\partial\tau\partial\xi^2)$, a higher derivative term multiplied by a small coefficient. This is precisely the situation encountered in Sections 8.3 and 8.4. To solve the constant profile problem for eqn (10.37), we therefore introduce an inner region with coordinates ϕ, ζ, given by

$$\phi = \xi - \tau, \quad \zeta = \tfrac{1}{2}\varepsilon\tau, \tag{10.41}$$

where the constant $\frac{1}{2}$ is included for algebraic convenience.

The inner region in ϕ, ζ coordinates must merge into the outer regions on either side where the solution is given by eqn (10.40). In the outer region behind, $\partial\eta/\partial\xi = 1$; in the outer region ahead, $\partial\eta/\partial\xi = 0$.

$$\frac{\partial}{\partial \xi} \equiv \frac{\partial}{\partial \phi} \quad \text{and} \quad \frac{\partial}{\partial \tau} \equiv -\frac{\partial}{\partial \phi} + \frac{1}{2}\varepsilon\frac{\partial}{\partial \zeta}, \tag{10.42}$$

so that eqn (10.37), correct to $O(\varepsilon)$ becomes

$$\left(1 + \varepsilon\frac{\partial \eta}{\partial \phi}\right)\frac{\partial^2 \eta}{\partial \phi^2} - \varepsilon\frac{\partial^3 \eta}{\partial \phi^3} = \frac{\partial^2 \eta}{\partial \phi^2} - \varepsilon\frac{\partial^2 \eta}{\partial \phi \partial \zeta}.$$

Note that it is in the neglect of higher-order trms in ε in the above equation that the small strain and 'small' viscosity hypotheses are used.

Divide through by ε and put

$$\hat{m} = \frac{\partial \eta}{\partial \phi}: \tag{10.43}$$

$$\frac{\partial \hat{m}}{\partial \zeta} + \hat{m}\frac{\partial \hat{m}}{\partial \phi} = \frac{\partial^2 \hat{m}}{\partial \phi^2}, \tag{10.44}$$

which is Burgers' equation (9.29) with $v = 1$.

The constant-profile solution of eqn (10.44) with the above stated limits on the inner region boundaries is given by eqn (9.35) with $\phi_1 = 1$, $\phi_2 = 0$, and $v = 1$ as

$$\hat{m} = \tfrac{1}{2}(1 - \tanh\tfrac{1}{4}(\phi - \tfrac{1}{2}\zeta)) \tag{10.45}$$

if $\hat{m}=\frac{1}{2}$ when $\phi-\frac{1}{2}\zeta=0$. By eqn (9.31), the profile travels with velocity $d\phi/d\zeta=\frac{1}{2}$ in ϕ,ζ-space. Since, from eqns (10.41),

$$\frac{d\xi}{d\tau}=1+\frac{\varepsilon}{2}\frac{d\phi}{d\zeta},$$

the velocity in ξ,τ-space is $1+\varepsilon/4$. This velocity is the same as the shock velocity in eqns (10.39).

The 'width' of the profile, see Chapter 9 p. 278 for definition, in ϕ,ζ-space is $W_{\phi\zeta}$, where by eqn (9.36), $W_{\phi\zeta}=12$. Since the width of the profile is measured at constant time and $\phi=\xi-\tau$, its width is the same in ξ, τ space, i.e. $W_{\xi\tau}=12$. In physical x,t-space, from the first of eqns (10.38),

$$W=W_{xt}=12\frac{\nu/(\alpha m_0)}{\sqrt{\rho(\lambda+2\mu)}}=6\sqrt{\frac{\lambda+2\mu}{\rho}}\frac{\nu}{\frac{1}{2}(\lambda+2\mu)\alpha m_0}. \tag{10.46}$$

The width of the profile is proportional to the viscosity coefficient ν and inversely proportional to the coefficient of the quadratic term in the stress–strain equation times the magnitude of the strain at the rear of the profile. In the limit as $\nu\to 0$, $W\to 0$ and the shock solution of the previous section is reproduced. The analysis of Section 9.4, starting from the Cole–Hopf solution, can be taken over to show that if a shock is fed in at the boundary of a semi-infinite non-linear Voigt half-space, then the constant profile of eqn (10.45) is eventually formed.

10.3. Non-linear thermoelastic constant profile

We now return to the thermoelastic solid. The governing equations are the equation of motion, from eqns (10.2) and (8.2),

$$\frac{\partial}{\partial x}\left(\frac{\partial U}{\partial m}\right)=\frac{\partial^2 u}{\partial t^2}, \tag{10.47}$$

and the energy balance equation, from $\partial Q/\partial t=T(\partial S/\partial t)$, eqns (8.9) and (8.21),

$$\rho T\frac{\partial S}{\partial t}=k\frac{\partial^2 T}{\partial x^2}, \tag{10.48}$$

with

$$T=\frac{\partial U}{\partial S},\ m=\frac{\partial u}{\partial x}\ \text{and}\ U=U(m,S)\ \text{given by eqn (10.1).}$$

In the constant profile solution, the dependent variables depend on x and t through ξ, where

$$\xi = x - Vt, \quad V \text{ a positive constant.}$$

The governing equations become

$$\frac{\mathrm{d}}{\mathrm{d}\xi}\left(\frac{\partial U}{\partial m}\right) = V^2 \frac{\mathrm{d}^2 u}{\mathrm{d}\xi^2} \tag{10.49}$$

and

$$-\rho V T \frac{\mathrm{d}S}{\mathrm{d}\xi} = k \frac{\mathrm{d}^2 T}{\mathrm{d}\xi^2}. \tag{10.50}$$

Equation (10.49) integrates to

$$\frac{\partial U}{\partial m} = V^2 \frac{\mathrm{d}u}{\mathrm{d}\xi} + C = V^2 m + C, \quad C \text{ constant.} \tag{10.51}$$

Since

$$\frac{\mathrm{d}U}{\mathrm{d}\xi} = \frac{\partial U}{\partial m}\frac{\mathrm{d}m}{\mathrm{d}\xi} + \frac{\partial U}{\partial S}\frac{\mathrm{d}S}{\mathrm{d}\xi} = (V^2 m + C)\frac{\mathrm{d}m}{\mathrm{d}\xi} + T\frac{\mathrm{d}S}{\mathrm{d}\xi},$$

substitution for $T\,\mathrm{d}S/\mathrm{d}\xi$ into eqn (10.50) gives

$$k\frac{\mathrm{d}^2 T}{\mathrm{d}\xi^2} = -\rho V T \frac{\mathrm{d}S}{\mathrm{d}\xi} = -\rho V \frac{\mathrm{d}U}{\mathrm{d}\xi} + \rho V (V^2 m + C)\frac{\mathrm{d}m}{\mathrm{d}\xi}.$$

On integration,

$$U - \frac{1}{2}V^2 m^2 - Cm - D + \frac{k}{\rho V}\frac{\mathrm{d}T}{\mathrm{d}\xi} = 0, \quad D \text{ constant.} \tag{10.52}$$

We consider a constant profile propagating into a medium at rest and unstrained with constant strain m^* at the rear of the profile. Then $m \to 0$, $S \to 0$, $\mathrm{d}m/\mathrm{d}\xi \to 0$ and $\mathrm{d}S/\mathrm{d}\xi \to 0$ as $\xi \to +\infty$ with the consequential limits, $U \to 0$ and $\mathrm{d}T/\mathrm{d}\xi \to 0$ as $\xi \to +\infty$. Substitution into eqns (10.51) and (10.52) gives $C = D = 0$ and back-substitution into those equations gives

$$\frac{\partial U}{\partial m} = V^2 m \tag{10.53}$$

and

$$U - \tfrac{1}{2}V^2 m^2 + \frac{k}{\rho V}\frac{\mathrm{d}T}{\mathrm{d}\xi} = 0. \tag{10.54}$$

The boundary conditions as $\xi \to -\infty$ are

$$m \to m^*, \quad S \to S^* \quad \text{and} \quad \frac{\partial T}{\partial \xi} \to 0, \tag{10.55}$$

where S^* is at present an unknown constant. If the profile is continuous, i.e. if it contains no shocks, then eqns (10.53) and (10.54) are valid for all ξ, $-\infty<\xi<+\infty$, and boundary conditions (10.55) require

$$\left(\frac{\partial U}{\partial m}\right)^* = V^2 m^* \tag{10.56}$$

and

$$U^* - \tfrac{1}{2}V^2 m^{*2} = 0. \tag{10.57}$$

Equations (10.56) and (10.57) are identical to eqns (10.10). Hence the velocity of a continuous constant profile is identical to that of the adiabatic shock between the same strain limits. Proceed as after eqns (10.10).

On eliminating V^2 from eqns (10.56) and (10.57),

$$U^* - \frac{1}{2}m^*\left(\frac{\partial U}{\partial m}\right)^* = 0. \tag{10.58}$$

Substitute from eqn (10.1) into eqn (10.58):

$$T_0 S^* - \frac{1}{2}\kappa m^* S^* + \frac{1}{2}\eta S^{*2} - \frac{1}{12}c_1^2\alpha m^{*3} = 0.$$

As $m^*\to 0$, $S^*\to 0$; hence the largest terms in the above equation are $T_0 S^*$ and $-(1/12)c_1^2\alpha m^{*3}$. Therefore

$$S^* = +\frac{c_1^2\alpha}{12T_0}m^{*3} + \mathrm{O}(m^{*4}) \quad \text{as} \quad m^*\to 0. \tag{10.59}$$

Substitute for S^* back into eqn (10.2) with m and S starred to give $(\partial U/\partial m)^*$ and use eqn (10.56) to give

$$\frac{V^2}{c_1^2} = 1 + \frac{\alpha}{2}m^* - \frac{\kappa\alpha}{12T_0}m^{*2} + \mathrm{O}(m^{*3}). \tag{10.60}$$

Note that κ only enters into the profile velocity in the coefficient of m^{*2}.

Substitute for V^2/c_1^2 from eqn (10.60) into eqn (10.53). Since U in eqn (10.1) is correct to $\mathrm{O}(m^3)$, only the terms $1+\frac{1}{2}\alpha m^*$ are needed in V^2/c_1^2.

$$c_1^2(m + \tfrac{1}{2}\alpha m^2) - \kappa S = c_1^2(1 + \tfrac{1}{2}\alpha m^*)m$$

or

$$S = -\frac{c_1^2\alpha}{2\kappa}m(m^* - m). \tag{10.61}$$

From $T = \partial U/\partial S$ and eqn (10.61),

$$T = T_0 - \kappa m + \eta S$$

$$= T_0 - \kappa m - \frac{c_1^2 \eta \alpha}{2\kappa} m(m^* - m), \qquad (10.62)$$

and

$$\frac{\mathrm{d}T}{\mathrm{d}\xi} = \left[-\kappa - \frac{c_1^2 \eta \alpha}{2\kappa}(\tfrac{1}{2}m^* - m)\right]\frac{\mathrm{d}m}{\mathrm{d}\xi} = -\kappa\left[1 + \frac{\alpha}{\varepsilon}(\tfrac{1}{2}m^* - m)\right]\frac{\mathrm{d}m}{\mathrm{d}\xi}. \qquad (10.63)$$

Correct to second order in m,

$$U - \tfrac{1}{2}V^2 m^2 = -T_0 S = -\frac{c_1^2 \alpha T_0}{2\kappa} m(m^* - m).$$

Since $\varepsilon = \kappa^2/\eta c_1^2$,

$$U - \tfrac{1}{2}V^2 m^2 = -\frac{\alpha \kappa T_0}{2\eta\varepsilon} m(m^* - m). \qquad (10.64)$$

Substitute from eqns (10.63) and (10.64) into eqn (10.54) to give

$$\left(1 + \frac{\alpha}{\varepsilon}(\tfrac{1}{2}m^* - m)\right)\frac{\mathrm{d}m}{\mathrm{d}\xi} + \frac{\alpha\rho V T_0}{2k\eta\varepsilon} m(m^* - m) = 0$$

or

$$(\varepsilon + \alpha(\tfrac{1}{2}m^* - m))\frac{\mathrm{d}m}{\mathrm{d}\xi} + \frac{V}{2\beta}\alpha m(m^* - m) = 0 \qquad (10.65)$$

since $\beta = k\eta/\rho T_0$.

The r.h.s. of eqn (10.65) is positive in the open interval $0 < m < m^*$. The coefficient of $\mathrm{d}m/\mathrm{d}\xi$ is positive at $m = 0$ and will be positive throughout the closed interval if it is positive at $m = m^*$, i.e. if $m^* < 2\varepsilon/\alpha$. Hence $\mathrm{d}m/\mathrm{d}\xi$ is negative for all m, $0 < m < m^*$, if and only if $m^* < 2\varepsilon/\alpha$. In that case m decreases continuously from m^* to 0 as ξ increases from $-\infty$ to $+\infty$. Integrate eqn (10.65) as follows:

$$\frac{V}{2\beta}(\xi - \xi_0) = \int^m \frac{m - \frac{1}{2}m^* - \varepsilon/\alpha}{m(m^* - m)}\,\mathrm{d}m$$

$$= \int^m \left(-\frac{\frac{1}{2}m^* + \varepsilon/\alpha}{m^* m} + \frac{\frac{1}{2}m^* - \varepsilon/\alpha}{m^*(m^* - m)}\right)\mathrm{d}m$$

$$= -\left(\frac{1}{2} + \frac{\varepsilon}{\alpha m^*}\right)\ln m - \left(\frac{1}{2} - \frac{\varepsilon}{\alpha m^*}\right)\ln(m^* - m). \qquad (10.66)$$

Equation (10.66) is the equation between m and ξ when the profile is continuous. The constant ξ_0 represents an arbitrary shift of the whole profile along the ξ-axis. Note that the profile is no longer anti-symmetric about the value $m^*/2$ as was the case for Burgers' equation.†

When $m^*=2\varepsilon/\alpha$, eqn (10.65) reduces to

$$\frac{\mathrm{d}m}{\mathrm{d}\xi}+\frac{V}{2\beta}m=0$$

or

$$m=\mathrm{e}^{-2V(\xi-\xi_0)/\beta}. \tag{10.67}$$

From eqn (10.67), $m=m^*$ when $\xi=\xi_1=\xi_0-(\beta/2V)\ln m^*$. If $\mathrm{d}m/\mathrm{d}\xi$ is to be non-positive, then a possible continuation of the profile to the left of ξ_1 is for m to equal m^* for $\xi<\xi_1$. The whole profile is illustrated in Fig. 10.8. The profile has a weak discontinuity at ξ_1 and so for this solution to be acceptable, the profile velocity $V=c_1\sqrt{1+\varepsilon}$ must equal the characteristic velocity at strain $m=2\varepsilon/\alpha$.

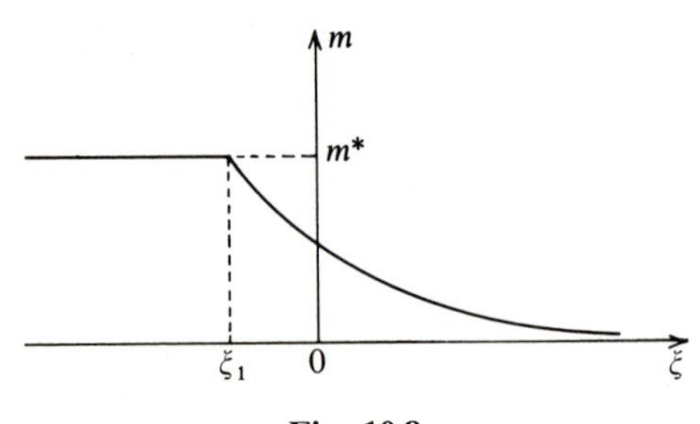

Fig. 10.8

The governing equations for u and S are given by eqn (10.47) with eqn (10.1) and eqn (10.48) with $T=\partial U/\partial S$ as

$$c_1^2(1+\alpha m)\frac{\partial^2 u}{\partial x^2}-\frac{\partial^2 u}{\partial t^2}=\kappa\frac{\partial S}{\partial x}$$

and

$$\rho T\frac{\partial S}{\partial t}-k\eta\frac{\partial^2 S}{\partial x^2}=-k\kappa\frac{\partial^3 u}{\partial x^3}.$$

† Further detail in Exercise 10.1 at the end of this chapter.

Elimination of S from the above two equations gives

$$-k\eta c_1^2(1+\alpha m)\frac{\partial^4 u}{\partial x^4}+k\eta\frac{\partial^4 u}{\partial x^2\partial t^2}+\rho k\kappa^2\frac{\partial^4 u}{\partial x^4}+\text{terms of lower order}=0.$$

The characteristic velocities of this equation are $\pm c_1\sqrt{1+\alpha m-\varepsilon}$ and ∞ (twice). When $m=2\varepsilon/\alpha$, the finite positive characteristic velocity is $c_1\sqrt{1+\varepsilon}$ as required.

If $m^*>2\varepsilon/\alpha$, the coefficient of $\mathrm{d}m/\mathrm{d}\xi$ on the l.h.s. of eqn (10.63) becomes negative for $m>\frac{1}{2}m^*+\varepsilon/\alpha$; the curve of m against ξ is sketched in Fig. 10.9. This solution is inadmissible, because m is single-valued in ξ. The same situation arose in traffic flow in Chapter 9, where the solution was to insert a shock to maintain single-valuedness. We now look for a solution of the same form for the thermoelastic constant profile when $m^*>2\varepsilon/\alpha$.

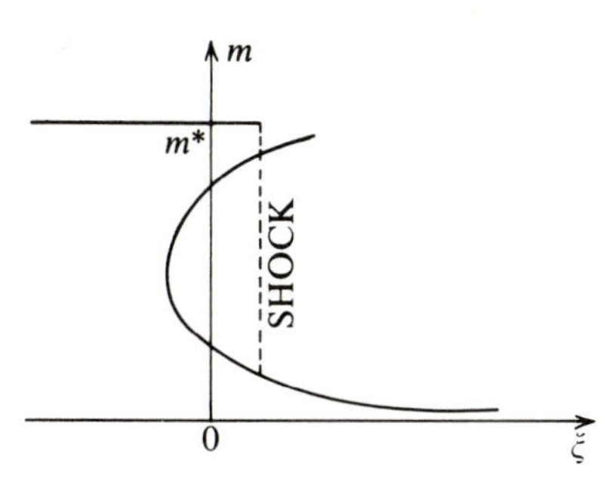

Fig. 10.9

It is first shown that $[T]=0$ across the shock. If the solid is acted on by a body force $X(x,t)$ per unit mass and contains a heat source $\theta(x,t)$ per unit volume, then the governing equations (10.47) and (10.48) are replaced by

$$\frac{\partial^2 u}{\partial t^2}=\frac{\partial}{\partial x}\left(\frac{\partial U}{\partial m}\right)+X=\frac{\partial}{\partial x}\left(\frac{\partial U}{\partial m}+\int^x X(y)\,\mathrm{d}y\right)$$

and

$$\rho T\frac{\partial S}{\partial t}=k\frac{\partial^2 T}{\partial x^2}+\theta=\frac{\partial}{\partial x}\left(k\frac{\partial T}{\partial x}+\int^x \theta(y)\,\mathrm{d}y\right).$$

When $X(x,t)=X(\xi)$ and $\theta(x,t)=\theta(\xi)$, $\xi=x-Vt$, then the same argument that led to eqns (10.51) and (10.52) now leads to

$$\frac{\partial U}{\partial m}+\int^{\xi} X(y)\,\mathrm{d}y = V^2 m + C \tag{10.68}$$

and

$$U-\frac{1}{2}V^2m^2-Cm-D+\frac{k}{\rho V}\frac{\mathrm{d}T}{\mathrm{d}\xi}+\frac{1}{\rho V}\int^{\xi}\theta(y)\,\mathrm{d}y \tag{10.69}$$

on replacing $\partial U/\partial m$ by $\partial U/\partial m+\int^{\xi} X(y)\,\mathrm{d}y$ and $k(\mathrm{d}T/\mathrm{d}\xi)$ by $k(\mathrm{d}T/\mathrm{d}\xi)+\int^{\xi}\theta(y)\,\mathrm{d}y$.

Let a constant profile contain a shock (Fig. 10.10). The zero of ξ is chosen so that the shock lies on $\xi=0$. $X(\xi)$ and $\theta(\xi)$ are now chosen to replace the profile from $-h$ to $+h$ by a continuous profile, the governing equations now being (10.68) and (10.69). Let the lower limit of the integral of θ in eqn (10.69) be $-h$, any constant so introduced being included in D. Integrate eqn (10.69) with respect to ξ from $-h$ to h and then let $h\to 0$. In the limit the shock is reproduced:

$$\lim_{h\to 0}\left(\int_{-h}^{h}(U-\tfrac{1}{2}V^2m^2-Cm-D)\,\mathrm{d}\xi+\frac{k}{\rho V}\int_{-h}^{h}\frac{\mathrm{d}T}{\mathrm{d}\xi}\,\mathrm{d}\xi + \frac{1}{\rho V}\int_{-h}^{h}\mathrm{d}\phi\int_{-h}^{\phi}\theta(y)\,\mathrm{d}y\right)=0.$$

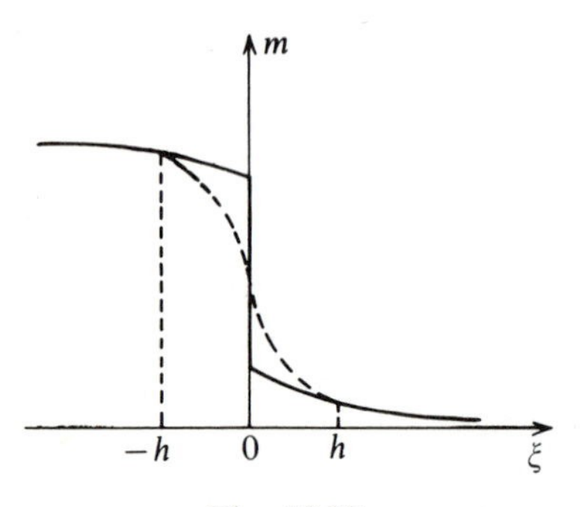

Fig. 10.10

Since the integrand of the first integral is finite, the limit of the integral as $h\to 0$ is zero.

$$\lim_{h\to 0}\int_{-h}^{h}\frac{\mathrm{d}T}{\mathrm{d}\xi}\,\mathrm{d}\xi=\lim_{h\to 0} T\Big|_{-h}^{h}=[T].$$

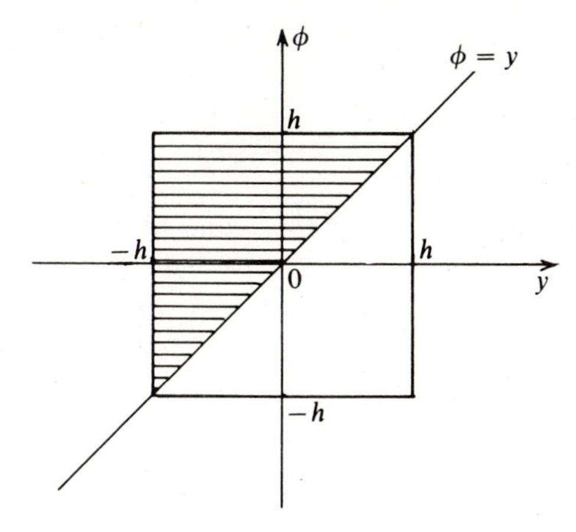

Fig. 10.11

With reference to Fig. 10.11,

$$\begin{aligned}\lim_{h\to 0}\int_{-h}^{h} \mathrm{d}\phi \int_{-h}^{\phi} \theta(y)\,\mathrm{d}y &= \lim_{h\to 0}\int_{-h}^{h} \theta(y)\,\mathrm{d}y \int_{y}^{h} \mathrm{d}\phi \\ &= \lim_{h\to 0}\left(h\int_{-h}^{h} \theta(y)\,\mathrm{d}y - \int_{-h}^{h} y\theta(y)\,\mathrm{d}y\right) \\ &= -\lim_{h\to 0}\int_{-h}^{h} y\theta(y)\,\mathrm{d}y.\end{aligned}$$

Hence

$$k[T] = \lim_{h\to 0}\int_{-h}^{h} y\theta(y)\,\mathrm{d}y. \tag{10.70}$$

But the r.h.s. of eqn (10.70) is zero in the absence of a surface distribution of heat dipoles. Hence

$$[T] = 0, \tag{10.71}$$

since, by definition for a heat-conducting solid, $k \neq 0$. Note that C and D need not be the same on both sides of the shock; the limits of their integrals as $h \to 0$ still vanish.

The position of the shock in the profile cannot be assumed *a priori*. The governing equations for shock propagation are (10.4), (10.5),

(10.16) and (10.71). Equations (10.4) and (10.6) can be combined to give

$$\left[\frac{\partial U}{\partial m}\right] = V^2[m]. \tag{10.72}$$

If there is a continuous part of the profile ahead of the shock, eqn (10.53) holds on this part of the profile and therefore

$$\left(\frac{\partial U}{\partial m}\right)_+ = V^2 m_+. \tag{10.73}$$

Adding this last equation to eqn (10.72),

$$\left(\frac{\partial U}{\partial m}\right)_- = V^2 m_-, \tag{10.74}$$

and therefore, by eqn (10.51), C is zero behind the shock and eqn (10.53) holds behind the shock as well as in front. In particular it holds as $\xi \to -\infty$, so that

$$\left(\frac{\partial U}{\partial m}\right)^* = V^2 m^*. \tag{10.75}$$

If the profile starts with a shock, apart from a constant strain ahead, them $m_+ = (\partial U/\partial m)_+ = 0$ and eqns (10.74) and (10.75) still hold.

Since $v = \partial u/\partial t = -V(\mathrm{d}u/\mathrm{d}\xi)$ and $m = \partial u/\partial x = \mathrm{d}u/\mathrm{d}\xi$, $v = -Vm$. Substitute $v = -Vm$, $\sigma = \rho(\partial U/\partial m)$, $q = -k(\mathrm{d}T/\mathrm{d}\xi)$ and eqns (10.73) and (10.74) into eqn (10.5), to give

$$[U] - \frac{1}{2}V^2[m^2] + \frac{k}{\rho V}\left[\frac{\mathrm{d}T}{\mathrm{d}\xi}\right] = 0. \tag{10.76}$$

From eqns (10.54) and (10.76) the same argument as in the last paragraph shows that eqn (10.54) holds behind the shock and therefore the limit as $\xi \to -\infty$ is

$$U^* - \tfrac{1}{2}V^2 m^* = 0. \tag{10.77}$$

Equations (10.75) and (10.77) are identical to eqns (10.56) and (10.57) and therefore S^* and V are given by eqns (10.59) and (10.60) in the presence of a shock. Also since C and D are zero everywhere, eqn (10.65) is valid everywhere; in particular, it is valid on any continuous section behind a shock. If m varies on such a section, then there exist values of m for which $\varepsilon + \alpha(\tfrac{1}{2}m^* - m) < 0$. But this implies

$dm/d\xi > 0$ over a finite interval. This is impossible if m is a non-increasing function of ξ. We conclude that the shock in a constant profile must come at the rear of the section in which m changes continuously, as shown in Fig. 10.12.

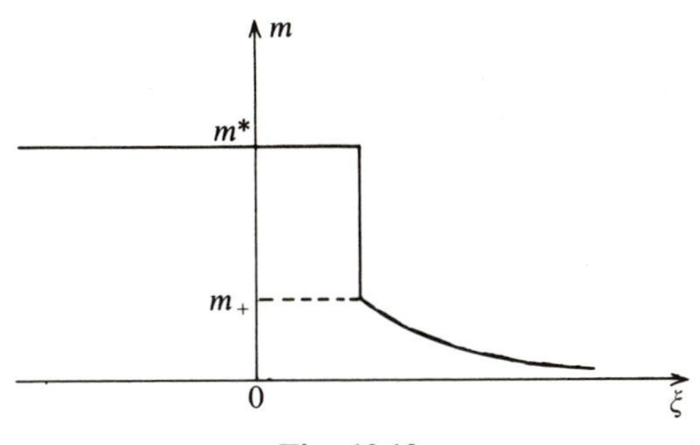

Fig. 10.12

The equation of the continuous changing section is given by eqn (10.66) for $m < m_+$, where m_+ is the value of m ahead of the shock; m_+ is given by eqns (10.71), (10.72) and (10.60). Note that the shock differs from the shock of Section 10.1 because it is isothermal not adiabatic.

$$c_1^2(1+\tfrac{1}{2}\alpha m^*)[m] = V^2[m] = \left[\frac{\partial U}{\partial m}\right] = c_1^2([m]+\tfrac{1}{2}\alpha[m^2]) - \kappa[S].$$

Since
$$[S] = \left[\frac{1}{\eta}T + \frac{\kappa}{\eta}m\right] = \frac{\kappa}{\eta}[m],$$

$$(\varepsilon + \tfrac{1}{2}\alpha m^*)[m] = \tfrac{1}{2}\alpha[m^2].$$

Now
$$[m^2] = m^{*2} - m_+^2 = (m^* - m_+)(m^* + m_+) = (m^* + m_+)[m];$$

therefore,
$$\varepsilon + \tfrac{1}{2}\alpha m^* = \tfrac{1}{2}\alpha(m^* + m_+)$$

or
$$m_+ = \frac{2\varepsilon}{\alpha}. \tag{10.78}$$

The strain just ahead of the shock is the same for all $m^*(>2\varepsilon/\alpha)$. The strain changes continuously from 0 to $2\varepsilon/\alpha$, then jumps to m^* and is constant and equal to m^* thereafter.

The solution is not yet complete. It remains to show that the entropy condition across a shock is satisfied. Since T is constant, the entropy condition $\delta S \geqslant \delta Q/T$ becomes

$$[S] \geqslant \frac{[Q]}{T}. \tag{10.79}$$

Apply the inequality to the small element from x to $x+\delta x$ in the small time interval $\delta x/V$ during which the shock passes through the element (Fig. 10.13). First, since the element has mass $\rho\,\delta x$,

$$[\rho\,\delta x Q] = (q_x - q_{x+\delta x})\frac{\delta x}{V} = [q]\frac{\delta x}{V},$$

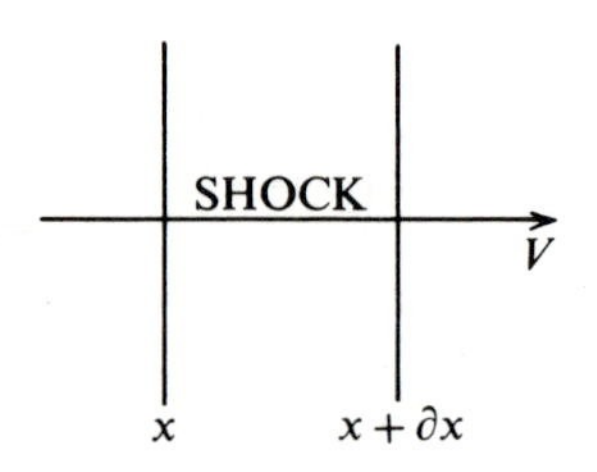

Fig. 10.13

or

$$\rho V[Q] = -\left[k\frac{\partial T}{\partial x}\right].$$

Now substitute for $[Q]$ into eqn (10.79). Then, for constant-profile shocks,

$$T[S] + \frac{k}{\rho V}\left[\frac{\mathrm{d}T}{\mathrm{d}\xi}\right] \geqslant 0,$$

or, on use of eqn (10.76) and $T = T^*$,

$$T^*[S] - [U] + \tfrac{1}{2}V^2[m^2] \geqslant 0. \tag{10.80}$$

Substitute

$$S = \frac{\kappa}{\eta}m + \frac{1}{\eta}(T - T_0)$$

into the last two terms of eqn (10.1):

$$U = T_0 S + \frac{1}{2}c_1^2\left(m^2 + \frac{1}{3}\alpha m^3\right) - \frac{1}{2}\frac{\kappa^2}{\eta}m^2 + \frac{1}{2\eta}(T - T_0)^2.$$

Hence,

$$[U]=T_0[S]+\frac{1}{2}c_1^2\left((1-\varepsilon)[m^2]+\frac{1}{3}\alpha[m^3]\right), \qquad (10.81)$$

since

$$[T]=0.$$

From eqns (10.80) and (10.81),

$$(T^*-T_0)[S]-\frac{1}{2}c_1^2(1-\varepsilon)[m^2]-\frac{1}{6}c_1^2\alpha[m^3]+\frac{c_1^2}{2}\left(1+\frac{\alpha}{2}m^*\right)[m^2]\geqslant 0.$$

Since, from eqn (10.62) and $[S]=\frac{\kappa}{\eta}[m]$,

$$(T^*-T_0)[S]=(-\kappa m^*)\left(\frac{\kappa}{\eta}[m]\right),$$

$$-\varepsilon m^*[m]+\frac{1}{2}\left(\varepsilon+\frac{\alpha}{2}m^*\right)[m^2]-\frac{1}{6}\alpha[m^3]\geqslant 0. \qquad (10.82)$$

From eqn (10.78) substitute $\varepsilon=\frac{1}{2}\alpha m_+$ into eqn (10.82). After some elementary algebra,

$$\alpha(m^*-m_+)^3\geqslant 0, \qquad (10.83)$$

which condition is satisfied.

The practical application of the results of this section is limited by two physical considerations. Some materials cease to be elastic before the strain reaches the value $2\varepsilon/\alpha$, and for strains much larger than $2\varepsilon/\alpha$ more terms will be needed in the expansion of U than are given by eqn (10.1). If terms of $O(m^4)$ are included in U, the shock may move to the front of the profile as m^* increases.‡

In the non-linear Voigt solid constant profile of Section 10.2, the strain tends asymptotically to its limiting values both at the front and the back of the profile. The same is true of the thermoelastic profile of Section 10.3, except that for strains $m^*\geqslant 2\varepsilon/\alpha$ the strain behind the shock is constant. If the width of a profile is defined as the distance between the points at which the strain is γm^* and $(1-\gamma)m^*$, where $\gamma\ll 1$ and constant,† then the width of the profile is proportional to the viscosity ν, eqn (10.46), or to the conductivity k, eqn (10.66) with $\beta=k\eta/(\rho T_0)$. In the limit as $\nu\to 0$ or $\beta\to 0$, the widths of the profiles tend to zero and the shock of adiabatic theory is reproduced.

---

† γ is frequently taken as 0.05, as it is in Chapter 9.

‡ See Bland, D. R. (1969). *Non-linear dynamic elasticity*. Chap. 5. Ginn, Aylesbury.

Conversely, one can say that the introduction of either small viscosity or small conductivity spreads out the adiabatic shock. For real elastic solids, the small dissipative effects of viscosity and conductivity are not often significant. However, these effects do cause shocks to have non-zero thickness and the detail of the spread-out shocks so produced is called 'shock structure'.

The adiabatic elastodynamic equations and the thermoelastodynamic equations are an example of a non-linear wave hierarchy in which the latter equations reduce to the former when the heat conductivity k is zero. Figure 8.11 illustrates how in the linearized theory the single positive velocity characteristic of the former equations lies between the two characteristics of the latter. The situation is more complicated in non-linear theory because the characteristic velocities are amplitude-dependent and shocks do not travel at the characteristic velocity. The single lines of Fig. 8.11 are replaced by wedge-shaped regions which can overlap (Fig. 10.14). The subject is treated in detail in chapter 10 of the first reference below. Suffice it for now to comment that the constant-profile solution of this section becomes discontinuous at the strain m^* for which the characteristic velocity of the thermoelastic equations becomes equal to the shock velocity of the adiabatic equations.

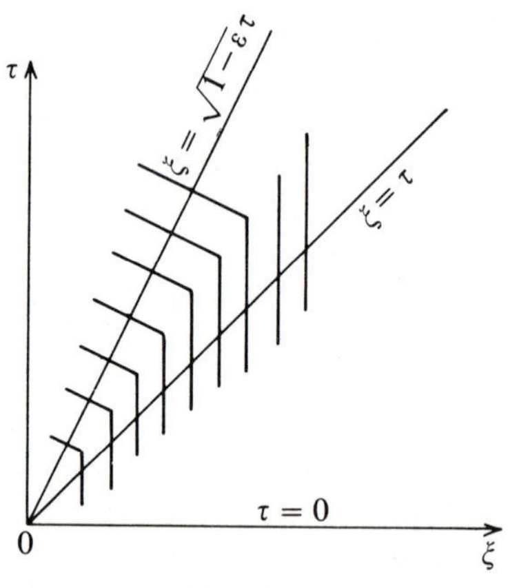

Fig. 10.14

The standard advanced work on waves is *Linear and non-linear waves* by G. B. Whitham. The standard text on waves in fluids is *Hydrodynamics* by H. Lamb; a more recent treatment is *Waves in fluids* by M. J. Lighthill. Two classical books are 'Theory of sound' by Lord Rayleigh and 'Propagation des ondes' by J. Hadamard. The reader can also refer to the books referenced elsewhere in this text.

Exercises

10.1. Equation (10.66) can be written

$$\frac{m^{-(\frac{1}{2}+\varepsilon/(\alpha m^*))}}{(m^*-m)^{-(\frac{1}{2}-\varepsilon/(\alpha m^*))}}=\mathrm{e}^{\frac{1}{2}V(\xi-\xi_0)/\beta}.$$

As $\xi-\xi_0\to+\infty$, $m\to 0$ and m^*-m can be replaced by m^* in the denominator of the l.h.s. of the above equation, so that

$$m\sim C_1\exp\left(-\tfrac{1}{2}V(\xi-\xi_0)\beta^{-1}\left(\frac{\frac{1}{2}+\varepsilon}{\alpha m^*}\right)^{-1}\right)\quad\text{as }\xi\to\infty.$$

As $\xi-\xi_0\to-\infty$, $m\to m^*$ and m can be replaced by m^* in the numerator, so that

$$m^*-m\sim C_2\exp\left(\tfrac{1}{2}V(\xi-\xi_0)\beta^{-1}\left(\frac{\frac{1}{2}-\varepsilon}{\alpha m^*}\right)^{-1}\right)\quad\text{as }\xi\to-\infty.$$

C_1 and C_2 are constants. Show from the above two equations that the rear of the profile approaches its limiting value faster than the front.

10.2. The equations

$$c_1^2\left(1+\alpha\frac{\partial u}{\partial x}+\nu\frac{\partial}{\partial t}\right)\frac{\partial^2 u}{\partial x^2}-\kappa\frac{\partial S}{\partial t}=\frac{\partial^2 u}{\partial t^2},$$

$$k\frac{\partial^2 T}{\partial x^2}=\rho T_0\frac{\partial S}{\partial t}$$

and

$$T-T_0=-\kappa m+\eta S$$

reduce to the governing equations of Section 10.3 when $\nu=0$, except that the heat conduction equation has been linearized. The additional term involving ν, known as an artificial viscosity, is added to spread out the shock for large strains. Find the constant-profile velocity, using $\xi=x-Vt$, for the above equations when m, S and their derivatives tend to zero as $\xi\to+\infty$ and $m\to m^*$ with the derivatives of m tending to zero as $\xi\to-\infty$.

Answer:

$$V^2=c_1^2(1+\tfrac{1}{2}\alpha m^*).$$

Author Index

Subject Index